Total Manufacturing Assurance

Second Edition

Total Manufacturing Assurance
Controlling Product Quality, Reliability, and Safety

Second Edition

Dr. Douglas Brauer and Dr. John Cesarone

CRC Press
Taylor & Francis Group
Boca Raton London New York

CRC Press is an imprint of the
Taylor & Francis Group, an **informa** business

Second edition published 2022
by CRC Press
6000 Broken Sound Parkway NW, Suite 300, Boca Raton, FL 33487-2742

and by CRC Press
4 Park Square, Milton Park, Abingdon, Oxon, OX14 4RN

© 2022 Douglas Brauer and John Cesarone

CRC Press is an imprint of Taylor & Francis Group, LLC

Library of Congress Cataloging-in-Publication Data
Names: Brauer, Douglas, author. | Cesarone, John, author.
Title: Total manufacturing assurance : controlling product quality, reliability, and safety /
Douglas Brauer and John Cesarone.
Description: First edition. | Boca Raton, FL : CRC Press, 2022. | Includes
bibliographical references and index.
Identifiers: LCCN 2021048082 (print) | LCCN 2021048083 (ebook) |
ISBN 9781032076362 (hbk) | ISBN 9781032076379 (pbk) | ISBN 9781003208051 (ebk)
Subjects: LCSH: Quality assurance. | Total quality control.
Classification: LCC TS156.6 .B75 2022 (print) |
LCC TS156.6 (ebook) | DDC 658.5/62—SSdc23/eng/20211109
LC record available at https://lccn.loc.gov/2021048082
LC ebook record available at https://lccn.loc.gov/2021048083

ISBN: 978-1-032-07636-2 (hbk)
ISBN: 978-1-032-07637-9 (pbk)
ISBN: 978-1-003-20805-1 (ebk)

DOI: 10.1201/9781003208051

Typeset in Times
by codeMantra

Contents

SECTION I   Introduction to TMA

SECTION II   Strategic Manufacturing Management

SECTION III   *Manufacturing System Control*

SECTION IV   System Improvement Monitoring

Foreword

Technology increased the standard of living to unprecedented levels for most of the world. The drive to further lift up the ability of people to afford and access products is underway. For consumers around the globe, key factors such as affordability, quality, reliability, and safety of products are of paramount importance. People expect the products to be low cost and very reliable, high quality, and safe. With the increased use of quality control technologies in design and manufacture of goods, including robotics and Internet of Things (IoT), the product quality across the board has made significant improvement over the past few decades. In the next industrial revolution, powered by IoT and Industry 4.0, all goods and services will be monitored from conception though manufacturing, operation, and disposition. This will lead to an exponential increase in product quality, reliability, and safety. Organizations that master the product quality will be the winners in the marketplace. That is possible with a holistic and strategic approach.

Total Manufacturing Assurance (TMA) is a holistic concept to address these issues that includes all aspects of quality control. The book provides guidance for professionals to develop long-term strategies for quality for a corporation with global footprint, as well as case studies that will be tutorial for young engineering students. The book is valuable for professionals and executives at corporate world, as well as students in management, mechanical, industrial, and manufacturing engineering. It outlines the strategic planning of global manufacturing processes with quality in mind; factories that can be controlled and monitored in real time from anywhere in the world 24/7.

This second edition of the textbook, by global experts Professors D. Brauer and J. Cesarone, is a much-needed addition to the foundational technical material in Total Manufacturing Assurance. It is a timely public service to diligently compile this material to educate the technologists around the world.

Sabri Cetin, PhD
Professor, University of Illinois at Chicago
President, Servo Tech, Inc.

Preface

The second edition of *Total Manufacturing Assurance: Controlling Product Reliability, Safety, and Quality* continues its objective to present an enhanced perspective for the innovative concept of Total Manufacturing Assurance (TMA) and the holistic means by which such assurance can be attained. In fulfilling this objective, the textbook discusses the management and engineering techniques and tools, and their practical implementation, necessary to achieve TMA.

The uniqueness of the textbook is derived from its presenting the concept of TMA. As with the original edition, TMA remains a highly relevant concept, as there remains no known other referenceable work which focuses on establishing a manufacturing environment and process that assures the attainment of manufactured products that exhibit, and retain, their designed-in levels of reliability, safety, and quality. This second edition is expanding on and integrating fundamental manufacturing, engineering, and management topics, which are holistically key in achieving TMA. Originally included as part of a quality and reliability textbook series, this new edition expands its reach as a seminal textbook as part of an engineering coursework and professional development focused on, for example, General Engineering, Industrial & Systems Engineering, Manufacturing Engineering, Mechanical Engineering, and Business Operations.

TMA comprises three primary elements: (1) Strategic Manufacturing Management, (2) Manufacturing System Control, and (3) System Improvement Monitoring. Each of these elements plays a key role in ensuring that products are manufactured in a manner which supports the profitability, sustainability, and growth strategic goals of a business. A major theme throughout the text is the strategic importance of TMA for companies competing for local, regional, and/or global market share. TMA is a strategic-driven, holistic pathway engaging management, engineering, and production personnel to maximize efficiency, effectiveness, and profitability.

The textbook is divided into four major sections: (1) Introduction, (2) Strategic Manufacturing Management, (3) Manufacturing System Control, and (4) System Improvement Monitoring. This enables the reader to focus on a particular aspect of TMA.

The Introduction section consist of two chapters. Chapter 1 provides an overview of the history of manufacturing detailing the global industrial revolutions; recognizing key innovations in management and manufacturing; highlighting seminal business, technology, and world events; and providing insight to the manufacturing advances seen with the ongoing advent of Industry 4.0. Chapter 2 introduces and defines the three major elements of TMA: strategic manufacturing management, manufacturing system control, and system improvement monitoring. Additionally, the TMA thread of *create, control, and critique* is presented.

The Strategic Manufacturing Management section consists of five chapters. Chapter 3 discusses strategic planning and the roll out of tactics. The use of key performance indicators is highlighted. Chapter 4 provides an overview of innovation and technology transfer. Technology development and the strategic implications of its commercialization are presented in conjunction with a detailed technology transfer checklist. Chapter 5 dives into engineering economic analysis, with a focus on financial proforma development and looking at a variety of financial analysis tools to make business decisions. This includes exploring make/buy decisions and the fundamental role of breakeven analysis. Chapter 6 discusses the role of management control to achieve TMA. This includes the integration of key business areas impacting TMA, project management methods, decision-making methods, leveraging of I4.0 technologies, implementation of computer-integrated manufacturing, and use of expert systems. Chapter 7 addresses organizational leadership. Several topics addressing the workforce are addressed, such as motivation, teaming, performance appraisal, workforce development and training, technical personnel hiring, and organizational succession planning.

The Manufacturing System Control section consists of four chapters. Chapter 8 provides insights into systems engineering in the manufacturing environment. This includes a comprehensive study of moving material through the factory, with special attention to simulation methods and advanced automation building blocks. Additionally, this chapter introduces manufacturing process engineering and methods (i.e., metal working, molding, fiber composites, assembly methods, and micromanufacturing tools), as well as information for industrial materials and their selection performance features and characteristics. Chapter 9 addresses product manufacturing degradation control during manufacturing. Engineering and control information and analysis methods are provided for areas of reliability, safety, and quality. Additionally, lean manufacturing, design for manufacturability, and manufacturing effectiveness design and assessment techniques are delineated. Chapter 10 provides a comprehensive insight to system maintenance covering maintenance program planning, reliability-centered maintenance, and maintenance strategy (specifically, preventative, corrective, and on-condition maintenance approaches).

The System Improvement Monitoring section consists of three chapters. Chapter 11 addresses big data system planning addressing data organization, system structure, and system operation. Chapter 12 addresses data recording and feedback to facilitate real-time collection, analysis, and feedback to optimize manufacturing operations. Data communications and FRACA are highlighted. Finally, Chapter 13 provides insight into manufacturing system performance analytics, with special emphasis on key performance indicators.

This textbook exhibits many special and innovative features. A key special feature is the inclusion of a case study at the end of chapters to facilitate keen understanding of the topical subjects addressed. While serving as a valuable teaching tool, the case studies illustrate the practical application of the material presented in each chapter and the benefits that can be realized. The case studies are standardized to provide an overview of the issue at hand, the strategic objective for resolution, the approach taken, results realized, and a summarizing conclusion. It will also help personalize the material for the readers so that they can more readily understand, embrace, and implement TMA as a strategic initiative. A second special feature is the highlighting and emphasis of Industry 4.0 technologies prominent in advanced manufacturing, but also broadly engaging the financial, logistics, and information technology business sectors.

The intended audience for the textbook is both working professionals and students in higher education. Working professionals include management, engineering, and others intimately involved in the manufacturing system. Students studying engineering management and mechanical, industrial, and manufacturing engineering, and business students (both university advanced undergraduate and graduate levels) will find the textbook an invaluable instructional resource, as will professors. The textbook's material will address the management and technical topics presented in the level of detail necessary for their full understanding and implementation to achieve TMA. PowerPoint slides and a solutions manual are also available to instructors for qualified course adoptions (http://www.routledge.com/9781032076362). In general, it is envisioned that the text will serve as a comprehensive and affordable fundamental reference and educational tool.

The authors thank everyone engaged for direct and indirect significant contributions to the writing of this second edition of TMA. A special acknowledgment goes to Ms. Laurie Brauer (MudTurtle Industries, LLC) for her work in preparing the graphic artwork. Additionally, the following people provided guiding insights for various sections of the textbook: Cheston Brauer (Manager Strategic Planning, Advocate Hospitals), Dr. Sabri Cetin (Professor, University of Illinois at Chicago & President, Servo Tech Inc.), Dr. Sheri Litt (Associate Provost, Florida State College at Jacksonville), Dr. Roh Pin Lee (Head of Technology Assessment, Technische Universität Bergakademie Freiberg, Germany), and Dr. Clifford Harbour (Professor, University of North Texas).

Douglas Brauer
John Cesarone

Authors

Dr. Douglas Brauer has over 30 years of global expertise in facilitating solutions to maximize organizational leadership, sustainability, profitability, and return on investment encompassing both and higher education. His scope of global strategic leadership activities involving education, engineering, and manufacturing operations has engaged organizations globally in the Americas, Europe, Africa, India, and the Pacific Rim. He has taught higher education undergraduate and graduate courses in finance, organizational leadership, mathematics, and manufacturing technology. His private business activities include founding Design Assurance Sciences in 1990 and MudTurtle Industries, LLC, in 2016. Degrees include PhD, Education & Human Resource Studies, Colorado State University; MS, Industrial Engineering, University of Illinois at Chicago; and BS, Industrial Technology, Illinois State University. Additionally, he is currently the Dean of Engineering & Industry at the Florida State College at Jacksonville.

Dr. John Cesarone has engaged in an engineering consulting practice since 1998. His research interests and areas of expertise include Intelligent Manufacturing, focusing on the following: (1) Computer-Integrated Manufacturing Technologies, (2) Simulation and Functional Modeling Methodologies, (3) Optimization Using Operations Research Techniques, (4) Automation and Robotics, and (5) Distance Learning. Degrees include PhD, Mechanical Engineering, Northwestern University; MS, Mechanical Engineering, University of Illinois-Urbana-Champaign; and BS, Mechanical Engineering, University of Illinois-Urbana-Champaign. He has taught industrial engineering courses for the University of Illinois and Northwestern University, and is currently a Senior Lecturer of Mechanical Engineering for the Illinois Institute of Technology.

Section I

Introduction to TMA

NEVER SHRINK FROM DOING ANYTHING WHICH YOUR BUSINESS CALLS YOU TO DO. THE MAN WHO IS ABOVE HIS BUSINESS, MAY ONE DAY FIND HIS BUSINESS ABOVE HIM.

—DREW

DOI: 10.1201/9781003208051-1

1 The World of Manufacturing

By the simplest definition, manufacturing is the process of making a finished product from raw material in accordance with an organized plan. A piece of cake. Anybody can do it. Manufacturing can be accomplished by anyone blindfolded and with one hand tied behind their back. Nothing could be farther than the truth.

All too often this seems to be the carefree attitude taken. With some raw materials, a couple of machines, and a vague product idea, it is only a matter of time before the products start rolling out and the money starts rolling in. Obviously, there is much more involved in manufacturing than merely shot-gunning hardware out the plant door.

There are three fundamental concerns of a manufacturing organization which must be integrated together in order to ensure success: (1) organizational management (which includes marketing, sales, finance, etc.), (2) engineering, and (3) production. Each of these concerns centers on people and each provides key contributions in the overall manufacturing effort. Yet without the other two, each is helpless. So, it must be reconciled that a "team" effort is necessary for organizational success.

The corporate structure and its ability to be successful is analogous to a needle and thread. There must be a common thread that binds everything, and, most importantly, that binds everyone, together to maximize the probability for business success. The thread itself is the organizational management approach and the corporate culture, or attitude, instilled in its team members.

So, what does all this mean? It means that the organization is, at the very least, a team. The manufacture of products is a result of many people working together to achieve a common goal. The organization comprises team members intimately linked together by the organizational thread...weak thread: weak team...strong thread: strong team.

Obviously, there must be a hierarchy of management in order to ensure that products do get manufactured correctly, efficiently, and in a timely manner. However, the full potential of any organization cannot be realized if there is a lack of mutual respect and caring between persons positioned on different levels of the corporate ladder. A strong binding thread must be in place.

1.1 INDUSTRIAL REVOLUTION

The current state of manufacturing is a result of much patient learning and heuristic applications of innovative concepts. This learning has by no means stopped and it must not. Global and national markets are becoming increasingly more competitive. Aside from the issue of cost, there is an ever-increasing consumer awareness of the right for high-quality products and services.

DOI: 10.1201/9781003208051-2

3

By taking a quick look at history, it becomes clear that much progress has been made. Corporations for the most part have evolved from a neanderthal-type management mindset to a more progressive, understanding, and compassionate style. (Unfortunately, this is not universally the case.)

A good place to pick up in history is the industrial revolution. But, which one? It is generally accepted that there have been four defined to capture significant turning points in time. It is important to note that each industrial revolution resulted an accelerated use of, and need for, advanced manufacturing systems. From the very first one, it may very well have witnessed the birth of the manufacturing engineer. The industrial revolutions are generally classified into four periods:

1. **Steam (circa 1765)**: highlighted by use of coal to produce steam resulted in more mechanization and movement of people and goods by the railroad.
2. **Electromechanical (circa 1870)**: highlighted by the emergence of electricity, gas, and the development for steel demand, chemical synthesis, and methods of communication such as the telegraph and the telephone, and, of course, the inventions of the automobile and the airplane.
3. **Digital/Computer (circa 1969)**: highlighted by the emergence of nuclear energy, electronics, telecommunications, and computers, all led to advancements in, for example, space exploration and biotechnology, as well as led to the invention of programmable logic controllers (PLCs) and robots to introduce the era of high-level industrial automation.
4. **Technology/Informatics (circa 2000)**: highlighted by a focus on renewable energies (e.g., solar, wind, thermal, and water), the connection of "operations" to "analytics" has facilitated real-time advanced decision-making using "big data," which simply means accelerated system optimization. The global institutionalization of the internet and computing power has led to routine applications of artificial intelligence (AI), virtual/augmented reality, machine learning, etc., to manufacturing optimization. This industrial revolution is often referred to as Industrial 4.0.

Some interesting characteristics of each industrial revolution are as follows: they typically coincide with an "energy" source, they result in iconic inventions and iconic entrepreneurs, and they drive and accelerate innovation and technology development to decrease the time to the next industrial revolution, that is, the "industrial revolutions" are coming faster. Throughout history, people have always been dependent on technology and leveraged its development in vast economic and social ways. Of course, the technology of each era did not have the same shape and size as today, but for each in their time technology advancements were certainly significant and helped to enrich lives.

There are many lessons to be learned by taking a deeper look at the first industrial revolution. It began in England in the early eighteenth century. It rapidly spread to other parts of the world during the remainder of the eighteenth century and into the nineteenth century. Although the first industrial revolution gained much worldwide momentum, many countries failed to witness the movement. Even today, there are many areas in the world which have remained essentially untouched by mechanized manufacturing system technology. However, during the latter part of the 1800s, several countries emerged as major industrial leaders. England, Germany, Italy, Japan and the United States became the most important industrial nations.

A key beneficiary of the great strides being made in mechanization was agriculture. As with agricultural processes, advances in engineering made it possible to improve and employ innovation in manufacturing processes. Machines and equipment began replacing inefficient systems driven by manually created motive power. A classic example is the steam engine. Steam power characterized the nineteenth century at its climax. Other sources of power, including electricity, petroleum, and nuclear power, proved to characterize the continuing industrial revolutions of the twentieth century.

The first industrial revolution also had far-reaching socioeconomic consequences. These varied in detail from country to country and are still present in countries undergoing technological changes. The most obvious and most direct consequences were a tremendous increase in the production of material goods. For example, mechanization enabled textile plants in England to produce billions of yards of cotton cloth a year. This increase in production was many times as much as that capable by people (i.e., men women, and children) working with old-fashioned spinning wheels and handlooms.

By 1900, the United States had become the greatest manufacturing nation in the world. Technological skill, government policies, business leadership, natural resources, labor supply, and large markets all contributed to the rapid progresses made. The combination of these advantages enabled the United States to become the world leader in industrial productivity.

While the total output of industry in the United States was increasing, the number of individual plants was decreasing. As plants grew fewer, and on the average larger, so did the business firms that operated them. Before the Civil War, the typical company was owned and run by an individual or partners, though a number of corporations existed. After the war, corporations became larger and more numerous.

For the owners of a company, its large size provided certain advantages. It enabled the company to lower its costs through mass purchasing and mass production. Often, it also served as a tool in weakening or eliminating competitors, and thus it enabled the company to maintain or even to raise prices. For consumers, large businesses were not necessarily advantageous. To the extent that low costs led to low prices, the consumer benefitted. However, in many instances, the benefits never came to fruition.

Significant events ongoing in the United States during this period created fundamental changes in demographics. People moved to large cities, medical advances caused people to live longer, and new means of transportation allowed people to move widely around the country. Also, at this time, a great insurgence of immigrants occurred.

A key side effect of the technological and economic changes ongoing was the formation of unions or organized labor groups. This in turn led to the formation of the classic industrial conflict between corporations and unions. As corporations became the owners of industries, organized labor became necessary to offset the power and consequent employee abuses present. The unions implemented strikes and boycotts.

Basic demands included improved working conditions, recognition of unions as bargaining agents, and the passing of federal laws to regulate various industries.

These early years were shaped by many well-known business persons. Names such as Cyrus McCormick (reaper works), Charles Goodyear (vulcanized rubber), Isaac Singer (sewing machine), Samuel Colt (gun mass production), John Rockefeller (oil), Andrew Carnegie (steel), Thomas Edison (electricity/inventions), George Westinghouse (electricity), J. Pierpont Morgan (finance/banking), Ransom Olds (automobile), and Henry Ford (automobile) all provided key contributions to our manufacturing management and engineering knowledge and growth [1]. However, as manufacturing capability, efficiency, and productivity increased, so did the ignorance of corporate management to the fundamental needs of the workforce regarding respect, honesty, caring, equity, and education/workforce development.

1.2   INNOVATION IN MANAGEMENT AND MANUFACTURING

Without a doubt, the list of contributors to manufacturing management, organization, and engineering innovation fills several pages. Their contributions enabled major strides to be made in technological and management advancements. Corporations such as Ford, Motorola, General Motors, International Business Machines, Xerox, Kodak, and many others continue to push the state of the art in both automated manufacturing and management concepts.

The automated manufacturing capabilities of today (e.g., computer-integrated manufacturing and flexible manufacturing systems) are far superior to the capabilities available to Henry Ford. But, many of the manufacturing philosophies in use worldwide today were seeded by people of that era. For example, in the early 1900s, the Just-In-Time approach to manufacturing was used in a crude form by Ford; Statistical Process Control was defined by Shewhart; and Design of Experiments was developed by Fisher. These manufacturing management/control tools are the subject of much discussion and use during the 1980s and will be well into the 1990s. With the re-discovery and use of many innovative manufacturing ideas (which were ahead of their time) to increase efficiency, productivity, and profitability, organizations are taking to heart the demand by consumers to get reliable, safe, and quality products.

This perhaps was a hard rediscovery and admittance on the part of business in the United States. It took a raising of the quality bar by Japanese corporations to wake up the quality unconscious organizations around the world, and particularly those in the United States. Names such as Deming, Juran, Crosby, Feigenbaum, Taguchi, and Shainin identify some of the more well-known quality gurus. Their concepts continue to be used in high demand and their ideas are common place in quality discussions and activities around the world. These concepts continue to be foundational for programs such as Six Sigma, Lean, Just-In-time, and Manufacturing System Effectiveness.

1.3   INDUSTRY 4.0

The world is in a rapid state of technology research, development, and implementation. The time between each industrial revolution decreases and will continue to do so as world technological events and consumer thirst for advanced technologies continue.

Industry 4.0 (I4.0), as an advanced manufacturing driver, necessarily must address the widening skills gap resulting from businesses redesigning and streamlining production lines through intelligent sensor technology, motors, robotics, and increasingly automating processes with the adoption of Internet of Things (IoT) technology and computers. As the Manufacturing Institute stated, "With smart manufacturing or Industry 4.0, manufacturers are moving towards a new level of interconnected and intelligent manufacturing systems which incorporate the latest advances in sensors, robotics, big data, controllers, and machine learning. The greater digital interconnectedness between parts of the supply and production chains, as well as higher reliance on automation in these smart factories, is going to make manufacturing ultra-efficient, ultra-sophisticated, and ultra-productive" [2].

As a result, companies have a heightened need for a comprehensive strategic model (i.e., TMA) to embrace advanced manufacturing and an advanced workforce to understand and apply complex science and engineering concepts along with analytical skills to address the I4.0 revolution of manufacturing. Because advanced manufacturing also includes workers' ability to analyze and manipulate big data, attaining TMA reinforces the strategic need to "create, control, and critique" to ensure business sustainability and profitable application of I4.0 technologies.

By 2020, it was predicted that IoT technology would be in 95% of electronics for new product designs. Already, smart factories have become a reality as the IoT connects any factory device to the internet, from industrial robots to 3D printers, enabling goods to be produced around the clock and managed remotely. Key components that allow Industry 4.0 to come to life are: Cyber-Physical Systems: electro-mechanical devices with connectivity and digital communication capabilities; IoT: the network of physical objects that use sensors to capture data through embedded connectivity to exchange it over the internet; Industrial Internet of Things (IIoT): the application of IoT to the manufacturing world; Internet of Services: refers to the usage and combination of IoT devices and applications to provide services to end users [3]. Augmented Reality (AR) technology allows technicians to watch real-time information from the work they are performing. Because manufacturing processes are becoming more autonomous, integrated, and intelligent, this leads to changes in requirements for the manufacturing workforce towards an increasing need in software, programming, and interdisciplinary skills [4].

As a global manufacturing industry growth example, according to Deloitte's 2020 Global Aerospace and Defense Industry Outlook [5], the global commercial space sector is likely to see steady investments in new and existing space technologies and services, with funding coming primarily from governments and venture capital. Additionally, demand for military equipment is expected to increase over the next 5 years as governments across the globe focus on military modernization, given increasing global security concerns. Global defense spending and investment is expected to grow at a compound annual growth rate of about 3% over the 2019–2023 period reaching US$2.1 trillion by 2023 [6].

It is important to recognize that I4.0 technologies are not limited to advanced manufacturing but also readily cross over financial services, logistics and transportation,

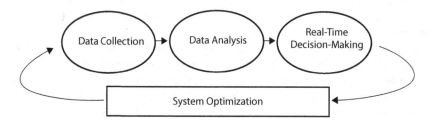

FIGURE 1.1   I4.0 manufacturing system intelligence.

and information technology. It is clear that advanced technologies, fundamentally identical, are used across all facets of the global economy. It really does not matter what industry or business sector one looks at, the result is the same in that advanced technologies are integrated into and present in everyday operations.

This indicates that not only must businesses be very strategic in implementing technology to achieve TMA but also workforce development must be driven by on-coming and next-generation I4.0 work issues and technologies that require enhanced skills and knowledge of topics such as mechatronics, composites, automated machining, AI, big data and analytics, IoT, virtual/augmented reality, advanced robotics, cybersecurity, supply chain visibility, block-chaining, additive manufacturing, and autonomous technologies. Broadly, the need is now for a talented and flexible workforce aligned with advanced manufacturing, information technology, logistics/transportation, and financial security to implement, use, and advance I4.0 technologies.

Where does society and manufacturing organizations, and businesses in general, go from here? There is little chance to escape the continuing advancements in technology, which drive advanced manufacturing. Computers play a major role in manufacturing today, and their role will broaden with advances in the development of AI. The use of AI is a field of computer science focusing on using computer technology to emulate the behavior of humans in solving problems. The use of AI magnifies the need for both a comprehensive data system and supporting data recording and feedback system. Figure 1.1 illustrates the conceptual framework for manufacturing system intelligence to attain ongoing optimization [3,7–13]. Three fundamental components for any AI-based application include data collection, data analysis, and real-time decision-making. Technology makes collecting data easy, probably too easy at times, but the trick is to be able to analyze and make decisions in a rapid manner. It still warrants a reminder that Henry Ford, for example, collected vast amounts of data using the technologies available in his day to continually improve the automobile assembly line. And, he likely would have collected more types of data if data collection technologies allowed. However, it is the strategic use of the "right" data, its analysis, and its use for decision-making that facilitates manufacturing success.

These AI-based systems draw upon their own store of knowledge and by requesting information specific to the problem at hand. Obviously, AI and the host of associated technologies will continue to increasingly become an integral part of manufacturing system design and implementation. Computer-integrated programmable logic controllers will even more effectively monitor and gather direct process information. Upon the identification of a problem, an expert system using AI will define the corrective action necessary to maintain process control. Ultimately, the expert system will make intelligent corrective action decisions on its own using the information it receives from the programmable logic controller. The expert decisions are then relayed back to the controller which in turn implements the decisions made.

The use of AI-based technologies will not only be limited to the manufacturing floor but also be applied to aid the making of corporate management decisions. Its use will continually enhance the soundness and correctness of management decisions made by looking at the same rules to make-a-decision as would a corporation's chief executive officer or president.

REFERENCES

1. Smith, G. & Dalzell, F. (2002). *Wisdom from the Robber Barons.* Edison, NJ: Castle Books.
2. Giffi.C, McNelly, J., Dollar, B., Carrick, G., Drew, M., Gangula, B. (2015). *The Skills Gap in US Manufacturing from 2015 and Beyond.* Deloitte Manufacturing Institute, Deloitte Development, LLC. www.themadeinamericamovement.com
3. Fourtane, S., (2020). *Top 10 Strategic Technology Trends.* Interesting Engineering. www.interestingengineering.com/Top-10-strategic-technology-trends-for-2020.
4. Haeffner, M. & Panuwatwanich, K. (2018). *Perceived Impacts of Industry 4.0 on Manufacturing Industry and Its Workforce: Case of Germany.* doi:10.1007/978-3-319-74123-9_21.
5. Lineberger, R. (2020). *Deloitte.* https://www2.deloitte.com/global/en/pages/manufacturing/articles/global-a-and-d-outlook
6. Lineberger, R., (2020). *Deloitte Estimates and Deloitte Analysis of Data from Stockholm International Peace Research Institute (SIPRI) Military Expenditure Database.* https://www. sipri.org/databases/milex, accessed October 1, 2019.
7. Boggess, M. (2019). *10 Trends That Will Dominate Manufacturing in 2019.* Hitachi Solutions.www.insights.csivendorinfo.com/view/content/QXwtV.
8. Wright, J. (2019). *Top 7 Manufacturing Trends for 2020.* www.advancedtech.com/blog/top-7-manufacturing-trends-for-2020.
9. Dowd, K. (2019). *5 Financial Services Tech Trends to Watch in 2020.* Business Intelligence, BizTech. www.biztechmagazine.com/article/2019/12/5-financial-services-tech-trends-watch-2020.
10. Marr, B. (2019). *The 7 Biggest Technology Trends to Disrupt Banking & Financial Services In 2020.* Enterprise Tech. www.forbes.com/sites/bernardmarr/2019/11/08
11. Insights Team. (2018). *The 4 Forces Transforming Logistics, Supply Chain and Transportation Today.* Forbes Insights.www.forbes.com/sites/insights-penske/2018/09/04
12. Transmetrics Blog. (2020). *Top 10 Logistics and Supply Chain Technology Trends in 2020.* Transmetrics.www.medium.com/@transmetrics.
13. Raiker, R. (2020). *Top Technology Trends for 2020.* www.towardsdatascience.com/top-10-technology-trends-for-2020-4a179fdd53b1.

2 Total Manufacturing Assurance

Total Manufacturing Assurance (TMA) is the ability to ensure that products are manufactured in a manner that enhances reliability, safety, and quality. TMA is the necessary objective of corporations seeking the benefits of market share leadership and high, long-term profitability. Attaining TMA is a function of three integral elements: (1) strategically planning manufacturing management activities, (2) controlling the manufacturing system, and (3) monitoring the manufacturing system for continual improvement.

These three elements in themselves sound pretty simple. However, TMA requires knowledgeable foresight on the part of corporate management in order to effect persistent and comprehensive efforts in planning, implementing, and monitoring manufacturing directives.

There are many well-founded management concepts documented such as Total Quality Control, Total Quality Management, Total Preventive Maintenance, Juran's Quality Trilogy, Deming's 14 Points For Management To Improve Productivity, Crosby's 14-Step Quality Improvement Program, Thriving On Chaos, Just-In-Time, Taguchi Experimental Design, Six Sigma Quality Belts, Lean Manufacturing, and 5S, to name a few. Each has their merits and each has undoubtedly numerous application success stories. It is terrific to see specialized innovative ideas created and applied.

TMA requires looking at the whole organizational picture, not just one segment of activity such as quality control. All corporate functions and activities support the ultimate objective of profitable manufacturing and product commercialization. This is what distinguishes TMA from other management concepts. Figure 2.1 illustrates the elements, and major activities therein, of TMA.

Notice that each of these elements is sequential and progressive. The first element (Strategic Manufacturing Management) feeds into the second element (Manufacturing System Control) which feeds into the third element (System Improvement Monitoring). For TMA, these activities must be close-looped. That is, the third element feeds back into elements one and two. This enables continual improvement in the TMA process.

Together, these major elements form a network to fuel the TMA process. TMA process; think about these two words for a moment. TMA is a process, not a program. Programs are generally activities that are planned to end upon completing a defined objective. Processes are dynamic entities that are forever ongoing and updated to reflect current innovative methods and concepts.

Again, it is emphasized that TMA is a process. If the organizational goal is to attain TMA and then the effort that made it possible is reduced or abandoned, you can bet your last dollar that TMA is not going to be maintained for very long. Every gain is lost and every market competitor is smiling from ear to ear.

DOI: 10.1201/9781003208051-3

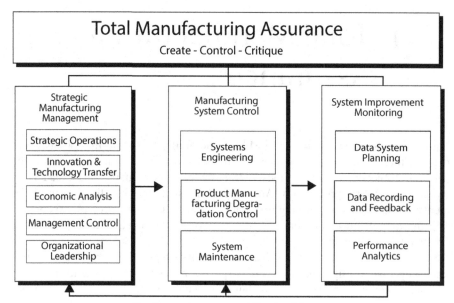

FIGURE 2.1 TMA network.

Obviously, it is not good news if market competitors are smiling. The way to make them frown is through defining and implementing, and subsequently revising as necessary, the corporate management and manufacturing objectives for launching a product into successful commercialization. This requires integrating and coordinating all of the groups (and their functions) in the company. This includes all levels of management, engineering, marketing, and production personnel. Each group works with the others and builds on the incremental corporate strides towards TMA.

The bottom line is that the TMA process leads to a shifting of many manufacturing assurance issues upfront into lower cost, lower-risk directives. This sets the stage for more effective and well-focused product manufacturing, subsequently leading to a reduction in organizational waste and productivity losses.

This enables us to define the three Cs of TMA:

1. **Create Vision**: Only those organizations having the ability to identify future opportunities are likely to experience successful long-term survival. Know where the organization is to be in the future….visualize it and strategically plan for it.
2. **Control the Manufacturing System**: Detour any possibility of being at the mercy of, and taking, whatever the manufacturing system provides. Always make the system provide high reliability, safety, and quality. The tactical strategies implemented are to attain the strategic goals and objectives defined in the strategic plan.
3. **Critique Performance**: Know what was done yesterday and compare it to what is happening today, and what is required tomorrow. Any strategic

plan is useless if results are not measured…all goals, objectives, and tactical strategies must have performance metrics defined. The results, both good and bad, are what keep the organization flexible and adaptable to global obstacles, challenges, and opportunities.

In moving the organization forward strategically and realizing desired results, the ability of the organization to adhere to the discipline of *create, control, and critique* is not without its challenges. But, it is much easier and productive to *create, control, and critique* with all organizational team members understanding their role as stakeholders in the organization's success.

The remainder of this chapter broadly defines each element of the TMA process and the specific activities required therein. Keep in mind that the overall management of these elements must emphasize the consolidation of the TMA activities. Consolidation meaning establishing a "team-oriented" relationship between all corporate groups. Isolation of activities and groups is not a part of the TMA process. Remember, consolidation, not isolation.

2.1 STRATEGIC MANUFACTURING MANAGEMENT

Strategic manufacturing management consists of five key activities. These include strategic operations, innovation and technology transfer, engineering economic analysis, management control, and organizational leadership, which fall under the umbrella of planning and control.

Broadly, manufacturing planning involves establishing the strategic direction of the organization in the near and long term. By examining internal and external environmental conditions current and projected, a corporate-wide strategy is defined and implemented. Obviously, attaining and maintaining TMA are an integral consideration in the manufacturing, as well as business, strategic segments of the overall corporate strategy. However, in addition to defining the manufacturing strategy, the tactical implementation of this strategy must be given careful consideration both in terms of its impact from capital investment and in terms of corporate culture standpoints.

Additionally, management control addresses the means by which the organizational strides towards TMA are achieved. Numerous management techniques, tools, and concepts are available as aids and should be open-mindedly used. If for some reason a management control technique fails to work, it should be replaced or revised. The hardest task is to change bad management habits and honestly make an effort to improve worker productivity and efficiency.

2.2 MANUFACTURING SYSTEM CONTROL

Manufacturing system control consists of three key activities. These are systems engineering, product degradation control, and system maintenance.

System engineering addresses implementation of a rational means for operating the manufacturing system(s). Efficient and effective manufacturing systems are not an accident. Through proper planning and evaluation of alternatives, it is possible to put in place the optimal system. The optimal system being the design which maximizes manufacturing assurance.

Product degradation control addresses the product as it is manufactured. A product is designed to provide various performance capabilities and levels. However, the manufactured product often exhibits performance levels lower than that designed-in, particularly in regard to reliability, safety, and quality. Care must be taken to ensure that the manufacturing system is designed to minimize the degradation which will occur.

System maintenance is a critical part of the manufacturing system. Not only must system availability be maximized, but if not properly maintained, product degradation is accelerated. The maintenance program defined for a manufacturing system must be considered as important as the product and system design themselves.

2.3   SYSTEM IMPROVEMENT MONITORING

System improvement monitoring consists of three key activities. These are data system planning, data recording and feedback, and performance analytics.

Data system planning addresses the breadth of data and information required to assess the current state of TMA. Common issues arise such as what data should be collected and how to mechanically collect it. Ideally, all information sources automatically feed into a central data system. Numerous technologies exist to make this happen. An important ground rule to keep in mind is that one should only collect that data and information which is practically useful, evaluated, and reacted to.

Data recording and feedback provides the means for collecting and evaluating pertinent data and information. This is achieved by having a closed-loop system in place which records data, looks for trends, and follows through with a proper response to highlighted issues. This is particularly important in regard to internal and external identified problems which endanger market performance. Such problems require timely evaluation and corrective action, as appropriate.

Performance analytics are the measures that the organization looks at to gage its performance against that defined and expected in the overall strategic plan. The vast amounts of data generated and capturable, particularly in light of Industry 4.0 initiatives, warrant a focused approach to looking at strategically "relevant" data.

2.4   SUMMARY

It is easy to develop an understanding of what TMA is all about. It focuses on the product and the corresponding manufacturing system(s). This focus begins with strategic manufacturing management, manufacturing system control, and system improvement monitoring.

All organizations in the corporation must participate in this focus and be committed to achieving TMA. TMA cannot be attained by going through the motions and pretending things are better. Nobody is fooled.

Remember the three Cs of the TMA process. *Create* a vision for the future. *Control* product design and the manufacturing system(s) to deliver what the customers expects. Finally, *critique* organizational performance in meeting and exceeding strategic requirements.

Section II

Strategic Manufacturing Management

THERE ARE IN BUSINESS THREE THINGS NECESSARY: KNOWLEDGE, TEMPER, AND TIME.

—FELTHAM

DOI: 10.1201/9781003208051-4

3 Strategic Operations

A fundamental task in any business is planning. Without a disciplined planning effort, it is impossible for an organization to effectively and efficiently respond to business opportunities and challenges. The result is knee-jerk reactions to the day-to-day dynamics of our global economies. Clearly, what is necessary is a strategic business plan which enables a corporation to direct itself towards maximizing return on investment and profitability.

A strategic business plan is a compilation of direction for organizational components. These generally include manufacturing, finance, management, marketing, and research and development. Each requires its own strategic plan, which is aligned and visibly linked to the strategic objectives of the organization as a whole.

This chapter addresses the manufacturing strategic component of the overall organizational strategy. Accordingly, the focus is on defining and implementing a TMA-based strategy to advance and control product reliability, safety, and quality. The intent is to eliminate product performance losses during manufacture, gain customer confidence and preference, capture market share, and enhance organizational profitability over the long term.

3.1 STRATEGIC PLANNING

Strategic planning involves several fundamental tasks: strategy formulation, strategy implementation, and evaluation and control [1]. Strategy formulation is the development of long-range plans to deal effectively with environmental opportunities and threats in light of organizational strengths and weaknesses. This consists of defining the organizational mission, specifying achievable objectives, developing strategies, and setting policy guidelines.

Strategy implementation is the process of putting strategies and policies into action through the development of various programs, budgets, and procedures. This is often referred to as operational planning, since it is concerned with day-to-day resource allocation problems. Finally, evaluation and control are the process of monitoring organizational strategic activities and results. From this, an assessment is made of actual performance relative to desired performance.

Strategic manufacturing planning is understandably a necessary and complex activity. A strategy has to be formulated, implemented, and then monitored to see if it works to a satisfactory level. But before a company can perform the latter two activities, it is necessary to have in place a comprehensive and complete strategic manufacturing plan.

This reference to a *strategic manufacturing plan* implies several key ideas. First, an organizational-wide strategy is defined based on input from some of the best innovative, visionary, and pragmatic minds in the company. Second, the focus is on manufacturing, and very likely some specific product line(s). Third, a detailed

DOI: 10.1201/9781003208051-5

FIGURE 3.1   Inputs to manufacturing strategy.

plan is developed reflecting realistic goals, objectives, timing milestones, and cost estimates.

The functional elements of a strategic manufacturing plan are depicted in Figure 3.1. Two major inputs are illustrated which impact the planning process. These are the business and tactical strategic considerations.

A third input is also depicted. This input, which is often overlooked, is the product(s) itself. Aside from traditional performance parameters addressed in design, the attributes of reliability, safety, and quality are critical attributes that can make or break the ability to successfully gain customer preference and market share.

It is acceptable that rational manufacturing decisions are made based on the defined strategic issues surrounding the corporation as a whole and, particularly, the product(s) to be manufactured. Several key issues include market share potential, probability of success, and return on investment (ROI).

The market share potential for a given product varies with the "growth" characteristics of the specific market to be entered (assuming successful product technology transfer occurs). Market share potential directly affects the probability of commercial success and the ROI that may be seen. Figure 3.2 depicts the relationship between ROI and probability of commercial success as a function of market share potential.

These strategic issues are a direct reflection of the consumers' impression of product reliability, safety, and quality. It therefore logically makes sense to develop and implement a strategic manufacturing plan which enhances product commercialization success potential and ROI.

From identification of the strategic issues, baseline product reliability, safety, and quality requirements are determined. The significance of these product attributes is evidenced in the extensive Profit Impact of Market Strategy (PIMS) database [2]. The PIMS data implies that the definition of product reliability, safety, and quality requirements is strongly dictated by market issues. This is extended to emphasize their driving positive influence in achieving high market share, ROI, and successful commercialization.

A key influence in structuring the manufacturing plan is the characteristic nature of the market intended to be entered. As shown in Figure 3.2, whether market pull or market push exists strongly affects the required product reliability, safety, and quality attributes. Market pull indicates the situation where the product is highly desired by the consumer and it will be readily received in filling a market void. Market push

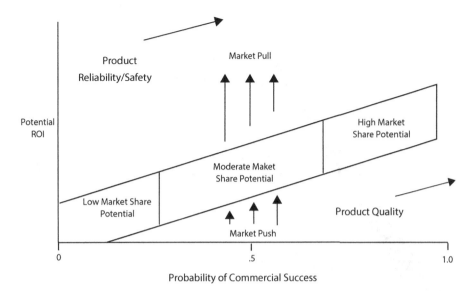

FIGURE 3.2   Product ROI/success relationship.

indicates the situation where the product is not highly desired by the consumer and, consequently, it must be made desirable to the consumer.

It was stated earlier that for a product to be successful it must be designed with the proper levels of reliability, safety, and quality. Accordingly, the manufacturing strategy must reflect, as one of its primary objectives, the need to control and maintain these inherent product performance attributes. The customer should get the product as designed and not some degraded version of the original design.

A sound framework for adequately addressing manufacturing strategy issues is the strategic market management approach [3]. Strategic market management is designed to help precipitate and make strategic decisions. A strategic decision involves the creation, change, or retention of a strategy. An important role is to precipitate, as well as make strategic decisions. Figure 3.3 shows the pyramid nature of the analyses that provide the input to strategy development and the strategic decisions that are the ultimate output.

As illustrated in the figure, there are two fundamental analysis areas aiding strategic formulation. These are external analysis and internal analysis.

External analysis consists of examining variables existing outside of the organization which typically are not within the short-term control of management. Specific analyses performed include (1) customer analysis to identify customer segments, motivations, and needs; (2) competitive analysis to identify other company strategies, objectives, cultures, costs, strengths, and weaknesses; (3) industry analysis to identify current size, potential growth, available distribution systems, entry barriers, and success factors; and (4) environmental analysis to identify factors of existing technologies, government controls, cultures, demographics, and economics.

Internal analysis consists of examining variables within the organization which, also, are not typically within the short-term control of management. Specific analyses

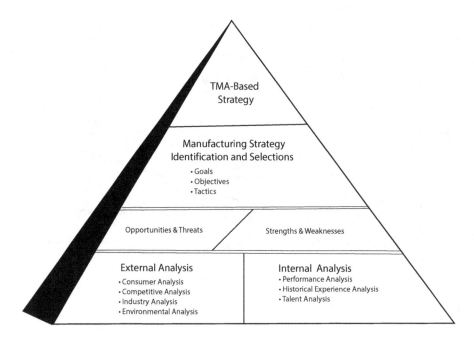

FIGURE 3.3   Strategic planning pyramid.

include (1) performance analysis to identify current and projected return on assets (ROA), market share, product value and performance, cost, new product activity, corporate culture, productivity, and product portfolios; and (2) historical experience analysis to identify past and current strategies, problem areas, organizational capabilities and constraints, financial resources available, and corporate flexibility.

Upon defining a strategic manufacturing plan, it should be evaluated relative to several issues. These include (1) internal consistency, (2) consistency with the environment, (3) capabilities of available resources, (4) acceptable level of risk, (5) appropriate time horizon, and (6) overall feasibility. Internal consistency refers to the cumulative impact of a strategy on organizational objectives. Consistency with the environment refers to strategy compatibility with what is going on outside the organization.

Capabilities of available resources refer to what an organization has to help it achieve its objectives. Balance must be achieved between strategies and resources. The acceptable level of risk refers to the uncertainty of strategies. Different strategies involve different risks (both internal and external) and management must establish its risk preferences. As time horizons increase, risk increases, particularly in unstable markets. Appropriate time horizons refer to the time period over which the strategies are to be followed. The overall feasibility refers to the workability of strategies on a practical level.

For a manufacturing strategic plan to be successful, it must be compatible with all the strategic components involved in the overall organization and with the particular product strategy. Typically, there are four widely recognized product strategies:

(1) market leader, (2) follow the leader, (3) applications oriented, and (4) production efficient [4].

1. **Market Leader Strategy**: An intense product R&D and engineering effort is implemented. Technical leadership is exerted through large financial investment(s). It involves a high degree of risk-taking, and usually only large organizations with substantial resources are able to adopt this posture. This is necessary since one must be able to absorb financial mistakes.
2. **Follow-the-Leader Strategy**: An effort involving minor product research but a strong development and engineering effort. It requires a rapid technical response to new products developed by market leaders. Manufacturing and marketing expertise are also vital to success. The market is generally entered with a competing product during the growth stage of the product life cycle.
3. **Applications-Oriented Strategy**: An effort involving minor product R&D but a strong engineering effort. The focus is on developing product modifications to serve a particular, specialized, or limited market segment. The market is generally entered during the maturity stage of the product life cycle.
4. **Production-Efficient Strategy**: An effort involving little product R&D and engineering effort but a strong manufacturing effort. It is based on achieving superior manufacturing efficiency and cost control. Competition in price and delivery is paramount. The market is generally entered during the maturity stage of the product life cycle.

A preference for manufacturing strategies with greater flexibility is desirable. Strategies that permit corrective action and adjustment are better than those that cannot be changed once they are implemented.

3.2 STRATEGIC PLANNING PATH

A fundamentally desired outcome of any strategic plan is to reduce "uncertainty" among technology development, financial and economic viability, and public acceptance. The development of a strategic business plan serves as a roadmap to quickly and cost-effectively get a desired outcome that provides a targeted ROI. That outcome is the result of collaboration of all stakeholders (industry, government, general public) and it starts with a "vision."

However, vision alone is not enough. All organizations (e.g., for profit, nonprofit, and governmental) quickly find that effective strategic planning is necessary to be a TMA-based provider of innovative, science-based systems engineering solutions that are developed in a way that inspires public/customer confidence (i.e., from technological, safety, and financial perspectives) and market demand. The path forward becomes one of well-developed strategy to effectively reduce technology development process uncertainty and inspire funding "customer" confidence through successful technology demonstration in the marketplace [5].

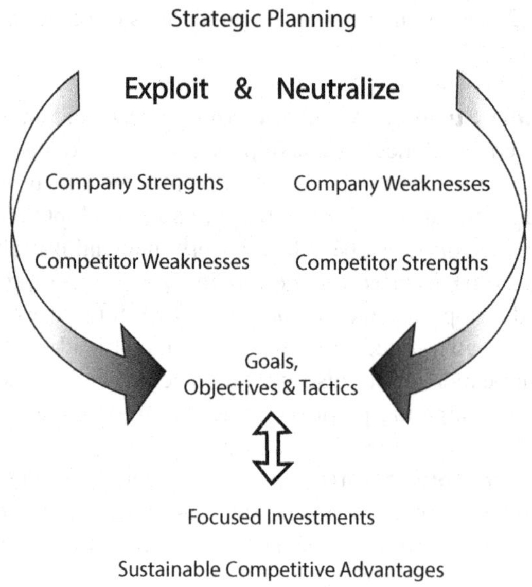

Strategic Planning

Exploit & Neutralize

Company Strengths   Company Weaknesses

Competitor Weaknesses   Competitor Strengths

Goals,
Objectives & Tactics

Focused Investments

Sustainable Competitive Advantages

FIGURE 3.4   Strategic planning fundamentals.

As depicted in Figure 3.4, strategic planning is fundamentally about exploiting strengths and competitors' weaknesses and neutralizing weaknesses and competitor's strengths. The GSOT approach (i.e., goals, strategies, objectives, and tactics) [6] becomes an integral part of creating a strategic plan that is meaningful and usable. This approach defines:

- A goal is a broad primary outcome
- A strategy is the approach taken to achieve a goal
- An objective is a measurable step taken to achieve a strategy
- A tactic is a tool or method used in pursuing an objective associated with a strategy

But, arguably the most beneficial part of strategic planning comes from performing a SWOT analysis (i.e., strengths, weaknesses, opportunities, and threats) that addresses the organization and its competitors. A competitor may be another company, but it may be another technology. Ultimately, this leads to implementing tactical actions, with accompanying performance metrics, that drive the decisions leading to sustainable advantages (e.g., financial, technological, and organizational). It is through the SWOT that one is able to exploit the organization's strengths and the competitor's weaknesses... and neutralize the organization's weaknesses and the competitor's strengths. Formal strategic planning is an integral element for engaging all stakeholder groups (i.e., team member, company, investor, and governmental) having a role in advancing the organization regionally, nationally, and globally.

Vision, mission, and strategy are all concepts that elicit a diverse set of emotions from people. For some it generates excitement, optimism, potential, for others it creates anxiety, confusion, apathy, and disillusionment. Developing a strategy is a journey that requires time, resources, awareness, and the ability to look philosophically at yourself as an organization and understand your purpose.

Beginning the strategic planning pathway is a strong, clear, concise, and accepted vision statement. While "mission" reflects what the organization does, the "vision" reflects where the organization wants to go. It is the vision statement that sets the tone and the direction of the strategic plan. The vision statement should be aspirational and provide identity and purpose. Without it, any GSOT are weakened and will do little to advance an organization. It is out of that vision that goals, strategies, objectives, tactics, and performance metrics are derived [7].

Of course, any strategic plan is useless if it is not implemented. That is, a project plan with tasks, timeline, milestones, and assigned responsibility needs to accompany the strategic plan. With disciplined implementation, the probability of realizing the envisioned opportunity and attaining the desired financial ROI is low.

The pathway steps include [8]:

Step 1: Define the Business Vision...The "business vision" identifies where you are going and reflects sustainable competitive advantages. It is what you always go back to reorient yourself...if you begin to go astray it will always point you in the right direction. As the fundamental business vision changes, let people know what has changed.

Step 2: Define your business characteristics...This is the company's/business' fundamental characteristic(s) or fingerprint. It reflects what are, or may not be, sustainable competitive advantages...that is, the way you compete, basis of competition (assets, skills, capabilities), where you compete, who you compete against.

Step 3: Look outside your organization

Consumer analysis

> Customer segments...commercial, consumer, military, etc.
> Motivations...what makes them buy or not buy
> Needs...delivery, purchase options, product options, etc.

Global market analysis

> What markets
> Current size
> Potential growth
> Available distribution systems
> Market share barriers
> Critical success factors
> Opportunities
> Threats
> Requirements to "play"
> Requirements to "win"
> Current posture...are you withdrawing, milking, maintaining, or growing

Competitor analysis

> Who
> Current and past strategies

Business objectives
Business/organizational culture
Size, growth, and profitability...costs/revenues
Major overall strengths and weaknesses

Industry environment analysis
Market drivers/trends
Technologies...maturing and new
Manufacturing
Product
Government...regulations, tax policies, stability
Cultures...lifestyles, fashions, opinions
Demographics...age, income, family formations, location
Economics...who can afford to buy what, health of countries, health of
    industries, interest rates, currency valuation
Product outlets/distribution mechanisms
Projected future environment
Market challenges
Market opportunities
Potential new competitors

Step 4: Look inside your organization

Performance analysis
Current/projected ROA
Market share(s)
Product value/performance
Costs
New product activity
Corporate culture
Productivity
Product portfolios

Historical analysis
Past/current strategies
Problem areas
Organizational capabilities/constraints
Financial resources available
Company flexibility

Step 5: Define your strengths and weaknesses...Look at the following areas
and others as appropriate.

Innovation
Customer services
Product characteristics and capabilities
R&D output
Technology use
Idea patentability/patent activity

Manufacturing
Cost structure
Production operations flexibility

      Equipment
      Raw material accessibility
      Vertical integration
      Workforce attitude and motivation
      Workforce availability/skills
      Capacity
      Product quality
Finance
      Profitability of operations
      Available assets
      Debt
      Investment back into business
Management
      Quality of workforce
      Knowledge of business
      Culture
      Strategic objectives, goals, and plans
      Entrepreneurial spirit
      Business planning/operation system
      Employee loyalty (turnover)
      Quality of decision-making
      Implementation of ideas and plans
Marketing/Sales
      Product reputation (cost, delivery, lead time, quality, reliability)
      Product performance characteristics/differentiation
      Brand name recognition
      Breadth of product line (total solution/system)
      Customer orientation
      Market focus
      Distribution mechanisms
      Retailer relationship
      Advertising/promotion
      Sales force
      Customer service/product support
Customer base
      Size and loyalty
      Market share(s)
      Growth of markets served
      Global presence

Step 6: Define strategic objectives...use the information developed from Steps 1–4 above to define SOs directed to get, maintain, or enhance sustainable competitive advantages that are meaningful.

Step 7: Define tactical strategies (typically major programs)...identify the tactical strategies that will achieve the SOs. Identify investments required in terms of capacity, cost, time, equipment, space, headcount, etc., to achieve the SOs.

Step 8: Define operational strategies (typically a day-to-day major activity)

Step 9: Define a performance measure(s) for each operational strategy... identify what metrics will be used to assess the progress and/or effectiveness of each operational strategy.

Step 10: Do a reality check...does it, as defined, really make sense?...is practical?...is achievable with acceptable risk?

Internal consistency
   Cumulative impact on the organization?
Environmental consistency
   Compatibility with rest of the world?
Capability of available or planned resources
   Investments needed?
Acceptable level of risk
   Success probability?
Time horizon for completion
   Timeframes to implement and complete?
Overall feasibility
   Workability in current organization?
Interdependencies/functional attachments
   Keeping intact key functional relationships...physically and/or in communication?

Step 11: Implement the strategic plan...Initiate ongoing status monitoring for all strategies...re-look/re-think any above issues as necessary to maximize success.

3.3 PRODUCT ASSURANCE STRATEGY

3.3.1 RELIABILITY, SAFETY, AND QUALITY REQUIREMENTS

The product reliability, safety, and quality requirements are a key part of the overall manufacturing plan. As stated earlier, this is supported by the PIMS database. These performance parameters provide a basis for defining the product assurance strategy necessary as part of the overall manufacturing strategy.

Developing a product assurance strategic plan involves balancing many interrelated variables and factors. These include:

1. Establishing quantitative product performance goals (mean time between failure, mean time to repair, service life, defectivity level, etc.) based on customer requirements.
2. Defining and implementing an effective management control program to guide the product development effort and provide timely outputs consistent with major design and program decision points.
3. Performing and implementing analyses and quality checks, audits, and controls to ensure that all tests, inspection, and screen data are complete and acceptable.
4. Performing a fully coordinated product development program which emphasizes failure analysis and corrective action, and provides growth/verification of specified design requirements.

5. Establishing a set of documentation through which conformance to the product assurance strategy is assessed and tracked.
6. Exposing design deficiencies and initiating corrective action in a timely manner.
7. Improving product operational availability to enhance the probability of commercial success.
8. Reducing the need for maintenance and logistic support (reduce life cycle costs).
9. Identifying, evaluating, and eliminating design hazards and risks.
10. Applying historical data, including lessons learned from other development projects, into the design process.
11. Minimizing unforeseen impacts on overall project cost and schedule.

The product assurance strategy provides an organized method of assuring that the above factors (and others) are adequately considered during a product development program. With such a strategy developed and documented, it is possible to initiate a properly planned program. That is, one that ensures critical project decisions involving significant investment of resources are keyed to the achievement of specific reliability, safety, and quality requirements, as well as other performance requirements.

The specific program objectives are dependent on the uniqueness of the technology under consideration, the nature of the product, the customer, and other factors as appropriate. Merely defining a product assurance strategy orients product development, and ultimately manufacture, towards a practical, serviceable, and affordable commercial product. Additionally, it is emphasized that ALL organizational components/groups are subject to strategic direction. This includes not only product development and manufacturing groups but also the service arms of the organization (e.g., customer service, employee training).

A product assurance strategy itself promotes and enables trade-off analyses between design and test engineering functions. Once implemented, it avoids duplication of effort and minimizes the probability of omitting essential elements. Always remember that there is an intimate relationship between the total product assurance effort and the TMA-based strategic plan.

TMA Case Study: Service Organization Strategic Plan

OVERVIEW

Products come in many forms. That is, services are as much a product as a piece of hardware. The fundamentals towards strategic planning do not change, but sources of data, benchmarks, and performance measures will have a slightly different orientation for every organizational application (e.g., product design, customer service, global expansion, or employee engagement). For this case study, the organization is developing and implementing a structured strategic business plan for delivering a unique museum experience for multiple generations of visitors. Missing has been a plan that the Board of Directors and staff follow to move the organization to financial security and long-term sustainability in the communities served.

ISSUE

The organization has not historically engaged in formal strategic planning. Additionally, there has not been any organizationally adopted performance measures to ensure accountability for delivering a product, which is the personal museum experience shared with children, parents, and grandparents.

STRATEGIC OBJECTIVE

The strategic objective is to develop the organizational strategic plan that provides foundational direction to guide the organization, overseen by a Board of Directors, through the next 3 years.

CASE BACKGROUND

A strategic plan was developed as follows:

Vision: Establish the Center as a premier science education museum. The Center will implement relevant science-based programs and exhibit gaining local and regional visibility, which attracts families resulting in (1) new enrollment/membership streams, (2) new programmatic/event venues, (3) proactive community support/engagement, and (4) financial sustainability.

MAJOR MILESTONES

A. Formalize operational policies and procedures
B. Initiate new/enhanced STEM K-9–12 student experiences
C. Establish new/enhanced local and regional partnerships
D. Pursue new revenue sources via grant submittals and foundations requests
E. Engage in aggressive center branding/marketing activity
F. Establish state-of-the-art facility
G. Become an accredited children's museum

SWOT (quick summary of strengths, weaknesses, opportunities, and threats)

What does it take to play (with your competition?)	What does it take to win (be recognized premier source?)
• Be known in the community as a valued educational resource. • Establish key partnerships (governmental and private) that bring resources to the Center (e.g., financial, in-kind, intellectual).	• Have differentiated programming/outreach. • Be perceived as a world class professional museum (facilities, programs, personnel).
Strengths: • Own facility. • Educational coordinator and engagement/events coordinator in place.	**Weaknesses:** • Limited local visibility as a differentiated educational resource. • Facility space and maintenance issues.

Opportunities:	Threats
• Exploit relationships with other local and regional educational and governmental partners. • Establish new educational partnerships with the regional university, community college, and K-12 school districts. • Exploit theater space event potential. • Exploit facility rental potential.	• Local programming from local educational institutions, park district, and private organization STEM programming. • Facility location and accessibility. • Facility presentation/condition.

Strategic Goals: (1) Maintain a Midwest recognized top tier science museum by 2020, (2) provide uniquely differentiated educational experiences by 2019, (3) be self-sustaining with significant financial reserves by 2020.

STRATEGIC OBJECTIVES

A. Ongoing sustainability
B. Advanced museum operations *(administrative development)*
C. Differentiated student experiential opportunities *(education & exhibits)*
D. Desired community engagement and outreach
E. Effective marketing and visibility

TACTICAL STRATEGIES

A. Sustainability
  a. Increase facility usage
    i. Increase family memberships
    ii. Increase student/family participation
    iii. Increase the onsite museum events
      1. K-12
      2. Teacher recertification
        i. Increase public access hours
        ii. Increase space rentals
  b. Optimize facility operations
    i. Establish world class museum facility
      1. Evaluate long-term building options
        a. Remain in current building
        b. Sell building and move to new location
        c. Lease/buy decision
        d. Move to new location and keep current building (for rentals)
        e. Build new museum
      2. Address missing museum amenities
        a. Food and drink
        b. American Disabilities Act (ADA)
    ii. Obtain exhibit corporate sponsorships
  c. Optimize personnel operations
    i. Increase staffing
    ii. Increase operational hours
    iii. Evaluate compensation baseline

      d. Attain grants

      e. Attain foundation contributions/endowments

B. Museum operations

      a. Implement clear public perception of organizational professionalism

      b. Develop administrative operations policy/procedure manual

      c. Develop Board of Directors in "Board role"

C. Student experiential opportunities

      a. Implement interactive, relevant, and technology-based exhibits

         i. Reflect diversity in museum user demographics in programmatic activities and exhibits

      a. Implement STEM activities with the integration arts and humanities

      b. Conduct educational exchanges with partners

      c. Implement after-school programs

D. Community engagement and outreach

      a. Establish partnerships with local and regional schools, governments, and communities

      b. Provide teacher recertification seminars and events

      c. Provide presentations for local community groups

      d. Conduct formal supporter recognition event(s)

         i. Member, business, volunteer, foundation

E. Marketing and visibility

      a. Brand the museum

         i. Keep Center mascot

      b. Implement and showcase clean exciting exhibits

      c. Implement consistent signage

      d. Become an accredited museum

      e. Participate in media outlets (television, radio, social platforms)

      f. Maintain/update website

PERFORMANCE METRICS

A. Enrollment/activities

      a. # students for programs/events

      b. # family walk-ins

      c. # outreach events

B. Quality

      a. # complaints

      b. Annual customer satisfaction survey

      c. Monthly exhibit appearance, safety relevancy assessment

C. Financial

      d. # family memberships

      e. Program $ revenue

      f. Program $ expenses

      g. $ Donations

      h. $ grants

RESULT/CONCLUSION

The following performance outcomes were derived from year one implementation of the strategic plan.

Tactical Strategies	Outcomes
Ongoing sustainability	
a. Increase facility usage	
i. Increase family memberships	(1) 2017 Annual Report, (2) 2018 Annual Report, (3) Charting for Center BOD in 2019
ii. Increase student/family participation	(1) 2017 Annual Report, (2) 2018 Annual Report, (3) Charting for Center BOD in 2019
iii. Increase the onsite museum events	(1) 2017 Annual Report, (2) 2018 Annual Report, (3) Charting for Center BOD in 2019
iv. Increase public access hours	No action
v. Increase space rentals	(1) 2017 Annual Report, (2) 2018 Annual Report, (3) Charting for Center BOD in 2019
b. Optimize facility operations	
i. Establish world class museum facility	(1) Following 2018–2021 BOD-approved strategic plan
ii. Obtain exhibit corporate sponsorships	(1) UI, (2) Phillips 66, (3) K Center, (4) B Bank
c. Optimize personnel operations	
i. Increase staffing	(1) Added three museum assistants
ii. Increase operational hours	No action
iii. Compensation review	No action
d. Attain grants	(1) IMA, (2) MSigma
e. Attain foundation contributions/ endowments	(1) General public contributions, (2) R Foundation, (3) B Bank
Advance Center Operations	
f. Implement clear public perception of organizational professionalism	(1) Participation in Chamber, CCP, City downtown plaza advisory committee, (2) Annual Reports for 2018 and 2019
g. Develop administrative operations policy/procedure manual	(1) Board Policy Manual in 2017
h. Develop Board of Directors in "Board role"	(1) Update Org Bylaws in 2018, (2) adhere to Bylaws in 2017, (3) approve annual budget for 2018 and 2019
Differentiated Student Experiential Opportunities	
i. Implement interactive, relevant, and technology-based exhibits	(1) UI engineering collaboration to accelerate new exhibit introduction
i. Reflect diversity in museum user demographics in programmatic activities and exhibits	(1) Emphasized serving the diverse "urban" population, (2) revised mission statement to reflect "diverse children of all ages"

j. Implement STEM activities with the integration of arts and humanities	(1) punkSTEM girls in 2018 (2) KArt Wall in 2018
k. Conduct educational exchanges with partners	(1) Corn Festival in 2018, (2) Illinois Marathon in 2018
l. Implement after-school programs	(1) Fall clubs, (2) summer camps, (3) K-12 collaboration
Desired Community Engagement and Outreach	
m. Establish partnerships with local and regional schools, governments, and communities (e.g., UI and City)	(1) UI College of Engineering in 2018, (2) UI College of Business in 2018, (3) K Center in 2018, (4) City in 2018
n. Provide teacher recertification seminars and events	No action
o. Provide presentations for local community groups	(1) Library
p. Conduct formal supporter recognition event(s)	
i. Member, business, volunteer, foundation	(1) "founder" thank you event in 2018, (2) donor wall in 2018, (3) 2017 Annual Report, (4) 2018 Annual Report
Effective Marketing and Visibility	
q. Brand the museum	
i. Keep mascot	(1) Marketing emphasized Center mascot in 2018
r. Implement and showcase clean exciting exhibits	(1) Astronomy exhibit spring 2018, (2) magnet wall in 2018, (3) exhibit Master Plan created in 2018
s. Implement consistent signage	(1) Signage on building windows in 2018
t. Become an accredited museum	No action
i. American Alliance of Museums	
u. Participate in media outlets (television, radio, social platforms)	(1) TV promotion in 2018 and 2019, (2) CM in 2019, (3) SP in 2019, (3) 94.5 in 2018
v. Maintain/update website	(1) Website redesign in 2018

Case Question: What changes are necessary to advance the Board of Directors' understanding, engagement, and use of the strategic plan?

3.4 SUMMARY

This chapter addressed strategic manufacturing planning to achieve TMA, as it evolves out the overall organizational strategic plan. Strategy planning plays a vital role in defining where an organization wants to be in the near and long term, and how they will get there.

A strategic plan consists of goals, objectives, tactics, and performance measures. The strategic planning process involves assessing the internal and external worlds impacting the organization. This is accomplished via analyses addressing areas such as the environment, competition, investment, potential, distribution, and customer base. Most importantly, the process takes the organization into an honest assessment of its SWOT. Fundamentally, the directive is for an organization to exploit its strengths and its competitors' weaknesses, and minimize its weaknesses and its competitors' strengths.

A key consideration in strategy development is the product itself. The product must be able to fill a market void and differentiate itself from the competition. Performance parameters key in this regard are reliability, safety, and quality. It therefore makes sense to develop a strategic manufacturing plan which optimizes these parameters to increase the probability of successful commercialization.

Tactical strategies address defining the methods for guiding, supporting, and performing the actual manufacturing activity. Broad tactical strategy components are discussed in Chapter 6 and include a variety of topics such as design for manufacture, computer-integrated manufacturing, material requirements planning, and Just-In-Time tools.

QUESTIONS

1. Identify a broad range of topics that could be addressed in a business strategic plan.
2. Business strategic plans typically begin with a mission statement. Develop a concise, explicit mission statement for your area of concern.
3. Identify the external and internal issues impacting your corporation or current activity.
4. Discuss the significance that environmental issues might have for some corporations in strategy formulation.
5. Discuss the differences between tactical and strategic planning.
6. What is the difference between vision and mission?
7. How does project management interface with the strategic plan?
8. Why is it important to understand "what it takes to play" and "what it takes to win?"
9. Who is responsible for the strategic plan implementation?
10. Look at your specific job and do a SWOT analysis.

REFERENCES

1. Buzzel, R. & Gale, B. (1975). *Market Share-A Key to Profitability.* Brighton, MA: Harvard Business Review.
2. Hunger, J. & Wheeler, T. (1986). *Strategic Management and Business Policy.* Boston: Addison-Wesley Publishing Company.
3. Aaker, D. (1988). *Developing Business Strategies.* New York: Wiley and Sons.
4. Tersine, R. (1980). *Production/Operations Management-Concepts, Structure, and Analysis.* New York: North Holland.
5. Beck, D. F. (2013). *Technology Development Life Cycle Processes.* Albuquerque, NM: Sandia National Laboratories. SAND2013-3933.
6. Belicove, M. (2016). *Understanding Goals, Strategy, Objectives, and Tactics in the Age of Social.* Forbes. https://www.forbes.com/sites/mikalbelicove/2013/09/27/understanding-goals-strategies-objectives-and-tactics-in-the-age-of-social/?sh=a624c984c796.

7. Savar, A. (2013). *Content to Commerce: Engaging Consumers Across Paid, Owned, and Earned Channels*. New York: Wiley.

8. Brauer, D., Lee, R. P. & Brauer, C. (2018). CarbonSAFE: Business Development Case Study; "Commercialization of Emerging Environmental Sustainability Technologies: CCUS Strategic Business Development." *Report prepared for U.S. Department of Energy and University of Illinois at Urbana-Champaign, Division of Illinois State Geological Survey.*

4 Innovation and Technology Transfer

Innovation is an outcome of the need to accomplish something, develop something, or, simply, survive. Generally, large organizations engage in broad innovation but typically not in the greatest depth. Smaller organizations engage in much more focused, limited innovation, but at a greater depth. The organizational resources available for research and development (R&D), and market outreach, typically dictate the innovation capabilities and spectrum of innovation that are possible.

The decision to invest in emerging technologies is generally considered to be associated with considerable investment risks, both financial and technological. In addition to dealing with multiple technological and cost-related challenges, organizations may also have to consider the issue of public acceptance and societal uptake to ensure the successful implementation of such technologies [1,2].

Insights obtained support the development of a holistic general commercialization framework to create a strategic vision and business plan for existing and, particularly, emerging technologies, including those that may have significant social impacts.

4.1 TECHNOLOGY DEVELOPMENT

Managing the overall technology development process often includes phase transitions taking on a particular focus, such as stakeholder management, as illustrated in Figure 4.1 [3]. Embedded in the development process within phases are the principal activities that have to be managed at different times in the technology life cycle (TLC) (e.g., technological feasibility, application feasibility, political feasibility, demand feasibility, market creation, and market share) [4]. Additionally, it is key to be aware that any developmental process intended to reach commercialization requires public acceptance at various points in the pathway.

However, new technology development is infrequently a straightforward process without problems and a systematic meeting of all development process milestones. In 1992, the Industrial Research Institute (IRI) [5] sponsored an investigation that tried to systematically identify causes of uncertainty in the R&D process. The underlying assumption that "application of appropriate solutions," which reduce these uncertainties (or risks), would "shorten project cycle times and improve the efficiency and productivity of the innovation process." [6].

Using common Total Quality Management tools, 45 major causes of uncertainty in R&D were identified and grouped into eight categories. The most frequently encountered causes of uncertainty throughout these categories were "customer requirements not defined" and "delays in decision." The categories include the following:

DOI: 10.1201/9781003208051-6

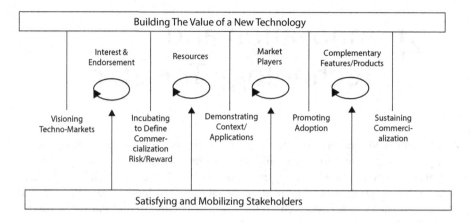

FIGURE 4.1 Emerging technology commercialization and stakeholder engagement.

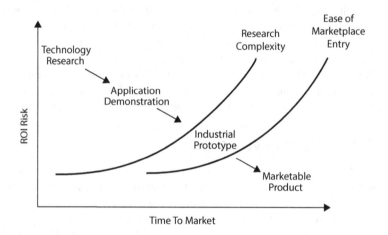

FIGURE 4.2 Technology development cycle.

1. Market
2. Competitor
3. Technical
4. Business processes
5. Management style
6. People/culture
7. Communication
8. External efforts

In large part, it has been revealed that human factors, and not technology, create the greatest uncertainty and risk in timely emerging technology development.

Figure 4.2 illustrates that the TLC is rooted in economics and emphasizes a focus on how quickly and completely new technologies are adopted within a consumer market.

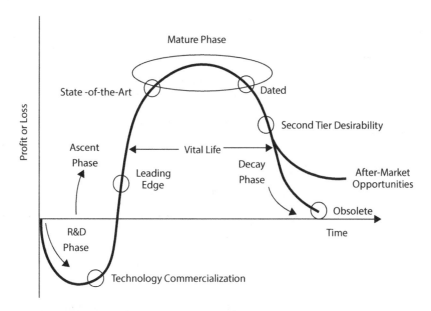

FIGURE 4.3 Management view of a technology life cycle.

That is, from a business perspective, with an overriding concern for return on investment (ROI), technology development and commercialization strategies focus on finding a way to enable early, rapid, and complete penetration of the market. Very simply, commercialization risk diminishes as the technology development process continues.

The research and marketplace lines depict the points at which the stakeholder focuses changes. The point at which the development pathway crosses these lines illustrates the amount of risk associated with the emerging technology development and time to get to the marketplace.

In addition to models of market share or consumer utility, emerging technology development and commercialization life cycles often reflect out of necessity the business model and profitability perspective. For example, Figure 4.3 illustrates a life cycle model concerned with R&D sunk costs, the timeline of recovering these costs, and the modes (e.g., types of licenses) of making the technology yield a profit proportionate to the costs and risks involved [7].

The use of advanced economic analyses techniques such as "discounted cash flow" and "compound options valuation" are tools to be of benefit in technology development as an aid to decision-making, risk assessment, infusion planning, probabilistic cost estimation, schedule uncertainties, and program-level decision tree analysis [8].

4.1.1 TECHNOLOGY S-CURVE & TECHNOLOGY LIFE CYCLE

Once a new technology comes into existence in the market, it shows a stable pattern which can be elucidated in terms of performance characteristic. The performance characteristic refers to an element of interest to a developer of a product or a user of a

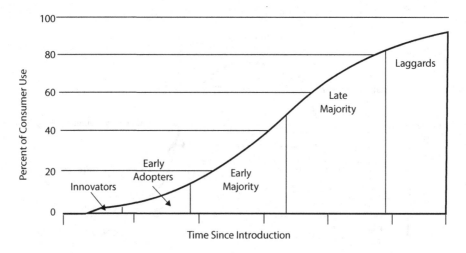

FIGURE 4.4   S-curve for emerging technological improvement.

specific technology. The performance of a technology has a recognized pattern over time that can be used in strategic planning of emerging technology management.

Technological innovation refers to the changes in performance characteristics of a specific technology over time. As depicted in Figure 4.4, the life cycle of innovations (or TLC) can be described using the S-curve, which depicts performance variables, such as growth of revenue or productivity, against time. In the early stage of a particular innovation, growth is slow as the new product establishes itself. At some point, customers generate demand and the product growth increases more rapidly. New incremental innovations or changes to the product allow growth to continue. Towards the end of its life cycle, growth slows and may even begin to decline.

There is a unique S-curve for each emerging technology performance characteristic, which changes uniquely for a specific technology over time. The performance characteristics show very little improvement in the early stages of the technology. This initial stage is followed by a second phase of very rapid improvement in the performance characteristic. During the third stage, the performance characteristic continues to improve, but the rate of improvement begins to decline. In the final stage, very little improvement is visible and the graph that charts the progress in the performance characteristic of a technology over time takes an S-shape [9].

It is argued that people have different levels of readiness for adopting new innovations and that the characteristics of a product affect overall adoption since customers respond to new products in different ways. The Diffusion of Innovations [10] classifies individuals into five groups: innovators, early adopters, early majority, late majority, and laggards. In terms of the S-curve, innovators occupy 2.5%, early adopters 13.5%, early majority 34%, late majority 34%, and laggards 16% as illustrated in Figure 4.4.

An innovation can be a product, a process, or an idea that is perceived as new by those who might adopt it. Innovations present the potential adopters with a new alternative for solving their problems, but they also present more uncertainty about whether that alternative is better or worse than the old way of doing things. The primary objectives of diffusion theory are to understand and predict the rate of diffusion of an innovation and the pattern of that diffusion. Innovations do not always get quickly adopted.

The S-curve is really a graph of the relationship between the effort put into improving a product or process and the results one gets back for that investment. As initial funds are put into developing a new product or process, progress is typically slow. As key knowledge necessary to make advances is put in place, the technology development rapidly speeds up. Over time, it becomes more difficult and expensive to make significant gains in technical innovation [11].

S-curves almost always come in pairs. The gap between the pair of S-curves represents a discontinuity reflecting the point at which one technology replaces another. Rarely does a single technology meet all customers' needs. There are almost always competing technologies, each with its own S-curve. Thus, in reality, there may be multiple technologies involved in the battle, some on defense and some on offense.

Deciphering a discontinuity's S-curves when all the development is happening is difficult. Companies that have learned how to cross technological discontinuities have escaped this trap. They invest in research in order to know where they are on the relevant S-curves and know what to expect from the beginning, the middle, and the end of these curves. A few develop very precise S-curves, but it is often enough just to know the general dimensions and limits and accept the implications.

Perhaps the most important lesson learned by organizations is that in order to quickly move to commercialize a technology, a company must invest in understanding the science that supports the base of the S-curve. Insights are still prevalent today for public/private projects; in 1981, the RAND Corporation, a government-sponsored think tank, looked into reasons for cost overruns of "pioneer process plants." Significant insights included the following [12]:

1. Severe underestimation of capital costs is the norm for all advanced technologies,
2. The factors that account for poor cost estimates and poor performance can largely be identified early in the development of the technology long before major expenditures have been made for detailed engineering, much less construction, and
3. Seventy-five percent of the cost variance can be attributed to insufficient technical information before the project began.

The TLC illustrates the way in which technological developments of products create commercial gain over a particular time frame. These gains are necessary to offset the R&D costs inherent in their creation. Moreover, varying life spans make it important for businesses to understand and accurately project the returns on these investments based on their potential longevity [13].

Due to the rapidly increasing rates of innovation, products such as electronics and computers in particular are vulnerable to shorter life cycles when benchmarks such as steel or paper are considered. Thus, the TLC is focused primarily upon the time and cost of development as it relates to the projected profits [14].

4.2 STRATEGIC IMPLICATIONS ON TECHNOLOGY COMMERCIALIZATION

In bringing products from development to market place commercialization, there is the need to formulate an analytical framework to draw strategic implications on technology transfer and commercialization. They combined the technology S-curve and technology adoption life cycle. From that they could identify the relationship between innovator groups and TLC statistics and showed the statistics (Q1, Q2 or median, and Q3) of the distribution illustrated in Figure 4.5. This figure enables us to identify strategic implications on technology strategies for firms by translating it in terms of TLC.

The technology S-curve shows the stages of a TLC. As discussed previously, a TLC starts with the R&D or introduction stage, progresses to the growth stage and the maturity stage, and finally winds up the cycle with the decline stage. In Figure 4.5, a technology begins to progress from the introduction stage that lasts until Q1, and after Q1, the technology enters the growth stage that lasts until Q2. After Q2, a technology begins its maturity stage, lasts until around Q3, and after Q3, it starts to decline. In view of characteristics of a TLC, Q1 can be considered as an approximate divider between the introduction stage and the growth stage, Q2

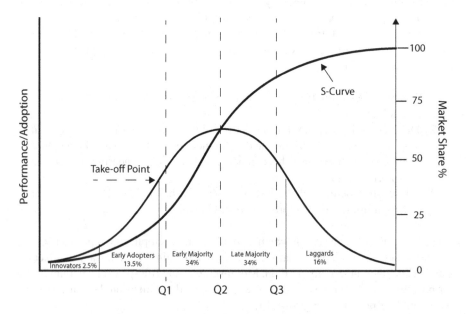

FIGURE 4.5 Analytical framework of technology management.

between the growth stage and the maturity stage, and Q3 between the maturity stage and the decline stage as follows:

Q1: R&D (introduction) stage – Ascent (growth) phase
Q2: Ascent (growth) stage – Maturity stage
Q3: Maturity stage – Decline (decay) stage

4.3 GLOBAL TECHNOLOGY TRANSFER

Technology transfer is a necessary activity to establish productive and efficient logistical pathways to customers. As companies grow and expand, their footprint and operational presence around the globe will necessarily increase to strategically support moving product and manufacturing technology between domestic and international facilities. Causes for a technology transfer effort include the following:

1. Localize manufacturing in the market region
2. Reduce logistics cost and time
3. Reduce/avoid import tariffs and value-added tax issues
4. Access supply chain partners

Another issue that may not receive the attention it deserves is for a manufacturing company to have contingencies in place to carry on operations should a situation occur that hurts or impacts production (e.g., fire, political unrest, and governmental retaliatory actions).

Typically, there is a procedure in place that an overall program manager (OPM) will use to direct all technology transfer activities to be performed. Often, the OPM will be from the facility transferring the technology with the receiving facility having a single person assigned to coordinate and manage the activities necessary at the receiving facility. The technology transfer procedure will include items, such as:

1. The OPM will be supported, as applicable, by appropriately assigned personnel from the following departmental areas at both the transferring and receiving locations: Quality, Training, Quality, Materials/Purchasing, Production, Manufacturing Engineering, Production Engineering, Toolroom, Engineering Laboratory, Product Management, Product Engineering, and Mold Design.
2. Obtain a master project number from the product line responsible home manufacturing site.
3. Use technology transfer checklists to guide activity completion (Table 4.1). Activity areas include schedule, training, quality, materials/purchasing, production, manufacturing engineering, production engineering, toolroom, engineering laboratory, product management, product engineering, mold design, and, of course, miscellaneous. Note that each transfer program must be tailored to the specific situation at hand. Not every item identified on a

transfer checklist may be applicable to the program at hand, and there may be others to be added.

4. Use a formal project management tool to facilitate communication to management and to illustrate the master program schedule, key milestones, and individual projects.

5. Initiate projects as necessary. Projects must be linked to the overall program master number so that cost and schedule monitoring can be done by the OPM.

6. Conduct technology transfer meetings, at defined intervals, with the appropriate team members at both the sending and receiving locations. Bring the right people into the program at the right time.

7. Scout the receiving facility prior to sending a technology transfer startup team to the receiving location. This enables identification of problems/ issues needing to be addressed prior to the arrival of a technology startup team.

8. Define the location (transferring or receiving) responsible for the product stocking quantity as part of the technology transfer.

9. Establish the OPM as the point of contact for all international communication. The OPM is responsible for making the program completed quickly, cost-effectively, and correctly. He/she must manage the program aggressively and proactively:

   a. Understand the culture where the manufacturing is going.
   b. Understand the people and players to work with to make this happen:
      i. Know who to trust to tell what is really happening, both good and bad. Spend work and leisure time with these people.
      ii. Be very gracious to people's need to show their culture. They are proud of it and want to show it off.
      iii. Show confidence and authority in facilitating company strategic directives and in giving clear direction.
   c. Understand the product and manufacturing technologies being transferred:
      i. Know the part numbers to be manufactured.
      ii. Know the Bill-of-Materials (BOM) items to be purchased and anticipated demand levels.
      iii. Know the Manufacturing System Effectiveness level before and after the transfer. In a perfect world, the system performance will be equal to or greater than that achieved before the transfer.
   d. Have the right people at the right time to make the implementation happen.

There are a myriad of issues surrounding technology transfer. Table 4.1 provides a list of topics to be considered. While this list is not all inclusive of every topic possible, it does provide a starting point for developing a strategically effective and efficient technology transfer effort.

TABLE 4.1
Technology Transfer Checklist

Schedule

1. Project numbers
    Master project number
    Subproject numbers
2. Key milestones
3. Status/progress meetings
4. Schedule tracking
    MS project
    Delays
5. Cost tracking
    MS project
    Transfer cost
        Purchase orders (PO)
        Manufacturers Statement of Origin (MSO)
        Engineer projects
        Leased assets handling
6. Startup incentives
    Tax
    R&D investment
    Market opportunities
    Competitor license agreements
7. Benefit–cost analysis for Manufacturing location
    Labor rate advantage
        USA versus Int'l
    Automation versus labor-intensive methods
8. Facility preparation
    Building
        Utilities
            Heating, ventilation and air conditioning (HVAC)
            Electrical
            Pneumatic
            Water
        Office areas
        Manufacturing areas
    License
    Government approval
    Dates for manufacturing start
    SOA of mfg. tech to be used

(Continued)

TABLE 4.1 (*Continued*)
Technology Transfer Checklist

9. Asset transfer
   Invoice
   Customs paperwork
   Inventory list – all items w/part number & value
   Container crate external ID for Panduit contents
   Int'l equipment transfer form (controller/tax dept.)

10. Operations monitoring
   Capacity usage
   Production volumes
   Profitability capability
   System effectiveness

11. Human resources
   Facility manager
   Mfg. manager
   Engineering
   Production workers
   Expatriate

12. Travel arrangements
   People
   Equipment

Training

1. Mfg. process/system product assy.
   Assembly
   Molding
   Stamping
   Plating
   Printing
   Coating
   Brazing
   Painting
   Machining
   Packaging
   Product point of display
   Cartoning
   Equipment/process ids (model #s, design level, etc.)

2. QA/process control
   Inspection instructions
   Gage use
   Colorimeter
   Calipers/blocks/pins

(Continued)

TABLE 4.1 (*Continued*)
Technology Transfer Checklist

	Height gage
	Continuity tester
	Electrical
	Optical
3.	Machine operation
	Troubleshooting
	Safety checks
	Maintenance (corrective maintenance, CM & preventive maintenance, PM)
	Startup
	Shut down
	Changeover
4.	Toolroom
	Repair
	Maintenance
	CM/PM
5.	Product applications and markets
6.	System effectiveness
7.	Training agenda
	Team members
	Schedule
	Operator videos
8.	Training grant application
	Schedule/training plan
	Training team
	Title
	Education
	Job experience
	Training responsibility

Quality Assurance

1.	Process control
	Sampling
	Mfg. parameters
	Number of samples
	Time internal (frequency)
	Correlation to product quality
2.	Finished incoming inspection
	Sampling
	Visual
	Functional

(*Continued*)

TABLE 4.1 (*Continued*)
Technology Transfer Checklist

3.   Incoming inspection
                 Sampling
                 Visual
                 Functional
4.   Test plan
                 Process control
                 Functional test
5.   Gages
                 Purchase
                 Manufacture
                 CAD files
                 Repeatability w/USA counterpart
6.   Supplier certification
                 New supplier
                 Existing supplier
7.   Data recording forms
8.   Control drawings
9.   Inspection aids
                 Illustrations
                 Samples
10.   Historical Data (customer complaints, production problems)
11.   FMECA

Materials/Purchasing

1.   Supplier qualification
                 Quality
                 Delivery
                 Financial stability
                 USA/foreign
2.   Materials management
                 Raw - identification/cost (supplier certifications,
                 receipt lead time)
                 Movement method
                 Method
                 Frequency
                 Receiving location
                 Source country
                 Order trigger date
                 Mfg. routings
                 BOM
                 Logistics handling

(*Continued*)

TABLE 4.1 (*Continued*)
Technology Transfer Checklist

       Costs/methods/durations/routes

       Truck/railroad

       Air

       Ship

       Volume versus weight

       Consolidation

3.   Purchasing

       Barcoded items

       Items to be barcoded

       Material ordering

       Materials purchased by sending location

       Materials purchased by receiving location

       Initial stock quantities

4.   Equipment shipment

       Packaging

         Clear

         Lube

         Mold saver

         Seal in plastic bag

         Desiccant use

         Crate for shipping

5.   Maintenance

       Spares to stock

       Annual usage

Production

1.   Operator instructions

       Process steps

       Equipment operation

       Quality

       Operator training

2.   Process certification (manufacturing system effectiveness)

       Quality capability (Cq)

       Achieved availability (Aa)

       Operational efficiency (Eo)

       Safety

       Product performance testing

3.   Maintenance (CM/PM)

       Failure history

       Trouble shooting

(Continued)

TABLE 4.1 (*Continued*)
Technology Transfer Checklist

4.   Production orders
                 Material requirements planning (MRP)
                     BOMs
                     Manufacture routings
                 Catalog items
                 Make-to-order
5.   Standard products
                 Parts
                 Kits
6.   Nonstandard products
                 Parts
                 Kits
7.   Outside mfg. services
                 Buyouts
                     Other division
                     Supplier
                 Manufacturing tasks
                     Other division
                     Supplier
8.   Inventory residence
                 USA warehousing vs Int'l warehousing
                 Service levels
                     a/b/c/d

                         Manufacturing Engineering

1.   Process performance benchmark
                 Manufacturing Se
                     Eo
                     Aa
                     Cq
2.   Maintenance
                 Sparing list
                 Sparing inventory
                 Initial stock
                 CM
                     Tasks
                     Frequencies
                     Videos
                 PM
                     Tasks

(Continued)

TABLE 4.1 (*Continued*)
Technology Transfer Checklist

        Frequencies

        Videos

3.   Machine design

        In house

        Supplier

        CAD

          Drawings

          Schematics

          Mechatronics

          Hydraulic

          Electronic

          Electrical

          Pneumatic

          Safety/reliability

4.   Equipment setup

        Complete system transfer (all hardware) versus transferring system   components being integrated with/into existing equip. (e.g., molds)

        Components needed to interface USA items with Int'l items

          Electrical

          Mechanical

          Power supply

          Electric

            Voltage

            Frequency

            Target location ability to make conversions

          Hydraulic

          Pneumatic

        Off-all

          Regrind

          Salvage/resale

          Disposal

        Auxiliary equipment

          Metal separator

          Conveyor

          Autobagger

          Label printer

          Grinder

          Robotics

(*Continued*)

TABLE 4.1 (*Continued*)
Technology Transfer Checklist

Heater

Chiller

Generator (to simulate other country's electrical conditions)

Material handling system/equipment

Material distribution systems/equipment

Production Engineering

1. Packaging method
   Autobagger
   Manual

2. System layout
   Floor layout
     Process flow
     Space requirements

3. Maintenance
   CM
   PM
   Sparing

4. Packaging
   Bag stock
   Carton
   Labels
   Product box
   Pallet stack pattern

5. BOM
   MRP
     Part number
     Kits

6. Mfg. system setup
   Changeover
   Startup
   Shut down

7. Equipment ordering
   Packaging
   Printing
   Cutting
   Inspection/functional test
   Tables
   Assembly equipment
   Molding machines
   Auxiliary equipment

(Continued)

TABLE 4.1 *(Continued)*
Technology Transfer Checklist

Toolroom

1.   CAM
     Fixtures
     Molds
     Other
2.   PM schedule
     Spares
3.   Machine setup
     Facilities
       HVAC
       Computer interface
       Power supply
       Layout position

Engineering Lab

1.   Product test plan
     Methods
     Sample sizes
     UL/CSA requirements
2.   Raw material qualification
3.   UL/CSA
       Products already UL approved
       Mfg. location UL approved status
       Coordinate with UL, CSA

Product Management

1.   Part numbers
     BOM
       Material sources
2.   Kit numbers
     BOM
       Direct sales impact
       Indirect sales impact
3.   Label
     Content
       Made in ...
       Assembled in ...
     Barcode
     Languages
4.   UL/CSA
5.   Customer instruction sheets
     Color

(Continued)

TABLE 4.1 (*Continued*)
Technology Transfer Checklist

	Content
	Graphics
	Text
	Languages
6	Sales forecast

Product Engineering

1.	Design
	CAD drawings
	Design performance
	Specifications
2.	Material
	Cross reference
	Formulation
3.	CAD file transfer to receiving location
	Type
	Conversion

Mold Design

1.	Mold handling
	Repair table
	Crane
2.	Mold acquisition
	Outside
	Supplier
	Manufacturing location
	Inside
3.	Mold design
	CAD
	Cores
	Inserts
	Blocks
	Mold setup
	Molding parameters
	Cycle times
	Mold heater/coolers
	Process control
	Critical performance measures (CPM)
4.	Mold maintenance
	PM

(Continued)

TABLE 4.1 (*Continued*)
Technology Transfer Checklist

	CM
	History log book
5.	Material handling
	Drying
	Coloring
	Mixing
	Pre-colored
	Regrind

TMA Case Study: R&D Strategic Test Planning

OVERVIEW

The methodology for optimized strategic testing (MOST) is applied for two competing R&D projects varying in the nature of the technology involved, the amount of prior experience at hand, the product cost and volume, and the level of safety and reliability concern.

In each example represented is the initial application of MOST at the beginning of the project planning process. Recognizing that a product R&D test program plan must be flexible/dynamic, the test plan should be updated at various points throughout the test program to reflect new information and insight gained as a result of completed tests and design refinements, as well as overall strategic forces.

STRATEGIC OBJECTIVE

The strategic objective is to select the best R&D test plan option for product development leading to successful commercialization. An effective test program plan is defined by the model in terms of how well it (1) identifies and removes design problems and minimizes technical risks; (2) assesses operational effectiveness, reliability, and suitability; and (3) provides a broad base of user acceptance.

BACKGROUND

For a given product R&D project, the test sequence, requirements, and decision milestones (i.e., proof of concept and technology transfer) are defined and integrated into a basic test plan via the MOST artificial-driven expert system. The test plan is then quantitatively addressed using an analytical evaluation model.

The MOST evaluation model is a benefit-cost-time-based model in which benefit is presented in terms of the effectiveness, E, of the test plan in contributing to successful technology transfer. This parameter, in turn, is a key factor in the overall achievement of successful product commercialization.

Test Plan Effectiveness Levels are defined as follows:

Level A – Highly Effective: A very high probability is required to contribute towards successful technology transfer. This corresponds to a range of 100–85.
Level B – Effective: A significant probability is required to contribute towards successful technology transfer. This corresponds to a range of 84–70.
Level C – Reasonably Effective: A moderate probability is required to contribute towards successful technology transfer. This corresponds to a range of 69–55.
Level D – Remotely Effective: A low probability is required to contribute towards successful technology transfer. This corresponds to a range of 54–30.

The issues of technology transfer and commercialization are broad concepts embodying, in addition to test program effectiveness and utility, such basic concepts as inherent product suitability, the quality of design execution, and the quality and integrity of the test specimens.

In particular, successful commercialization is ultimately dependent on the product having met or exceeded its design performance goals, including reliability and safety. A product lacking with respect to some of these fundamental characteristics may undergo a very successfully test program only to be shown to be lacking in commercial viability. However, the approach taken in strategically defining the test program plan should minimize this risk.

The evaluation model focuses on optimizing the overall effectiveness level of a test program plan in regard to its two constituent test segments (i.e., laboratory and field). The effectiveness level of each test segment is subsequently allocated to the individual tests. Test program plan effectiveness optimization is performed with respect to its estimated cost and time. This is achieved by estimating the cost and time requirements of a defined R&D test plan and then quantitatively evaluating this data against the known cost and time constraints impacting the project.

In general, the major elements contributing to test program cost and time are as follows:

1. **Test Unit**: The cost and time associated with the fabrication of the necessary test unit(s) taking into account any potential "manufacturing learning" when several identical test units are to be fabricated,
2. **Test Setup**: The cost and time associated with the installation of a test unit(s) at the test site or into a test bed taking into account test procedure preparation, chamber installation, instrumentation setup, thermo/vibration survey, and test infrastructure including logistic factors,
3. **Test Monitoring**: The cost and time associated with conducting and monitoring a test(s) and compiling all required test data,
4. **Test Facility**: The cost and time associated with the use of laboratory test facilities including necessary environmental test chambers and standard test equipment and instrumentation, and
5. **Special Test Equipment**: The cost and time associated with the development and use of any special or nonstandard test equipment, instrumentation, or fixturing that may be required.

In optimizing test program effectiveness relative to cost and time, the model provides a measure of the cost and time involved so that rational choices can be made in an ROI-based context for planning and implementing a test program tailored specifically for a subject product R&D project.

RESULT/CONCLUSION

PROJECT 1: This product development project was for a large complex system having the following characteristics:

1. High technology at state-of-the-art (SOA)
2. High product complexity
3. Residential/commercial application
4. Large potential market volume
5. Significant potential safety consequence
6. High availability requirements
7. Available schedule: 60 months
8. Available funding: $2 million
9. Level A reliability and safety program requirements
10. Level E1 QA program requirements
11. Market leader product R&D strategy.

The R&D objective for the test program plan was to commit adequate resources to ensure product R&D success. From the MOST expert system, the test program derived included (1) experimental (one prototype), (2) design verification (three early engineering test units, ETUs), (3) design approval (three advanced ETUs), (4) reliability (four advanced ETUs), and (5) field (10 pilot units).

The cost of implementing the test program plan was estimated from standard cost factors for each test from historical data. The total estimated cost of the test program defined was $1,084,800 compared to the $2 million available.

The time associated with conducting the test program defined was derived from a test time network diagram. The total time period defined for the test program was 53 months compared to the 60-month schedule limit set in order to meet market introduction factors.

The effectiveness level of the subject product R&D test program plan (minimal effectiveness) was specified to be 85 for Level A. The optimal effectiveness level for laboratory testing was 1.0 and 1.0 for field testing. The product R&D test program plan was considered acceptable in terms of its ability to meet the product and project strategic issues; therefore, no subsequent revisions were generated and a formal test program plan was prepared. The application of MOST resulted in a test program effectiveness level of 100% having an estimated cost of $1,084,800 and an estimated time of 53 months.

PROJECT 2: This product development project was for a small handheld instrument having the following characteristics:

1. Off-the-shelf electronic technology
2. Moderate product complexity

3. Residential application
4. Large potential market volume
5. Mild safety consequences
6. Normal availability requirements
7. Available schedule: 36 months
8. Available funding: $200,000
9. Level C reliability and safety program requirements
10. Level E3 QA program requirements
11. Application-oriented product R&D strategy.

The R&D objective for the test program plan was to perform minor research and emphasize product development and engineering effort. From the MOST expert system, the test program derived included experimental (one prototype), design approval (three advanced ETUs), and field (20 pilot units). It was determined that design verification testing was not applicable. This was due to there being no complex individual components in the design. Also, it was determined that reliability testing was not applicable. This was due to the readily available reliability experience data for the parts comprising the product. A very high product mean time between failure (MTBF) was predicted, and it was expected that any reliability problems would be exposed during design approval testing.

The cost of implementing the test program plan was estimated from standard cost factors for each test. The total estimated cost of the test program defined was $264,800 compared to the $200,000 available.

The time associated with conducting the test program defined was derived from a test time network diagram. The total time defined for the test program was 28.5 months compared to the 36-month schedule limit set in order to meet market introduction factors.

The effectiveness level of the subject product (minimal effectiveness) was specified to be 55 for Level C. The optimal effectiveness level for laboratory testing was 1.0 and 0.7 for field testing. The test program plan was considered acceptable in terms of its ability to meet the product R&D strategic issues; therefore, no subsequent revision was performed and a formal test plan was prepared. The application resulted in a test program effectiveness level of 85% having an estimated cost of $200,000 and an estimated time of 28.5 months.

Case Study Question: *Which product development test program should receive priority when considering technology transfer and commercialization?*

4.4 SUMMARY

This chapter addressed the move of products out of R&D and into commercialization, as well as the need to move products and manufacturing capabilities to cost-effectively meet customer demand and revenue expectations. The TLC is ultimately rooted in economics, which indicates a strong dependency on how quickly and completely new technologies are adopted within a consumer market. From a business perspective, with an overriding concern for ROI, technology development and commercialization strategies focus on finding a way to enable early, rapid, and complete penetration of the market. Comfort comes from knowing that during the TLC, commercialization risk diminishes as the technology development process continues.

Technology transfer is a necessary activity to establish productive and efficient logistical pathways to customers. As companies grow and expand, their footprint and operational presence around the globe will necessarily increase to strategically support moving product and manufacturing technology between domestic and international facilities. Causes for a technology transfer effort typically include the following: (1) localize manufacturing in the market region, (2) reduce logistics cost and time, (3) reduce/avoid import tariffs and value-added tax issues, and (4) access supply chain partners. Another issue that may not receive the attention it deserves is for the need to have contingencies in place to carry on manufacturing operations should a situation occur that hurts or impacts production (e.g., fire, political unrest, and governmental retaliatory actions).

QUESTIONS

1. What are the principal activities of the innovation TLC?
2. What happens by taking actions to reduce technology risk?
3. What is the variable causing the greatest innovation developmental risk? Why?
4. Why are economic considerations so prevalent in the technology development and implementation life cycle?
5. What are the four management perspective phases occurring during the TLC?
6. What is the S-curve?
7. What is the diffusion of an innovation?
8. Does each innovation have a unique technology S-curve? Why or why not?
9. Discuss the reasons for technology transfer.
10. Is training a critical component of technology transfer? Why or why not?

REFERENCES

1. Brauer, D., Lee, R. P. & Brauer, C. (2018). "Commercialization of Emerging Environmental Sustainability Technologies: CCU&S." *International Freiberg Conference on IGCC & Xtl Technologies, Coal Conversion, and Syngas*; Berlin, Germany.
2. Lee, R. P. (2016). "Misconceptions and Biases in German Students' Perception of Multiple Energy Sources: Implications for Science Education." *International Journal of Science Education*, Vol. 38, No. 6, pp. 1036–1056.
3. Jolly, V. K. (1997). *Commercializing New Technologies: Getting from Mind to Market.* Boston, MA: Harvard Business School Press, p. 5.
4. Wilson, M. (2001). "ICT Technology Lifecycles," World Wide Web Consortium (W3C), http://www.w3c.rl.ac.uk/pasttalks/tech_lifecycles.pdf.
5. Beck, D. F. (2013). *Technology Development Life Cycle Processes.* Albuquerque, NM: Sandia National Laboratories, SAND2013-3933.
6. Burkart, R. E. (1994). "Reducing R&D Cycle Time." *Research Technology Management*, Vol. 37, No. 3, pp. 27–32.

7. United Nations Industrial Development Organization (UNIDO). (1996). *Manual on Technology Transfer Negotiation: A Reference for Policy-Makers and Practitioners on Technology Transfer.* Vienna: UNIDO.
8. Tralli, D. (2004). "Valuation of technology development using a novel workflow approach to compound real options." *2004 IEEE Aerospace Conference*, Big Sky, Montana.
9. Narayanan, V. K. (2001). *Managing Technology and Innovation for Competitive Advantage.* Englewood Cliffs, NJ: Prentice Hall.
10. Rogers, E. M. (2003). *Diffusion of Innovations* (5th ed.). New York: Free Press.
11. Foster, R. (1986). *Innovation: The Attacker's Advantage. Fire Side Book.* New York: Simon and Schuster.
12. Merrow, E., Phillips, K., Myers, C. (1981). *Understanding Cost Growth and Performance Shortfalls in Pioneer Process Plants*, RAND. www.osti.gov/biblio/6207657
13. Park, H. W., Sung, T. E. & Kim, S. G. (2015). "Strategic Implications of Technology Life-Cycle on Technology Commercialization." *International Association for Management of Technology, IAMOT 2015 Conference Proceedings*, Cape Town.
14. Park, H. W., Sung, T. E. & Kim, S. G. (2014). "Technology Life Cycle," Boundless. https://fdocuments.in/document/strategic-implications-of-technology-life-implications-of-technology-life-cycle.

5 Engineering Economic Analysis

Decision-making is a required activity of any organization. The "best" decisions come from having the ability to distill all data available and arriving at the correct course of action through an intuitive process, weighing the tangibles and intangibles to arrive at the right, correct, or best conclusion.

As discussed in Chapter 3, strategic planning is a tool for establishing the foundation for achieving future organizational success. Financial analysis and budgeting are fundamentally the strategic plan expressed in financial terms, which is also necessary for achieving organizational and TMA success.

5.1 ECONOMIC ANALYSIS

All tactical manufacturing planning decisions have economic consequences which must be considered. Naturally, decisions are slanted towards those alternatives which maximize profits. An understanding of basic economic principles is therefore essential to sound strategic planning.

The fundamental arithmetic of economic planning is known as "engineering economics" and focuses on the time value of money. Note that the intent here is not to present a comprehensive discussion of accounting, finance, or engineering economics, there are many valuable resources available to guide the reader seeking more detail [1–4]. In this section, several key topics are addressed that are seminal to strategic financial decision-making.

5.1.1 TIME VALUE OF MONEY

The foundation of engineering economics is the concept that money has a time value. That is, there is a present value and future value of money, which provides significant guidance to financial decision-making. Present value is the value today of a monetary amount (e.g., $1 in a savings account today has a present value of $1). Future value is the value, at some later time, of a monetary amount today (e.g., $1 in a savings account today earning interest has a future value greater than $1).

Fundamentally, the time value of money refers to a discounted cash flow associated with projects that are adjusted to allow for the timing of the cash flow and the potential interest on the funds involved. Such allowance for timing is important, since most investment projects have their main costs or cash outflows in the first year or so, whereas their revenues or cash inflows are spread over many future years.

Consider, using an interest rate at 10%, a company could invest $100 today and see the investment grow at a compound interest of $110 at the end of year 1 and $121 at the end of 2 years. So, the $100 today is worth (future value) the same as $110 in

1 years' time or \$121 in 2 years' time (and this pattern continues for a specified time frame). Reversing this cash flow outlook, the \$110 receivable in 1 years' time, or \$121 in 2 years' time, has a present value of \$100.

To aid decision-making, it is possible to calculate the present value of the estimated stream of future cash outflows associated with an investment project and the present value of the estimated stream of future cash inflows from the project, and compare the two. If the present value of the cash inflows from the projects exceeds the present value of outflows when both are discounted at the same rate (e.g., 5%, 10%), the net present value is positive. This suggests that it would be worthwhile for a company to use its own money or borrow money at the applicable discount rate and undertake the project since it is projected to earn a return in excess of its financing costs.

To facilitate calculations, financial texts have tables for the future value of \$1 invested at an interest rate and for the present value of \$1 received in the future. Such tables enable comparison of cash inflows and outflows over time in order to make financially sound decisions confronting the manufacturing enterprise.

5.1.2 SELECTION OF ALTERNATIVES

The task of engineering economics is to evaluate various production alternatives and to determine which one gives the most advantageous time value for the money involved. The issues commonly considered are:

1. **Initial Cost**: How much money is required up front to implement a specific alternative or plan?
2. **Interest Rate**: How much "rent" is acceptable to be paid on the money that will be used to implement the plan?
3. **Annual Cost or Benefit**: What will the monetary consequences of the plan be on a yearly basis? That is, by how much will it increase profits? Or, alternatively, how much will it cost each year?
4. **Economic Life**: How long will this plan or investment last? How long can it be expected for the profits to keep coming? When will there be a need to re-invest to keep the manufacturing enterprise going?
5. **Salvage Value**: After the economic life of the investment has passed, will it be possible to sell the used equipment for any significant amount?
6. **After-Tax Cash Flow**: How does the investment affect our income tax status? Is the investment depreciable? Is this the best time to implement it?

Each of these values must be determined as closely as possible for a reasonable economic analysis to be made of a candidate plan. The results of the analysis will tell if the plan is profitable, and if several alternatives are available, which is the most profitable.

In terms of manufacturing planning, each of the factors listed above can be complex and involve many subproblems. Some of these problems are discussed in the following subsections.

5.1.2.1 Initial Cost

This is usually a fairly straightforward value to determine. While considering purchasing a piece of production equipment, for example a lathe, several vendors will probably offer acceptable items. The prices will be well defined but may be negotiable.

If a vendor gives a trade-in allowance on old equipment being replaced, this amount may be deducted from the cost of the new equipment in the economic analysis. Since trade-in values may vary from one vendor to another, it is important to determine the amount separately for each alternative.

It is also important to consider startup costs. These include the costs associated with installing and debugging the new equipment, hiring and/or retraining employees, and production time lost during the changeover.

5.1.2.2 Interest Rate

The interest rate to be paid on an investment is also simple to determine, but first one must determine the actual source of the capital funding to be used. In the case of a bank loan, the interest rate will be explicitly stated. However, many capital expenses will be paid for out of a company's cash reserves. In this case, the cost of the funds is the loss of the interest the cash is currently earning. Obviously, it would be unwise to remove funds from an investment portfolio which is yielding a 10% return, just to invest it in equipment which will return 5%. This would only be justified when the equipment is absolutely necessary to stay in business and protect or support other, more profitable, investments.

Another factor to consider is variability in interest rates. When an adjustable rate loan is used to finance the new equipment, a best estimate must be made of the interest prevailing in each future year. These varying values must then be used in the economic analysis.

5.1.2.3 Annual Cost or Benefit

The impact of an investment on annual cash flow can be difficult to determine. A new piece of machinery will impact overhead costs in terms of energy consumption, raw material use, insurance, maintenance, floor space, and work in process. It may also impact manpower requirements in either a positive or negative manner. Effect on overall factory efficiency will be difficult to determine, unless the new equipment is replacing an older machine that was performing precisely the same function.

Other effects on annual cash flow can be even more difficult to determine. These include such concepts as product quality, customer satisfaction, corporate image, factory safety, employee satisfaction, and environmental impacts.

5.1.2.4 Economic Life

It is also difficult to predict how long an item of equipment will continue to function. Even if equipment life could be determined accurately, there would be the problems of annual cost and benefits changing (generally getting worse) as the equipment aged and wore out.

The term economic life is preferred to "useful life," because an item of capital equipment is rarely used until it is no longer useful. It is generally used only so long as keeping it is the most economical alternative. As a piece of equipment gets older, its maintenance costs almost always increase. Further, the depreciation allowance will eventually have been used up. At some point, it is advantageous to purchase a newer piece of machinery, even though the old one still has some life left in it.

The point at which this replacement becomes the preferred alternative is very difficult to determine, especially at the beginning of the life cycle of the original equipment. The future maintenance costs are difficult to anticipate, as are the future costs of replacement equipment.

The moral of this discussion is that any estimate of the economic life of a capital investment must be considered a very rough approximation. The only exception is when you have already been through many life cycles of the same type of item and have extensive experience to draw upon.

5.1.2.5 Salvage Value

This is another value that is difficult to determine at the beginning of a machine's life cycle. It will vary with economic life of the equipment, final condition of the equipment, and prevailing economic, technological, and market conditions at the time of replacement.

5.1.2.6 After-Tax Cash Flow

Once all of the above values have been estimated, as closely as they can be, an economic analysis is possible. However, this will be a before-tax analysis, which does not necessarily reflect the actual cash flow that will be felt.

In general, money spent in the course of doing business is tax deductible, but there are several different ways of deducting it. If the items purchased will be consumed within a year, they are considered ordinary expenses. These include office supplies such as pencils and paperclips, as well as rent, salaries and wages, and insurance. The costs of these items are deductible in the year that they were purchased.

Other items last more than 1 year. These are called capital expenses, and their costs cannot be deducted in a single year. They include such items as machinery, computers, buildings, vehicles, and land. The costs for these capital items must be deducted over the course of several years, and each year's deduction is called a depreciation allowance. A large number of laws and regulations spell out detailed rules for determining how many years it takes to fully depreciate each type of equipment, and how much depreciation is allowed in each year. These laws are complex and tend to change from year to year.

In conducting a complete economic analysis of any investment under consideration, it is necessary to calculate the tax consequences, including depreciation allowances. The net effect of taxes is to minimize both profits and losses, since payment of the taxes will remove some of your income, and depreciation allowances will return some of your expenses. These after-tax rates of return are the numbers which must be compared between the various alternatives to actually determine the most economically attractive course of action for an investment.

5.2 PROFORMA DEVELOPMENT

There are several components of the financial plan that should be developed to provide the forward look at financials. Together, these components provide a proforma of operations and indicate whether or not the strategic plan is do-able within the current operating and financial structure. Additionally, this provides an essential management tool to measuring actual financial performance. Without the forward-looking strategic proforma, budget development is done blind and lends comparison to a report of actual results that really becomes a comparison to a series of abstract numbers.

A proforma typically consists of the following components:

1. Operating Budget (or P&L, profit and loss, spreadsheet)
   a. Sales plan (by product and/or territory)
   b. Cost and expenses (by product and/or department)
   c. Gross profit projection
   d. Production inventory plan
   e. Hiring/manpower plan
   f. Incoming order forecast
2. Capital Budget
   a. Asset purchases (what and when)
   b. Major repairs or maintenance
3. Cash Flow Forecast
   a. Source and use of funds
   b. Working capital changes and needs
   c. Long-term capital needs
   d. Borrowing or other financing activities

The oversimplified characterization of the proforma operating budget is to view it as a "forecasted P&L." This forecast typically runs out 5–10 years. However, it is important to remember that the longer the projected time frame, the more error and need for adjustment is built in. With the preparation of the proforma, a model has been constructed for the organization's anticipated operations element by element. In addition to establishing "how much it will cost," additional questions include "why should it cost $" and "is the organization getting fair value?" That is, need for scrutiny is necessary to ensure that it aligns with the dynamic organizational strategic business plan.

There are four main basic situations that lead to financial performance missing financial proforma projections/forecasts [5]:

1. Failure of the forecast/budget to have anticipated or allowed for the general business conditions or environment of the organization
2. Failure to have recognized a problem in the organization's products, processes, or personnel that existed at the time the forecast/budget was prepared
3. Failure, during the proforma period, to modify and adjust the forecast/budget in light of any sudden problems or changing conditions
4. Failure to have created a unity of purpose, teamwork, and an attitude of caring among the strategic human resources or the organization

In analyzing actual financial performance to proforma budgeted, there are profit and loss fundamentals to recognize that (1) all costs do not change in the direct proportion to changes in revenue, (2) all costs do not change at the same rate or as a result of the same stimuli, and (3) profits do not change in direct proportion to either costs or revenue. Only by analyzing and understanding the causative factors and relationships that can affect individual costs and revenues will the organization be able to strategically anticipate, plan, and exercise some control of profit.

All costs fit into one of four categories: fixed costs, variable costs, semivariable costs, or discretionary costs. Understanding the internal and external forces that cause a specific cost to increase or decrease is a first step to being able to exercise control over that cost. The proforma provides the all-important tool to strategically plan forward and make the "best" decisions possible.

The information needed by team members for making decisions and for planning and performance evaluation are generally the costs of carrying out the organization's activities. It is important to distinguish cost, as used in managerial accounting, from expense, as used in financial accounting. That is, a cost is a sacrifice of resources and an expense is a cost that is used up in a particular accounting period.

A useful way of classifying costs for decision-making is by cost behavior. While there are many ways to breakdown and describe costs, a common fundamental approach is to classify them as "variable" or "fixed." If total costs vary with activity, then they are variable costs. If not, they are fixed costs. With this classification, total cost is expressed as:

$$TC = F + VX$$

where TC is total cost for a particular period of time

F is the fixed costs for the period

V is the unit variable cost

X is the product volume for the period in units

The profit equation comes from adding average sales price (P) per unit to the equation to provide:

Operating Profit = Total Revenue – Total Variable Costs – Fixed Costs

$$OP = PX - VX - F$$

The point at which total cost equals total revenue is the break-even point. From this, the actual number of units defining the break-even point is derived:

n = Fixed Costs / (Sale Price per unit – Variable Cost per unit)

$$n = F/(P - V)$$

Price minus the variable cost is also known as the contribution margin. The contribution margin describes how the selling of each unit contributes to fixed costs and the earning of profit. The break-even point lends itself to graphical display for cost-volume-profit (see Figure 5.1). The graph lends itself to being a heuristic decision and analysis tool as individual costs and production volumes are strategically adjusted.

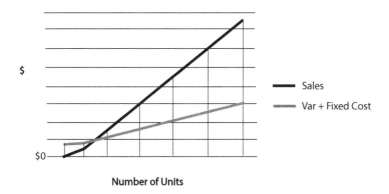

FIGURE 5.1 Break-even graph.

5.3 INVESTMENT DECISION MATRIX

The more information available the better the decisions that follow. The investment decision matrix (IDM) is a structure tool for collecting relevant financial information to support strategic decision-making. It consists of nine financial parameters to provide multiple data points. If all point in the same direction, then the best decision is readily evident. It they do not all point in the same direction, then a closer look at the financial assumptions used is warranted.

5.3.1 TIME TO MARKET

Time to market (TTM) is defined as the period from the conception of a new idea until it is released to the marketplace. Figure 5.2 illustrates the relationship of TTM to investment expense and revenue over time. That is, it is the time between when the team starts work and when the first unit is sold. Research has shown that new market entrants enjoy clear advantages in terms of market share, revenue, and sales growth when TTM is one of the critical new product introduction metrics. Many product development strategies depend on being first to market.

5.3.2 INITIAL INVESTMENT

Initial investment is the monetary amount required to start a business or a project. It is also called initial investment outlay or simply initial outlay. It equals capital expenditures plus working capital requirement plus after-tax proceeds from assets disposed of or available for use elsewhere.

5.3.3 EXCESS PRESENT VALUE INDEX

The excess present value index (EPVI) is computed as the present value of future cash flows divided by the initial investment. The EPVI indicates the number of net present value dollars generated per dollar of investment.

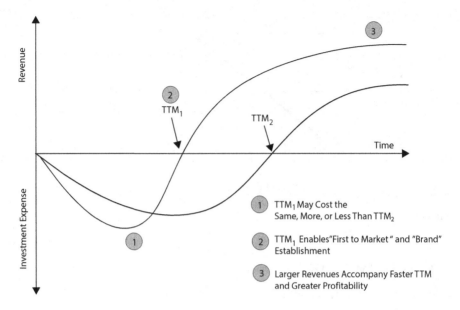

FIGURE 5.2   Relationship of TTM and $ over time.

$$EPVI = \frac{\text{Present Value of Future Cash Flows}}{\text{Initial Investment}}$$

Strategic initiatives with an EPVI value greater than 1.0 are generally acceptable. Initiatives with an EPVI value less than 1.0 are generally not acceptable. The EPVI enables ranking initiatives based on net present values and/or EPVI. Any difference in the rankings is caused by the fact that in the EPVI method the excess present value is related to the dollars of the initial investment required, rather than to the absolute size of the net present value.

5.3.4   BENEFIT-COST RATIO

Benefit-cost ratio (BCR) analysis provides a value that summarizes the overall value for money of a project or proposal. A BCR is the ratio of the benefits of a project or proposal, expressed in monetary terms, relative to its costs, also expressed in monetary terms. All benefits and costs are expressed in discounted present values. The net present worth of benefits (i.e., NPWB) includes the initial investment.

Benefit-Cost Ratio $= ($ Net Present Worth of Benefits $)/($ Present Worth of Costs $)$

BCR = NPWB / PWC

The BCR is interpreted as the higher the value, the more attractive the risk-return profile of the project/asset. The value generated by the BCR indicates the dollar value generated per dollar cost.

5.3.5 INVESTMENT PAYBACK

The payback period refers to the amount of time it takes to recover the cost of an investment. Simply put, the payback period is the length of time an investment reaches a break-even point. That is, payback period is the period of time required for the profit or other benefits of an investment to equal the cost of the investment. The desirability of an investment is directly related to its payback period. Shorter paybacks indicate more attractive investments.

There is no set rule for payback time frames, each organization will have their own requirement for decision-making. For example, a payback period less than 3 years is a common benchmark for engineering and manufacturing departments.

5.3.6 INTERNAL RATE OF RETURN

The internal rate of return (IRR) of a stream of cash flows is defined as the discount rate that equates the net present value of that stream to zero. That is, it is the rate that discounts the future cash flows to the present so that the present value of future cash flows is equal to the initial investment. It is often a method of trial and error to find the rate that makes the present value stream equal to the initial investment.

$$\text{Initial Investment} = \frac{\left(\text{in flow}\right)_1}{\left(1+r\right)^1} + \frac{\left(\text{in flow}\right)_2}{\left(1+r\right)^2} + \frac{\left(\text{in flow}\right)_n}{\left(1+r\right)^n}$$

$$\text{IRR} = r$$

Organizations set their own IRR value that apply to their specific strategic initiatives. The IRR acts as a hurdle for initiates, and an IRR of 15% is common hurdle benchmark used.

5.3.7 OWNERSHIP

Ownership is straightforward. The organization owns the asset or it does not own the asset. There may be numerous reasons for not owning, which typically lead to performing make/buy analysis.

5.3.8 ASSET VALUE

Asset value is defined as the value of an asset in the books of records of an organization at any point in time. For companies, it is calculated as the original cost of the asset less accumulated depreciation and impairment costs.

Asset Book Value

= Total Asset Value – Depreciation – Other Directly Related Expenses

where Total Asset Value is the original purchase value
Depreciation is the periodic reduction in the value of the asset amortized as per standards.

5.3.9   PURCHASE INVESTMENT ANALYSIS

The purchase investment analysis is a summary of the present value of cash flows. The summary table includes (1) present value factor (%, periods), (2) outflow, (3) revenue inflows, and (4) NPV inflows. The present worth data support the EPVI, BCR, payback, and IRR calculations. Present and future value tables are readily available in a variety of texts and internet portals [1–4].

TMA Case Study: Carbon Capture and Storage Financial Investment

OVERVIEW

The organization is making a business development assessment to identify the requirements for economic viability and evaluate the business climate and financial resources necessary for a CO_2 storage complex development, contractual agreements, state incentives and policies, and revenue sources.

ISSUE

The decision to invest in emerging environmental sustainability technologies requires an understanding of the ROI risks. Accordingly, numerous variables provide challenges for emerging carbon capture and storage (CCS) technologies including technological, financial, governmental policy, storage potential and permanence, and public/societal acceptance.

OBJECTIVE

The objective is to understand the stakeholder obstacles that may impact the decision to invest in CCS technologies and to bring them through technology transfer and into commercialization.

CASE BACKGROUND

While some countries have implemented explicit carbon prices in the form of a tax or cap-and-trade system that applies to CCS, that is not the case in the United States [6]. It is the specific combination of these factors for each individual source/reservoir pair that will determine the actual transport and storage costs and contribution of each to a total estimated combined cost [7]. With the wide range of variability for transport and storage from one potential project to the next, the cost of deploying CCS systems will vary widely.

The U.S. Environmental Protection Agency has used a cost of $15/tCO_2$ to represent the cost of transport and storage of CO_2 once it has been captured from a large anthropogenic CO_2 source, such as a power plant. This value has been used within EPA's macroeconomic modeling of the cost of complying with various proposed CO_2 emissions policies. This $15/tCO_2$ cost is based on previously published research to provide an assumed breakout cost of CO_2 transport and cost of CO_2 storage (including monitoring, verification, and accounting, MVA) for the overall $15/tCO_2$ value.

This $15/tCO_2 cost for storing CO_2 is characterized as a long-term average price that seems to provide for the site-specific circumstances likely to be encountered by anthropogenic CO_2 point sources. However, it is important to note that this value is neither the exact cost for a given source (i.e., customer) nor the cost likely to be experienced by all anthropogenic CO_2 sources. Despite using a general statement that a significant fraction of CCS projects in the U.S. will fall in the $15/tCO_2 range, the cost of deploying CCS systems will vary significantly from business to business, depending upon a host of source- and site-specific conditions.

However, for this very broad business financial analysis, a weighted-average CO_2 price of $15/tCO_2 will be used. While likely inaccurate for any given specific business situation, the $15/tCO_2 cost assumption is a useful, robust, and useful estimate for transport and storage costs likely to be encountered across a significant fraction of potential future commercial CCS deployment scenarios

BUSINESS PROFIT AND LOSS PROFORMA

The business profit and loss proforma and investment/depreciation schedule information are provided in Tables 5.1 and 5.2.

BREAK-EVEN ANALYSIS

Breakeven occurs at approximately 2.7 MtCO_2, as illustrated in Figure 5.3, based on using proforma year 1 total operating expense as the consistent representation of fixed cost.

RESULT/CONCLUSION

KEY FINANCIAL INDICATORS

Table 5.3 provides the IDM based on the financial proforma created for this business case example. The TTM needs to keep in mind that the investment clock actually starts ticking in 2020 time frame, as this example case assumes (theoretically) that infrastructure is in place to start commercial injection in 2025. The initial investment is manageable and reasonable to draw in serious potential investors. The net present value (7% required return) provides a good return over the 11-year proforma period. The EPVI and BCR values are greater than one and favorable. Investment payback is longer than typically desired. The internal rate of return is favorable.

From the IDM, all financial indicators point in the direction of making the investment in the CCS technology. It should be noted that past CCS projects have experienced significant cost overruns and have been troubled by lack of (1) investor development, (2) productive collaboration with governmental agencies regarding the establishment of policy and legislative action, and (3) long-term liability unknowns. These additional data points should be considered in conjunction with the strategic business plan and the IDM.

Case Study Question*: What are the topical thoughts when considering risk reward regarding investment in a project of this magnitude?*

TABLE 5.1
CO_2 Storage Facility Proforma

	2025	2026	2027	2028	2029
Million metric tons of CO_2	1,000,000	2,000,000	3,000,000	4,000,000	5,000,000
Gross sales	$15,000,000	$30,000,000	$45,000,000	$60,000,000	$75,000,000
Cost of Services					
Direct labor	$3,000,000	$3,150,000	$3,307,500	$3,472,875	$3,646,519
Equipment lease	$1,200,000	$1,320,000	$1,452,000	$1,597,200	$1,756,920
Freight incoming	$2,000,000	$2,400,000	$2,880,000	$3,456,000	$4,147,200
Insurance	$7,000,000	$7,350,000	$7,717,500	$8,103,375	$8,508,544
Misc	$2,000,000	$2,200,000	$2,420,000	$2,662,000	$2,928,200
Utilities	$1,500,000	$1,650,000	$1,815,000	$1,996,500	$2,196,150
Maintenance	$2,500,000	$2,750,000	$3,025,000	$3,327,500	$3,660,250
Pore space lease	$300,000	$309,000	$318,270	$327,818	$337,653
Total	$19,500,000	$21,129,000	$22,935,270	$24,943,268	$27,181,435
Gross profit	$(4,500,000)	$8,871,000	$22,064,730	$35,056,732	$47,818,565
Operating Expenses					
Selling Expense					
Misc	$100,000	$110,000	$121,000	$133,100	$146,410
Advertising	$1,500,000	$1,650,000	$1,815,000	$1,996,500	$2,196,150
General and admin expense	$3,000,000	$3,300,000	$3,630,000	$3,993,000	$4,392,300
Research expense	$2,000,000	$2,200,000	$2,420,000	$2,662,000	$2,928,200

(Continued)

TABLE 5.1 (Continued)
CO_2 Storage Facility Proforma

	2025	2026	2027	2028	2029
Legal	$1,000,000	$1,200,000	$1,440,000	$1,728,000	$2,073,600
Total	$4,600,000	$5,060,000	$5,566,000	$6,122,600	$6,734,860
EDITDA	$(9,100,000)	$3,811,000	$16,498,730	$28,934,132	$41,083,705
Depreciation					
Building & infrastructure	$4,615,385	$4,615,385	$4,615,385	$4,615,385	$4,615,385
Truck	$571,429	$571,429	$571,429	$571,429	$571,429
Amortization	$17,118,600	$17,118,600	$17,118,600	$17,118,600	$17,118,600
Total	$22,305,414	$22,305,414	$22,305,414	$22,305,414	$22,305,414
Total operating expenses	$26,905,414	$27,365,414	$27,871,414	$28,428,014	$29,040,274
Income from operations	$(31,405,414)	$(18,494,414)	$(5,806,684)	$6,628,718	$18,778,291
Net interest expense	$6,861	$6,900	$12,741,332	$12,424,899	$12,085,591
Income before taxes	$(31,412,275)	$(18,501,314)	$(18,548,016)	$(5,796,181)	$6,692,700
Provision for Taxes					
Sales	$1,650,000	$3,300,000	$4,950,000	$6,600,000	$8,250,000
Real estate	$3,956,000	$3,956,000	$3,956,000	$3,956,000	$3,956,000
	$5,606,000	$7,256,000	$8,906,000	$10,556,000	$12,206,000
Net income	$(37,018,275)	$(25,757,314)	$(27,454,016)	$(16,352,181)	$(5,513,300)
Cumulative		$(25,757,314)	$(51,514,628)	$(78,968,644)	$(95,320,826)

(Continued)

TABLE 5.1 (Continued)
CO$_2$ Storage Facility Proforma

		2025	2026	2027	2028	2029
Million metric tons of CO$_2$	5,000,000	9,000,000	1,000,0000	11,000,000	12,000,000	15,000,000
Gross sales	$75,000,000	$135,000,000	$150,000,000	$165,000,000	$180,000,000	$225,000,000
Cost of Services						
Direct labor	$3,828,845	$4,020,287	$4,221,301	$4,432,366	$4,653,985	$4,886,684
Equipment lease	$1,932,612	$2,125,873	$2,338,461	$2,572,307	$2,829,537	$3,112,491
Freight incoming	$4,976,640	$5,971,968	$7,166,362	$8,599,634	$10,319,561	$12,383,473
Insurance	$8,933,971	$9,380,669	$9,849,703	$10,342,188	$10,859,298	$11,402,262
Misc	$3,221,020	$3,543,122	$3,897,434	$4,287,178	$4,715,895	$5,187,485
Utilities	$2,415,765	$2,657,342	$2,923,076	$3,215,383	$3,536,922	$3,890,614
Maintenance	$4,026,275	$4,428,903	$4,871,793	$5,358,972	$5,894,869	$6,484,356
Pore space lease	$347,782	$358,216	$368,962	$380,031	$391,432	$403,175
Total	$29,682,910	$32,486,379	$35,637,091	$39,188,059	$43,201,498	$47,750,540
Gross profit	$45,317,090	$102,513,621	$114,362,909	$125,811,941	$136,798,502	$177,249,460
Operating Expenses						
Selling Expense						
Misc	$161,051	$177,156	$194,872	$214,359	$235,795	$259,374
Advertising	$2,415,765	$2,657,342	$2,923,076	$3,215,383	$3,536,922	$3,890,614
General & admin expense	$4,831,530	$5,314,683	$5,846,151	$6,430,766	$7,073,843	$7,781,227
Research Expense	$3,221,020	$3,543,122	$3,897,434	$4,287,178	$4,715,895	$5,187,485

(Continued)

TABLE 5.1 (*Continued*)
CO$_2$ Storage Facility Proforma

		2025	2026	2027	2028	2029
Legal	$2,488,320	$2,985,984	$3,583,181	$4,299,817	$5,159,780	$6,191,736
Total	$7,408,346	$8,149,181	$8,964,099	$9,860,509	$10,846,559	$11,931,215
EDITDA	$37,908,744	$94,364,440	$105,398,810	$115,951,433	$125,951,942	$165,318,245
Depreciation						
Building & infrastructure	$4,615,385	$4,615,385	$4,615,385	$4,615,385	$4,615,385	$4,615,385
Truck	$571,429	$571,429	$571,429	$571,429	$571,429	$571,429
Ammortization	$17,118,600	$17,118,600	$17,118,600	$17,118,600	$17,118,600	$17,118,600
Total	$22,305,414	$22,305,414	$22,305,414	$22,305,414	$22,305,414	$22,305,414
Total operating expenses	$29,713,760	$30,454,595	$31,269,513	$32,165,923	$33,151,973	$34,236,629
Income from operations	$15,603,330	$72,059,026	$83,093,396	$93,646,019	$103,646,528	$143,012,831
Net interest expense	$12,085,591	$12,085,591	$12,085,591	$12,085,591	$12,085,591	$12,085,591
Income before taxes	$3,517,739	$59,973,435	$71,007,805	$81,560,427	$91,560,937	$130,927,240
Provision for taxes						
Sales	$8,250,000	$14,850,000	$16,500,000	$18,150,000	$19,800,000	$24,750,000
Real estate	$3,956,000	$3,956,000	$3,956,000	$3,956,000	$3,956,000	$3,956,000
	$12,206,000	$18,806,000	$20,456,000	$22,106,000	$23,756,000	$28,706,000
	$(8,688,261)	$41,167,435	$50,551,805	$59,454,427	$67,804,937	$102,221,240
Net income	$(100,834,126)	$(109,522,387)	$(68,354,952)	$(17,803,147)	$41,651,280	$109,456,218

TABLE 5.2
Investment Costs and Depreciation Schedule

Investments			Annual Depreciation	
Pore space lease for 300 acres surrounding center	$ 300,000		0	
20 trucks & pressurized shipping containers	$ 4,000,000	$ 571,429	7 years	
Land purchase 300 acres	$ 7,500,000	$ 192,308	39 years	
Operations center	$ 5,000,000	$ 128,205	39 years	
Ten above ground tanks	$ 30,000,000	$ 769,231	39 years	
Dehydration center	$ 10,000,000	$ 256,410	39 years	
Compression center	$ 20,000,000	$ 512,821	39 years	
Class VI well	$ 30,000,000	$ 769,231	39 years	
Power substation	$ 25,000,000	$ 641,026	39 years	
Above ground piping	$ 7,500,000	$ 192,308	39 years	
Capture & storage tank at each customer site	$ 45,000,000	$ 1,153,846	39 years	
	$ 184,000,000	$ 4,615,385		

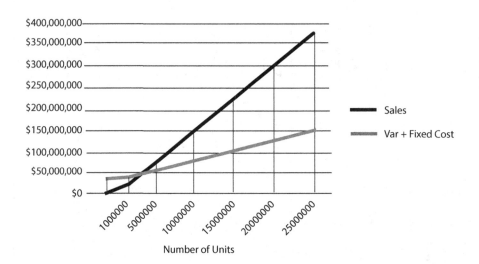

FIGURE 5.3 $MtCO_2$ injection breakeven.

5.4   SUMMARY

This addressed engineering economic analysis as it relates to strategic decision-making and maximizing TMA ROI. Financial analysis and budgeting are fundamentally the strategic plan expressed in financial terms, which is also necessary for achieving organizational and TMA success. For all projects, engineering based or other, evaluating alternatives involves identifying the most advantageous ROI time

TABLE 5.3
Investment Decision Matrix

Investment Decision Matrix

Time to market			5 years
Initial investment			$184,000,000
NPV	7%, 10 years		$549,605,000
Excess present value index			4
Benefit-cost ratio			2.99
Investment payback			9.4 years from net income, 4.2 years from gross sales
Internal rate of return			30%
Facility ownership			Yes
Asset value		yr1	$184,000,000

Purchase investment analysis

PV factor	7%, 10 years	NPV	Inflows
0	1	$184,000,000	
1	1	$15,000,000	$15,000,000
2	0.935	$28,050,000	$30,000,000
3	0.873	$39,285,000	$45,000,000
4	0.816	$48,960,000	$60,000,000
5	0.763	$57,225,000	$75,000,000
6	0.713	$53,475,000	$75,000,000
7	0.666	$89,910,000	$135,000,000
8	0.623	$93,450,000	$150,000,000
9	0.582	$96,030,000	$165,000,000
10	0.544	$97,920,000	$180,000,000
11	0.508	$114,300,000	$225,000,000
		$549,605,000	$1,155,000,000

value for the money involved. The issues commonly considered are (1) Initial Cost, (2) Interest Rate, (3) Annual Cost or Benefit, (4) Economic Life, (5) Salvage Value, and (6) After-Tax Cash Flow.

An essential part of strategic planning is to develop a proforma of operations that will indicate whether or not the strategic plan is do-able within the current operating and financial structure. Additionally, this provides an essential management tool to measuring actual financial performance. The IDM is a structure tool for collecting relevant financial information to support strategic decision-making. It consists of several financial measures, which provide multiple data points to support decision-making. If all point in the same direction, then the best decision is readily evident.

QUESTIONS

1. What is the difference between inflation and the "time value of money"? Are they related at all?
2. What is ROI applicable for throughout the organization? Why?
3. You place $1,000 into a savings account that pays 3% interest. How much will you have in 5 years?
4. An investment is set to pay back $5 million in 2 years. What is its present worth if the interest rate is 6%?
5. An investment of $1,000 paid back $600 after 1 year and another $600 after 2 years. What is its IRR?
6. A new product has fixed costs of $1M and variable costs of $10 per unit. It will sell for $20 per unit. What is the minimum sales volume before you make an overall profit (i.e., the break-even point)?
7. A depreciable asset had an initial cost of $15,000. It will be depreciated over 5 years by the "Straight Line" method. What will be its book value (or asset value) after 2 years?
8. What is the difference between EPVI and benefit cost?
9. How is the business financial proforma a part of the overall organizational strategic plan?
10. If fixed costs increase or decrease, what happens to the product volume break-even point for the business?

REFERENCES

1. Newman, D. G., Lavelle, J. P. & Eschenbach, T. G. (2013). *Engineering Economic Analysis*. New York: Oxford Press. ISBN: 0199740089.
2. White, J. A., Grasman, K. S., Case, K. E., Needy, K. L. & Pratt, D. B. (2013). *Fundamentals of Engineering Economic Analysis*. Hoboken, NJ: Wiley. ISBN: 9781118414705.
3. Van Horne, J. (2010). *Fundamentals of Financial Management*. Financial Times/ Prentice-Hall. ISBN: 9780273738015.
4. Cooke, R. (2004). *Finance for Nonfinancial Managers*. New York: McGraw Hill. ISBN: 0071425462.
5. Weiss, M. (1989). *Survival Plus: A Management Handbook*. Radnor, PA, Chilton Book Company.
6. Brauer, D., Lee, R. P. & Brauer, C. (June 2018). "Commercialization of Emerging Environmental Sustainability Technologies: Case Study based on CCU&S," *International Freiberg Conference on IGCC & Xtl Technologies, Coal Conversion, and Syngas*, Berlin, Germany.
7. Brauer, D. (2018). *CarbonSAFE: Commercialization of Emerging Environmental Sustainability Technologies: CCS Strategic Business Development*. Morgantown, WV: US Department of Energy, NETL, and UIUC, ISGS Division.

6 Management Control

Management control and leadership are critical factors in the success of a corporation. Accepting this, both control and leadership are recognized as an integral part of the whole Total Manufacturing Assurance (TMA) objective. It must further be accepted that TMA is not intended to be a trendy or fashionable activity that makes a grand debut and then quickly disappears but rather a long-term commitment the to continual improvement as a world leading corporation.

The control is demonstrated by maintaining a direct and expedient course towards corporate strategic goals and objectives. Leadership is demonstrated through the desire to ensure that products are insightfully and optimally designed upfront and subsequently manufactured and provided to the customer without any loss of intended design integrity.

This chapter outlines several concepts and tools to make management control of TMA activities easier. These concepts and tools are not presented as new and innovative techniques. They are presented as underused and very effective and efficient means to maintain management control.

6.1 INTEGRATED MANAGEMENT CONTROL

There are major organizational components that must come together to support and enable the ongoing TMA manufacturing environment and culture. In order to be effective, these individual elements must work together, communicate, feed on each other's strengths, and shore up each other's weaknesses. This requires that they be implemented via an integrated management control approach.

For this integrated structure to exist, management must take a very strong hand in control of all areas of endeavor, including engineering, manufacturing, marketing, and product assurance. Management must have a wide-ranging vision and avoid mistakes of shortsightedness or narrowness of focus. Far more than in traditional manufacturing environments, the TMA-oriented corporation must be glued together by the integrating powers of its management function.

This is not to say that successful TMA requires more management than traditional organizations. In fact, the truly integrated company often has fewer layers of management and fewer people employed at each level. This is not the contradiction that it seems to be at first glance. Integration of function requires that all hands know what each of the other hands is doing, and this becomes more natural when there are fewer hands involved. A large management structure tends to specialize and lose broad focus; a lean management structure necessitates a broad overview on the part of all participants.

The motto of the TMA management structure is "no walls." All departments and functions must cooperate and share information, responsibility, and decision-making functions. Interdisciplinary teams must be a standard working paradigm at all levels of control. Allowing "walls" between various functional groups encourages the

DOI: 10.1201/9781003208051-8

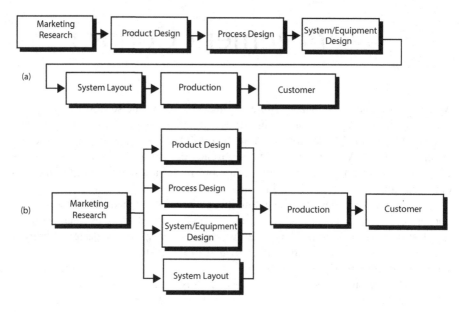

FIGURE 6.1 Operation and communication structure.

dangerous practice of the "throw it over the wall" style of interactions. In this mode, marketing will decide what product is needed, write it in a memo, and "throw it over the wall" into the design engineering department. Design will come up with what it deems to be an acceptable product design, and toss the blueprints over the wall to process engineering, which will come up with its vision of a good process plan. This will be lobbed over another wall to manufacturing, and so on.

What the organizational walls actually do is to force a sequential or "serial" operation and communication structure (see Figure 6.1, a level). This limits interactions, inhibits learning from the experiences of other groups, and generally keeps everyone's focus entirely too local for TMA.

A more appropriate structure is "parallel" operation and communication structure, where multiple departmental tasks are carried out more or less at the same time (see Figure 6.1, b level). Certain functions must naturally lead into other functions with their output, so that complete parallel structure is not usually feasible. But the more cooperation and interaction that is possible can only strengthen the entire organization.

6.1.1 MANUFACTURING ENTERPRISE FOCUS

As mentioned above, an integrated management approach requires a broad focus on the manufacturing enterprise. Certain management guidelines make this philosophical strategy more concrete:

- **Focus on Broad, Rather than Narrow, Performance Measures**: Avoid judging manufacturing tactics on short-term and short-range indices of performance. These short-sighted measures include such ideas as machine utilization, man-hours per part, floor space, and inventory levels. This does not mean that these goals are bad or even unworthy; they are quite good. But they are not proper goals for management to focus upon.

More fitting goals for high-level management to pursue are competitiveness in the industry, corporate image, customer satisfaction, adaptability to changing market conditions, and other issues that provide long-term staying power in relation to competitors. These high levels goals must be the guiding principles. When they are taken care of, the narrow focus measures will take care of themselves.

- **View Change as a Process, Not an Event or Program**: In keeping with a broad focus, it is necessary to recognize that change is the only constant in the manufacturing business, as in life. It is a mistake to look at a new production line or new organizational structure as a sudden change, dividing the old from the new in a quick, clear cut delineation. Rather, see the gradual implementation of new techniques and equipment as a way of life which, when constantly followed, ensures constant improvement.
- **View Reliability, Safety, and Quality (RSQ) as a Solution, Not as a Problem**: Too often, a management philosophy is to see acceptable performance by their products as a goal and lack of RSQ as an impediment to that goal. This leads to a manufacturing environment that implements only enough RSQ to avoid missing the given specifications. It also delays the injection of quality into products and processes until after they have both been designed and implemented. It results in a case of "too little, too late" and undermines the competitive edge. The correct philosophy is to see quality as the only goal and satisfaction with meeting specifications as the problem. When all efforts are tuned towards maximization of RSQ in all respects (including design, process, testing RSQ), the best possible, and most competitive, output will result.
- **Use Shorter Time Frames than Seem Comfortable**: Typical time frames for corporate goal programs run in the range of 3–5 years. The goals for these programs may be the implementation of a new customer service department, or the complete automation of an old process, or perhaps improvement of some performance index by a given percentage. Do not be afraid to attempt the same goals in half the usual time or even less. John Kennedy set the seemingly impossible goal of putting a man on the moon before the end of the 60s, and the goal was met. Forcing an organization to react as quickly as possible can motivate it to react as efficiently as possible.
- **Develop a Preemptive Attitude towards Problems**: Do not merely wait for problems to occur, thinking that they can be fixed when they crop up. This attitude is tempting, since it only requires action on problems that do, indeed, develop. But, it is self-destructive and anti-TMA in the long run. The broad focus approach is to develop an organizational structure which automatically senses any process that is beginning to deviate by even a slight amount and forces a solution before it can develop into a full-fledged problem. Just-In-Time (JIT) is an example of a preemptive approach to production control.

- **Be Willing to Take Risks**: There is an entire world full of competitors out there, and many of them are going to be taking risks. Some of those that do will fail, and fall behind, but some of them will succeed, and leave competitors to fall behind. To remain competitive, you must take risks, and you must make them pay off. Refusing to take risks is like the poker player who always antes but never sees a bet. The player will get nickeled and dimed to death and will never take home a pot.

These philosophical concepts may help develop broadly focused viewpoints for top management to follow, but they do little to help organize a corporation that is capable of following them. The successful organization is one with the proper interlinking of functions and lack of walls to allow these concepts to be implemented.

Guidelines exist for proper structure, some of which have already been discussed:

1. **Structure Like a Matrix, Not a Pyramid**: The typical organizational chart is based on a two-dimensional pyramid, with large numbers of low-level functions, each feeding upward to smaller and smaller levels of control. The top level is completely isolated from the lower levels, and no communication is possible in a sideways direction.

   For elimination of walls, this structure should be rearranged into a multidimensional matrix. The pyramid should be flattened so that there are fewer layers from the top to the bottom. The size of the bottom layer should be decreased, and the top layer increased, so that workers are less specialized and top managers are less abstract. There should also be a large number of horizontal lines of communication, so that all functions interact appropriately in all phases of a project. This fits in with the parallel, rather than serial, structure discussed earlier.

2. **Develop Interdisciplinary Knowledge in All Personnel**: Just as manufacturing functions are "integrated" by computer linkages and nontraditional organizations, people must also be integrated. This should be accomplished at two levels: at the level of the individual and of the group, or department.

   Encourage, or even require, individual employees to learn as much as possible about other departments and their functions, responsibilities, and problems. Publicize open positions within the company and encourage employees to apply for them. Institute training periods for new employees or newly promoted managers which rotate them through a variety of functions.

   On a group level, organize multidisciplinary teams to attack problems. Draw members for key committees from a large variety of departments. Invite people to meetings from different functional areas of the company. Whenever a group is organized, try to get a cross-section of the corporation's expertise involved.

3. **Eliminate Barriers to Communication**: Often, different departments have trouble communicating because of differences in vocabulary, viewpoint, or educational background. Try to standardize how information is stored and presented throughout the organization.

Vocabulary can be a significant problem. Each department may have a different term for the same concept or may use the same term for different concepts. Try to root out these sources of confusion and develop a standard set of definitions.

Time frames are another area where separate functions have trouble communicating. Top managers tend to think in terms of years, while sales and marketing personnel tend towards months and quarters. First-line managers think in terms of weeks and months, while workers often focus directly on individual shifts or even hours. While it is probably not possible to get everyone to agree upon a single unit of time, at least translate each individual piece of information into the appropriate time frame for the target audience.

4. **Avoid the NIH Syndrome**: This is the "Not Invented Here" attitude. This attitude wastes incredible amounts of time, as each functional unit attempts to reinvent its own wheel. Try to cultivate a team attitude; all departments and divisions are on the same side.

Management must not only prevent the NIH syndrome from wasting resources within the company, it must also resist the urge itself. Evaluating and possibly enhancing ideas derived from other companies may seem repugnant to some, but it is essential to retaining ompetetiveness.

5. **Do Not Allow Stagnated Benchmarks**: Posting a long-term stationary goal for a specific function can only lead to complacency and stagnation. Keep goals moving and constantly updated. Better yet, set meta-goals, such as increasing the amount of improvement each month or going longer without a defect than the last time.

6.2 DECISION-MAKING USING QUALITATIVE AND QUANTITATIVE TOOLS

Throughout the daily operations of a manufacturing organization, there are a multitude of management decisions being made in conjunction with numerous ongoing activities. This is true for corporations working for TMA, as well as corporations not striving for excellence. It is imperative that informed decisions are made, and that decisions are in accordance with a defined strategic theme. To ensure that this is the case, it is essential that formal project management exist as part of the corporate culture.

The key to successful project management is upfront planning. For a given project, the activity sequence, requirements, and major milestones (e.g., proof-of-concept and technology transfer) are defined and integrated into a basic project management plan. Various qualitative and quantitative analysis techniques are then applied to adjust the plan as necessary. The result is a plan which reflects the optimal approach to achieving the end goal(s) or objective(s) in support of TMA.

The fundamental objective of management discipline is to enhance the benefit-cost-time potential of a project. Without a defined project management activity, there exists the danger that little or no coordination between all the participating groups occurs. There is nothing worse than thinking a project is completed only to find out that fundamental elements of the project slipped through the cracks or that the elements completed fail to fit together.

For example, consider a product development project. It is known that successful commercialization depends on the product meeting or exceeding its design performance goals, including reliability, safety, and quality. Suppose that the product was lacking with respect to some of these fundamental performance characteristics but in general underwent a very successfully development program. However, at the end of development the product was lacking in commercial viability due to some missing performance characteristics required by the market. Somebody failed to get marketing involved to determine exactly what the targeted customer wanted. The preparation of a project management plan which identified all tasks and communication interfaces would have eliminated, or greatly reduced, the risk of such an error.

Sound project management optimizes the overall effectiveness of a project in regard to its constituent activities. With this approach, one minimizes the risks associated with a large number of activities and participants and maximizes their effectiveness contribution. Project management also assures that the proper activities are planned, that the activities are properly timed, that adequate resources (e.g., personnel, test items, test facilities, funding) are available, and that no unnecessary or redundant activities are conducted.

As a minimum, a project activity schedule should be developed. This involves defining the major project tasks and subtasks that will be performed. Also, timeline and milestones are defined for each task. Figure 6.2 illustrates a typical format for documenting this information.

Typically, the effectiveness of a project management plan is optimized with respect to estimated cost and time. This enables rational choices to be made as to overall project content and structure based on a return-on-investment approach.

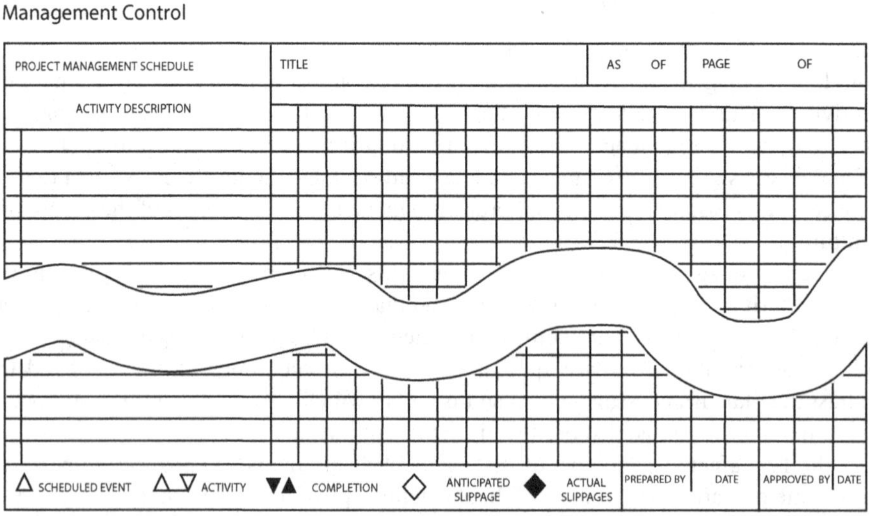

Management Control

FIGURE 6.2 Project management schedule form.

There are many techniques available to aid optimal management. Some of these include:

- Networking
- Linear Programming
- Inventory, Production, and Scheduling
- Econometric, Forecasting, and Simulation
- Integer Programming
- Dynamic Programming
- Stochastic Programming
- Nonlinear Programming
- Game Theory
- Decision Theory
- Optimal Control
- Queuing
- Difference Equations

All of these techniques have their place in practice. Their common objective is to aid the making of the best or optimal decision. The top two in the list are frequently used industry and are discussed in further detail in the following subsections.

6.2.1 PROJECT MANAGEMENT NETWORKING

A good tool for defining and guiding the project management plan is the use of a networking technique [1]. Networking is an enhanced graphical application of the Program Evaluation Review Technique (PERT) and the Critical Path Method. The network provides a view of the interactions and interconnections of various project activity time constituents. It also provides a powerful tool for identifying and recognizing relationships among interdependent activities.

This is particularly important since completing projects on time and within budget constraints is often difficult at best. The fact that certain activities must be completed before others can be initiated complicates matters. Therefore, the ability to effectively deal with interdependency between activities provides a great advantage in achieving success.

A network itself consists of a series of ovals, or nodes, connected by lines. The nodes represent the completion of activities and attainment of a project milestone. The nodes themselves contain several pieces of information. These include (1) the activity reference number, (2) the earliest start time (EST), (3) the latest start time (LST), (4) the earliest finish time (EFT), (5) the latest finish time (LFT), and (6) the activity duration time. Figure 6.3 illustrates the components of a network node.

The line between each node represents the activity in progress. The length of each line has no correlation to the actual duration time of the activity and is merely a function of convenience for the user.

The network illustrates how all the tasks of the project are tied together. The saying "a picture is worth a thousand words" certainly applies to project management networks. The network identifies which activities must be performed before others,

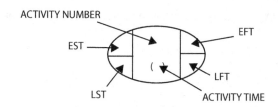

FIGURE 6.3 Components of a network node.

which occur in parallel, and which are unrelated. In addition, the time to complete individual activities, as well as the overall project, is clearly identified.

An important early step in preparing a network is to define the activities required in the project and to establish the proper order of precedence. Errors or omissions at this step will produce a faulty management plan and possibly have a catastrophic effect on the project at some point.

Proper project management networking involves identifying each activity, its description, its immediate predecessors, its estimated duration, and the resources available. The immediate predecessors of an activity are those activities that must be completed prior to the start of the activity in question. The estimated duration time is defined via discussion with the various persons participating in the project. Typically, both the overall time and resources available are well-defined project constraints and reflect corporate priorities.

As an example of project management networking, consider the project of conducting a design verification test for a hardware which is part of an overall system. The following test activities are defined as essential parts of the corresponding network diagram presented in Figure 6.4:

1. Test Unit Fabrication,
2. Special Test Equipment Development/Fabrication,
3. Test Procedure Preparation,
4. Support Requirements Definition/Establishment,
5. Test Configuration Setup,
6. Test Conduct/Monitoring,
7. Test Data Compilation,
8. Failure Reporting, Analysis, and Corrective Action,
9. Test Facility Use, and
10. Test Data Analysis/Reporting.

The network depicts the activities in their order of precedence. Note in the diagram that some of the project activities occur simultaneously as would be expected.

Once the project manager develops a network diagram, it is then filled out by estimating the time required of each activity. For all network diagrams, the earliest and latest start and finish times are defined in order to determine the "critical path" [2] and, subsequently, the total calendar project time required.

The EST is the earliest time an activity can begin when all preceding activities are completed as rapidly as possible. The EFT for each activity is the EST plus the

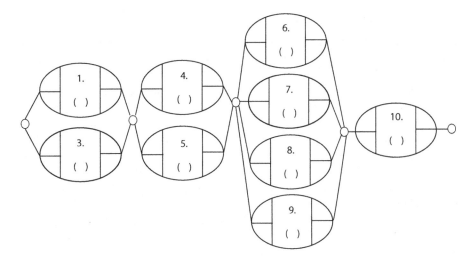

FIGURE 6.4 Example project management network.

activity time. When two or more parallel activities are immediate predecessors to another activity, the EST for the activity is the largest of the EFTs of its predecessors. The EST and EFT are defined by starting with the first activity and working through the network to the final test activity.

The LFT is the scheduled time when the test program must be completed. This time is fixed independently of the network diagram and, typically, corresponds to market introduction factors. The LST is the latest time at which an activity can be started if the test program schedule is to be maintained. The LST is LFT minus the activity time. These are defined by starting with the final activity and working backward through the network. To determine the LFT for an activity that has two or more parallel successors, the smallest LST of its successors is selected.

The EST and EFT times are developed by a forward pass through the networks. Identification of the LSTs and LFTs is done by a backward pass through the networks. The total slack is the difference between the two start times or the two finish times (these times are the same for each test activity). All activities along the critical path have the same amount of slack. Slack will either be zero or some minimum value for all the activities. The critical path activities are crucial to the on-time completion of the test program. These activities obviously receive the greatest amount of management attention.

In some cases, it may be found that the critical path is too long and must be shortened. To do this, two basic approaches are used. The first is the strategic approach. This consists of questioning the defined order of precedence of activities; particularly those on the critical path. It may be possible to make arrangements to complete some activities in a manner that removes them from the critical path (i.e., in parallel with some other activity).

The second is the tactical approach. This consists of determining if an increase in or acceleration of funding resources reduces the time of certain activities in the critical path. Initiating the paid working of overtime is an example of this. Obviously, this approach requires making a trade-off between project time and cost.

6.2.2 Optimization Modeling

Decisions are typically made in light of some definable constraint(s). Constraints many be self-imposed or dictated by others. Most frequently constraints are a function of time and money.

In order to make the best decision, it is necessary to balance, or trade-off, project variables (e.g., time and money) in order to make optimal decisions. Constrained Optimization Modeling (COM) is used to make management decision which achieves the best possible results considering restrictions. The whole area of making optimal decisions is encompassed in the field of management science. One of the most popular management science tools used for COMs is linear programming.

Linear programming-based COMs are used in defense, health, transportation, energy planning, and resource allocation. In purely private sector applications, uses include long-term and short-term scheduling of activities. Long-term planning activities include topics such as capital budgeting, plant location, marketing strategies, and investment strategies. Short-term planning activities include production and workforce scheduling, inventory management, and machine scheduling.

In all applications, the objective is to make optimal decisions which minimize or maximize some desired outcome. The trick for each application is to satisfy the defined set of restrictions or constraints in arriving at an optimal solution. Note the continual reference to the "optimal solution." A COM provides a solution that is called the optimal, or best, possible answer. Remember that these optimal solutions are produced relative to a mathematical problem posed by a model. Therefore, the optimal solution is only as good as the model defined to describe the problem.

Keep in mind that "optimality" is a theoretical concept, as opposed to a real-world concept. Although it is advantageous to have quantitative data to support and guide decisions, this alone should not form the basis for decisions. Management and engineering intuition must play a vital role in making sound, well-rounded decisions. Likewise, COMs are of great help in making final decisions; particularly, in the area of identifying the numerous variables involved in making a specific decision. However, the solutions generated must not be used in a vacuum excluding intuition.

6.2.3 Linear Programming

In structuring a COM, one must identify all the mathematical variables required in making the decision. The more complex the model, the more difficult it is to solve. However, increasing complexity also makes the model's solutions more accurate and credible. The bottom line is that models which realistically describe a situation are most apt to provide the best answers.

Look at the structure of a COM solved via linear programming. Every COM consists of an objective function, decision variables, constraints, and parameters. The objective function describes a decision function. In order to derive an optimal solution, the objective function must be either maximized or minimized relative to a set of constraints. Comprising the objective function are the decision variables. The decision variables represent actions or activities to be undertaken at various levels. Any selection of numerical values for the decision variables indirectly assigns values to the constraint functions. The COM requires that each of these constraint function values satisfies a condition expressed by a mathematical inequality or equality. The information serving as limiting values are the parameters.

There are five key steps in defining a COM [3].

1. Express each constraint in writing noting whether the constraint is a requirement (of the form >), a limitation (of the form <), or exact (of the form =).
2. Express the objective (i.e., maximize or minimize something) in writing.
3. Identify the decision variables.
4. Express each constraint in terms of decision variables.
5. Express the objective function in terms of decision variables.

In structuring the COM, it is important not to read more into a problem than precisely what is known.

As is obvious, linear programming-based COMS can rapidly become very complex. Such problems are conveniently solved using a computer to arrive at the optimal solution. (Numerous commercial software programs are available.). However, it is possible to solve simple linear programming models manually.

Simple COMs are comprised of two decision variables. This enables a graphical solution and enhances the conceptual visualization of a particular management problem. Graphical solutions provide key advantages in evaluating different problems. One visually sees what happens by adjusting the constraints. This might include adding a constraint to a problem, tightening or relaxing a constraint, or unintentionally leaving out a constraint or including a constraint in the model in the wrong way.

Linear programming makes use of the fact that equations of the form $ax + by > c$ are linear inequalities. This means that all constraints or restrictions in the COM are linear inequalities. When the above equation is changed to $ax + by = c$, it becomes a linear equation. In general, equations in which one side is of the form $ax + by$ and the other side is a variable value that depends on the choice of x and y is called a linear function.

Knowing that the graph of any linear equation is a straight line, this line is drawn by merely locating two points in the plane that lie on the graph. This is accomplished by finding the points of the line which intersect the x and y coordinate axes. To do this, a zero value is substituted for x to find the value of the y coordinate and vice-a-versa. This gives a set of two points (i.e., $(0, y)$ and $(x, 0)$) which graphs the linear equation.

This is similarly done for each constraint. The graph of all constraints together define a domain, or feasible region, which isolate all the possible combination of variables satisfying the objective function (see Figure 6.5).

The domain for the COM is determined by the relevant side of each constraint's equality line. This relevant side is determined by selecting a trial point (e.g., 0, 0) which is not on the constraint line. If the trial point satisfies the original inequality, then the linear line plotted and all points on the same side of the lines as the trial point satisfy the inequality. If the trial point does not satisfy the original inequality, then the linear line and all points not on the same side as the trial point satisfy the inequality.

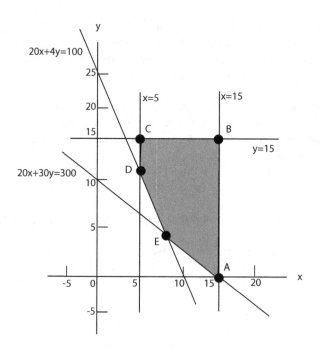

FIGURE 6.5   Graphing of constraints.

Depending on whether it is desired to maximize or minimize the objective function, the optimal variables are identified as the extreme points of the domain. These points are represented by a corner point of the domain; that is, a point where two edges of the domain come together. Note that the objective function is also a linear function and its graph passes through the extreme points of the domain for the optimal solution.

To sum up, linear programming optimization problems involving two unknowns are solved via three fundamental steps.

1. Construct a geometric representation of the domain, or feasible solution set of the optimization model.
2. Identify the extreme points of the domain.
3. Calculate the value of the objective function satisfying the criterion for maximization or minimization.

Keep in mind that optimization problems involving only two variables are unique. It becomes difficult if not impossible to solve a COM involving several variables, via geometric methods.

A popular alternative to geometric solution methods is the Simplex Method. The Simplex Method is a technique of matrix manipulation which automatically finds pairs of linear equations with graphs that intersect at extreme points, solves for variable values that satisfy both all constraint equations, and calculates the values of linear functions for the variable values found. In addition, the Simplex Method identifies the variable value combinations that solves the linear programming problem and, in general, does not need to examine all the extreme points in order to do this.

The most important feature of this method is its ability to work for linear programming problems in any number of unknowns, not just two. Also, it is well-suited to computer implementation in solving large linear programming problems. As with all decision techniques, there are some applications not suited for the Simplex Method. However, most COM-based linear programming problems arising in practice are solved using this method.

COM Example

Consider the need to establish the most effective product development test program plan. To do this, it is necessary to find the optimal solution via geometric representation of a linear programming-based COM. A graphical method is used since the natural segmentation of a product development test structure (i.e., laboratory testing and field testing) allows for two-dimensional analysis. It is desired to determine the maximum effectiveness levels for laboratory testing and field testing based on various project constraints.

The following objective function is defined:

$$\text{Maximize: } 70\,E_L + 30\,E_F$$

where E_L is the effectiveness level for laboratory testing and E_F is the effectiveness level for field testing.

The objective function is derived based on a maximum overall program effectiveness limit of 100%.

The following constraints are defined:

1. $E_L, E_F \geq 0$,
2. $E_L, E_F \leq 1.0$,
3. $70\,E_L + 30\,E_F \geq E$,
4. $t_L\,E_L + t_F\,E_F \leq T$, and
5. $c_L\,E_L + c_F\,E_F \leq C$.

where the decision variables are defined as follows: E is the minimum acceptable effectiveness level defined for the overall project (see Table 6.1) and the 70 and 30 represent effectiveness contributory weights; t_L is the estimated laboratory test time required, t_F is the estimated field test time required, and T is the actual test program time available; and c_L is the estimated laboratory testing cost, c_L is the estimated field testing cost, and C is the actual R/D test program funding available.

The COM defined states that the problem is to make the value of the objective function as large as possible, provided that the constraints are satisfied. The value of the objective function is measured in effectiveness (i.e., 0–100). Constraint one defines the non-negativity conditions. Constraint two defines the upper bound of laboratory and field-testing effectiveness. Constraint three defines the test segment effectiveness level possible based on the lower acceptable bound of overall test program effectiveness and the baseline test segment effectiveness weights. Constraint

TABLE 6.1

Test Plan Effectiveness Levels

Level A – Highly Effective:
A very high probability is required of the test program plan to contribute towards successful technology transfer. This corresponds to a range of 100 to 85.

Level B – Effective:
A significant probability is required of the test program plan to contribute towards successful technology transfer. This corresponds to a range of 84 to 70.

Level C – Reasonably Effective:
A moderate probability is required of the test program plan to contribute towards successful technology transfer. This corresponds to a range of 69 to 55.

Level D – Remotely Effective:
A low probability is required of the test program plan to contribute towards successful technology transfer. This corresponds to a range of 54 to 30.

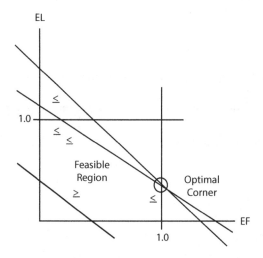

FIGURE 6.6   Optimality determined by a constraint set.

four defines the test segment effectiveness level possible based on the maximum test program time available and the program time identified for each test segment. Finally, constraint five defines the test segment effectiveness level possible based on the test program funding available and the estimated cost of each test segment. The model requires that each of these constraint values satisfies the condition expressed by the mathematical inequalities defined.

The established constraints are solved and plotted to define a feasible region from which the optimal corner found, and, consequently, the optimal effectiveness levels for laboratory and field testing. Figure 6.6 illustrates a possible feasible region formed by a constraint set. The optimal corner determined by the constraint set is that which maximizes the objective function defined previously.

6.3   TACTICAL OPERATIONS

The implementation leg of strategic goal and objectives, tactical operations, and activities encompasses those activities that are outcomes of the strategic planning pathway discussed in Chapter 3. Tactical operational strategies are the actions taken to achieve the TMA strategic goals and objectives and include the implementable details, in terms of priority, resource commitment, and timing.

The time for strategic planning is roughly after answering the following questions:

"What should we build?"
"Why should we build it?"
"When should we start building it?"
"How many should we build per day/month/year?"

At this point, tactical issues must be resolved. A new set of questions arises, including:

"How should we design it?"
"Should we make or buy the parts?"
"How should we manufacture it?"
"How should we assemble it?"
"How should we inventory it?"
"What types of machines should we use?"
"How should we arrange our factory?"
"How should we control our manufacturing processes?"

The activities/essential techniques for addressing these and related issues are presented in the remaining sections of this chapter.

6.3.1   DESIGN FOR MANUFACTURING

An important tactical tool is the philosophy of Design for Manufacturing (DFM). While this is considered a tactical tool from the perspective of manufacturing planning, it is not incorrect to consider it a strategic tool (which is to say, a far-sighted or long-range tool) from a product design viewpoint.

The idea of DFM is an example of TMA at its best: functions which are traditionally considered separate and sequential are now linked into a simultaneous and symbiotic process. In this case, the two functions are those of product design and process design. Rather than designing a part merely on the basis of its intended function, it is designed with an eye towards the issues of how to best manufacture it [4].

For example, imagine that several designs have been suggested for a part. Traditionally, the choice of the best design is based upon such issues as the fulfillment of the intended function of the part and costs of materials. Naturally, these are good and proper considerations, but they ignore the manufacturing consequences of the decision. What are the process costs of the selected design? How many machining steps are required? How many assembly steps, testing steps, and packaging steps are required? How many jigs and fixtures will be needed? How many man-hours?

Is it automatable? Can it be assembled from standard parts? The trick is to think of the product not as a standalone item, but as a cog in the entire business enterprise of the business.

One approach that improves DFM competency is the restructuring of design teams. A product development team should include one or more members from process or plant engineering who will have direct input into all design decisions. At the very least, a process representative should be consulted at regular intervals during the design phase.

This approach is especially important when a major new product line is being introduced and new facilities are being planned. The functions of product design, process design, tool and fixture design, plant layout, vendor selection, and equipment selection can and should be performed simultaneously.

6.3.1.1   DFM Guidelines

What are the rules to live by in designing for manufacturability? Certain guidelines have been established by various industries. Some are more applicable than others in different situations, so that these must be adapted to your particular product line and industry. Still, the DFM philosophy is well understood by studying these basic rules.

1. **Reduce the Number of Parts**: With fewer parts in a product, there are fewer drawings required, fewer assembly steps, and less paperwork. The results are lower error rates in both manufacturing and assembly, quicker production time, and increases in both quality and reliability (because there are fewer parts to fail).
2. **Reduce the Number of Part Numbers**: Not only should each product have as few components as possible, but the entire plant should require as few different types of components as possible. Always use a standard or "preferred" part that is already designed and on hand if it can do the job. This reduces inventory, paperwork, and chances for assembly errors.
3. **Reduce the Number of Vendors Used**: Try to select vendors who can deliver as many components as possible. This simplifies quality validation procedures, reduces number of shipments needed, and minimizes paperwork and chances of backorders.
4. **Design for Robustness**: Robustness is defined as insensitivity of output to variations in the input. The product should be designed to function correctly and consistently even in the face of variability of component characteristics (introduced by the manufacturing system).
5. **Reduce the Number of Adjustments Needed**: Try to eliminate, if possible, the use of adjustable components such as set screws, tunable electrical components, tensile, or compressive fits. Attempt to make all assembly steps of the "positive fit" variety, so that as soon as a component is in place, it is ready to function properly. This eliminates the possibility of adjustment errors, increases assembly quality and reliability, and vastly decreases assembly time and cost. This principle is worth following, even if it results in an increase in component cost.

6. **Use Foolproof Assembly Steps**: Use components that only fit one way – the correct way. Do not use force fits. Try to avoid use of threaded fasteners; snap-fit parts are quicker, cheaper, and easier to automate. Use chamfers and other location features to aid in assembly. Sequence assembly steps so that component "b" cannot be inserted until component "a" has been correctly inserted. The fewer the number of possible assembly errors, the fewer will be the number of actual assembly errors.

7. **Design for Vertical Stack Assembly**: Ideally, all components should be assembled along a single axis, like stacking up a sandwich or a pyramid. This reduces the need for reorientation during assembly, makes the process easier to automate, and helps make assembly foolproof (see previous guideline).

8. **Design Testing Procedures into the Product**: Testing should be accomplished with a minimum of disassembly, reorientation, and measurements. The quicker and cheaper testing is, the more testing you can do.

6.3.2 Computer-Integrated Manufacturing

The integration of computer technology to manufacturing has been increasing since the beginning of the digital industrial revolution. Computers have so permeated the manufacturing field in every aspect that much research and resulting texts are devoted to the various topics included within the general heading of Computer-Integrated Manufacturing (CIM).

The concept of CIM can be arbitrarily broken down into a number of overlapping areas: computer-aided design (CAD), computer-aided manufacturing (CAM), computer-aided process planning (CAPP), computer process control (CPC), and computerized business operations [5]. This section will focus on CAD, CAM, and CAPP and how they interact with computerized business operations in a general CIM system.

6.3.2.1 Computer-Aided Design

CAD is, obviously, the application of computer technology to the product design function. In the most trivial sense, CAD replaces the traditional drafting tools of drawing board, vellum, and pencil with a computer graphics workstation. This innovation alone supplies several advantages:

- Increased efficiency
  - A draftsman or designer can easily double his productivity by proficient use of a computer-based drafting system. The investment required to convert a drafting department to computer technology is roughly $20,000–$100,000 per workstation and 1–2 weeks of training per operator. This investment rapidly pays for itself, particularly if equipment is used for multiple shifts.
- Automatic documentation
  - A computer-generated document is easily stored and backed up on a file server system and can be readily distributed to the departments requiring it. It will not degrade with time or with multiple generations of copies nor will it distort when being reproduced.

- Ease of modification
  - If several similar parts are to be drawn, a single file can be easily modified to represent each version with no duplication of effort.

However, this list of advantages ignores the fact that a CAD system is far more than merely an automated drafting board. It is a computer-based system, with all of the programmability inherent in the power of the computer. The following new capabilities become available:

- Automatic design
  - Many CAD systems have been developed which automatically draft a component based on a few simple input parameters. For example, a software interface can be easily written by a CAD department which will query the draftsman or designer for number of poles, coils, and field magnets in a generator, and some basic dimensions, and will then automatically draw it. Time required for drafting of standard types of parts like this can be reduced to a mere percent or two of the time to draw it manually.
- Automatic analysis
  - Similarly, a design that has been created in a CAD system can also be linked to an analysis package to quickly and accurately calculate volume, mass, center of gravity, etc. Finite element packages can also be linked to CAD systems to do stress analysis, heat dissipation, etc.
- Automatic database
  - The CAD system can also be used to perform automatic feature extraction and database management. Such features as holes, fillets, finish surfaces, etc., can automatically be counted and stored in files to be sent to manufacturing, accounting, or bill of materials' packages elsewhere in the company.

6.3.2.2 Computer-Aided Manufacturing

CAM is a more complex topic than CAD and covers a broader range of functions. It can be thought of as the link between CAD and CPC. That is, it takes the database that was generated and stored by the CAD system and formats, transmits, and prepares the information to be sent to the numerically controlled machine tools on the factory floor.

Many incarnations of CAM exist, and each must be tailored for a specific factory and its equipment. Unlike CAD systems, which can be bought "off-the-rack" by any manufacturing company, a CAM installation is a complex system of many items from many vendors, carefully selected, installed, and orchestrated to implement the automation that the company requires.

CAM is best defined by example. Consider a fairly simple installation for the drilling of holes in circuit boards. A designer in the CAD department has just completed a drawing of a new circuit board, which consists of a new arrangement of holes and circuits on a standard-sized board. A CAM system for this application might start with a custom-written program which queries the CAD database and compiles a list of holes to be drilled. The list includes three numbers for each hole: the x and y location of the center of the hole from some reference point and the hole diameter. The list also includes the part number of the standard board from which the part will be produced. This list is transmitted via some network to the shop floor.

An operator, watching a human–machine interface (HMI) screen, sees that a board has been ordered. The required board is received from inventory and placed it in a CNC drill press. When it is ready, the operator presses a button, and the hole list proceeds to the CNC press, which then drills the required holes. No paper has been generated, and there have been very few opportunities for error. The hole file could even have specified that multiple parts be produced from the same file, which the operator would then have produced.

A more complex CAM system could have automated many more of the steps in the production process and could have reached more areas of the factory. Parts could be ordered from inventory automatically and delivered by an automated-guided vehicle (AGV) to the appropriate machine tool. Robots could manipulate the parts onto the machines and deliver them to the final assembly area.

As the role of CAM expands in a factory, the overall concept of CIM is approached. Other roles that the CAM system can support CIM is to include CAPP, cost estimating, part programming, inventory control, material requirements planning (MRP) and JIT implementation (discussed below), automated testing and inspection, and development of work standards.

6.3.2.3   Computer-Aided Process Planning

CAPP is one of the most vital links between the design and manufacturing functions in a CIM factory. By definition, process planning is the sequencing of the manufacturing steps to convert raw materials into products, in accordance with the design specifications. It is often thought of as something of an art form and can be very time consuming. For this reason, and because of its connection with many factory functions, CAPP is an important and far-reaching concept.

CAPP, and process planning in general, addresses the following questions:

1. What raw materials?
2. What machines should be utilized?
3. In what order should the parts be machined?
4. In what order should the components be assembled?

In a non-computerized process planning system, these questions are answered by manufacturing personnel and documented on a route sheet. This sheet is duplicated and distributed to all stations involved, and a large paper-trail results.

CAPP improves the process in two ways: reduction of the paperwork and removal of the human element from the planning process. The first task is a simple matter of information technology and data distribution, but the second is much more difficult. Development of the process plan is a task of synthesis, where various options are weighed, and an overall plan is created that will result in the finished product. It is a process that requires intelligence and as such is not automated without a great deal of effort. Two approaches are commonly used by a CAPP system to accomplish the task of generating a process plan, known as generative and derivative.

A derivative CAPP system, also called a retrieval system or variant system, seeks to create a new process plan out of an old process plan. It takes as input the CAD drawing of the new part and analyzes the part in terms of general shape, types of features, and complexity. This analysis results in a group technology (GT) code number, which describes the part in a general, feature-oriented manner. The CAPP system then searches the company's part database for a previous part with a similar GT code. When an old part is found with a sufficiently similar code, the route sheet for that part is retrieved and modified into a route sheet for the new part. The new part and its route sheet are then added to the database for use in the future. If no closely matching GT code can be found, the CAPP system either signals for an operator to develop the process plan by hand or defaults to a generative mode.

A generative CAPP system takes a "from the ground up" approach, generating a process plan completely from scratch. This system is an example of an artificial intelligence (AI) program and is usually created using expert system programming techniques (discussed later in Chapter 6). Basically, the program contains "rules" which relate product features to process steps. For example, one rule may state that if the overall shape of the part is cylindrical, the process plan should start with round bar stock, cut to the proper length. If it is cylindrical with varying diameters, a lathe step will be included. If it is cylindrical with holes, a drill press will be used, and so on.

Regardless of the type of CAPP system used, a route sheet is prepared and stored and routed to the necessary departments. It can be printed out and sent to each recipient on paper in a less-integrated factory or sent electronically in a more fully CIM-oriented factory.

The use of a CAPP system has other advantages besides reduction of paperwork and savings in time. One is standardization. While two human process planners might come up with two different process plans for the same part, a CAPP system will always provide the same plan. This can standardize factory operations and reduce time wasted in re-fixturing and re-tooling. Further, two similar parts will generally have similar process plans, which might not be the case with manual process planners.

Finally, since CAPP necessarily runs on a computer, an interface is automatically provided to other parts of the CIM operation of the factory. The CAPP system is easily interfaced to an MRP system, inventory system, customer billing system, etc.

6.3.2.4   Flexible Manufacturing Cells and Systems

Flexible manufacturing is the concept of using capital equipment such as machine tools, robots, and controllers for the production of a variety of products whose mix is not known in advance. In a factory which produces large volumes of a small number of parts, there is no need for flexible manufacturing. A factory which produces a variety of parts, but in well-defined and pre-planned proportions, has no need to adapt to changing demand. These types of factories are still well-advised to make use of TMA techniques but need not worry about the material in this section.

However, many industries do not have the luxury of advance warning as to their production requirements. In the quest for ever-decreasing lot sizes, setup times, and inventories, flexibility of production capability is becoming increasingly attractive to more and more companies.

What is this flexibility, and how is it achieved? The flexibility of interest involves the ability to produce:

1. A variety of products,
2. A variety of product mixes, and
3. New parts with minimal lead time.

This implies that it is not known, in detail, what will be called upon to be produced, or how much, or when. The production equipment must be flexible enough to perform whatever tasks are necessary, within reason, to meet the demands.

This flexibility is achieved by the use of a flexible manufacturing system (FMS). This is a group of equipment, generally under computer control, which has the capability of producing a variety of products with a small setup time between products. Ideally, the system should be able to produce one of each product at a time and produce it profitably.

A term often used to describe this type of system is flexible manufacturing cell (FMC). The difference between a FMC and a FMS is largely arbitrary, but in general, a system is thought of as a larger and more flexible installation than a cell. A FMC usually contains a small number of machine tools (say, a lathe, a mill, a band saw, and a robot) under direct control of one computer and designed to produce one part "family" of very similar parts. The term FMS generally designates a larger collection of machine tools, along with material handlers, some sort of storage and retrieval system, several controllers, and a central supervisory computer, all geared towards the production of several different part families.

Possible components of a FMS or FMC include all of the equipment used in any type of manufacturing. However, the most useful items to be included are the multi-function and programmable versions of these devices, such as CNC milling machines, machining centers, and turret lathes, along with the appropriate tool changing devices. Since FMSs are often run with minimal supervision, automated inspection equipment, such as automated coordinate measuring machines, are often included as well.

There is no need for the functions of an FMS to be limited to machining. They may also include tooling for sheet-metal forming, injection molding, forging, and joining processes. Larger systems require an extensive amount of material handling equipment such as conveyors and AGVs for moving material from one machine to another and robots for loading and unloading parts to and from each machine.

Another possible component for an FMS is one or more human operators. While the concept of FMS usually implies a high level of automation, that is not a requirement. A skilled human operator who is familiar with the process plans of all types of parts to be produced can replace the material handling equipment and greatly reduce the cost of the FMS.

How is an FMS designed? Obviously, there is an optimal selection of machine tools and material handlers for any one specific application, based upon the products to be produced, and a best guess as to the overall product mix required. Upon determining that an FMS is the appropriate approach, it is necessary to select the equipment, arrange it in the most efficient manner, and be able to control it for maximum throughput and machine utilization.

The first step, therefore, in FMS design is to identify the requirements for the system as closely as possible. Since the system will be, by definition, flexible, these requirements will naturally be ranges and approximations rather than exact numbers. The questions to ask are:

1. **How Many Different Products Will Be Produced and How Different Will They Be?** For our purpose, these are really the same question; they seek to determine the amount of flexibility needed in the FMS. The general rule is that efficiency decreases with increasing flexibility. When the need for flexibility becomes too large, the loss of efficiency makes the FMS approach infeasible. On the other hand, if very little flexibility is needed, a dedicated manufacturing setup will be more efficient than an FMS could ever be.

   The moral of this trade-off is that the FMS implementation is most attractive when a moderate amount of flexibility is required. If only one or two very similar parts are to be produced, the FMS approach should be abandoned in favor of a less flexible but more efficient installation. If there are a large number of widely varying products, a sufficiently flexible single system will end up with poorly utilized equipment. In this case, the operations would be best broken up into separate cells, some flexible, some dedicated, as circumstances dictate.

   What amount of flexibility is appropriate for FMS operation? How many products should there be, and how different should they be, to require sufficient but not excessive flexibility? Unfortunately, there is no general rule, as the actual numbers depend upon the nature of the industry in question.

2. **What Volume of Production Is Required?** This must be estimated from the product mix you intend to produce, the market conditions that you expect to prevail, and the level of production that you wish to attain. The production volume does not have a particularly strong influence on whether or not to use FMS but must be known to determine capacities and amounts of equipment to purchase.

3. **What Is the Anticipated Product Mix?** That is, of the different products the FMS will produce, approximately what percent of the production volume will be devoted to each product? Obviously, this question cannot be answered with precision. If it could, we would have no need for flexibility at all; we want the flexibility so that we can respond to changes in demand for each product. But we must decide what range of variation is likely for each product, so that we will be prepared to handle it.

Once these questions are addressed to the best of our knowledge, detailed specifications for the cell or system can begin. There is no general rule for FMS design, as distinct from design of any other production system, with regard to selection and layout of the equipment. The best approach is to use standard industrial and systems engineering techniques, as presented in Chapter 8, to develop several candidate manufacturing system layout designs. These various alternatives should then be evaluated using simulation techniques to determine which best meets the requirements for throughput, machine utilization, efficiency, and adaptability to the range of product mixes that are anticipated.

Once the FMS has been designed, specified, and built, it must be controlled. This is generally accomplished by one or more computers, operating at several levels of supervision. In the case of a FMC, with a small number of machines in one localized area, a single dedicated computer of midrange capability is usually be sufficient to handle all control tasks. A larger FMS, composed of several cells and an extensive material handling system, will require an additional level of computer supervision and will usually need several autonomous controlling computers with one supervisory mainframe to keep track of them all.

Whatever the arrangement of computer architecture, there are several levels of control function that must be met.

1. **Shop Floor Control**: This level is only necessary for a large FMS and is responsible for keeping track of the activities at each individual cell and location within the system. It is usually performed by a mainframe computer with connections to each of the individual cell controller computers. It tells each cell what to produce, when, and how many. It controls the AGVs and conveyors and keeps track of the parts and raw materials in the automated storage and retrieval systems. It uses an optimization procedure to allocate tasks for maximum throughput of the system as a whole. It also keeps track of the performance of the individual cells and may include some AI programs for adapting to changing conditions and improving performance. It will probably be connected to the company's financial and business records, as well as the CAD system.

2. **Cell Control**: In the case of a Flexible Manufacturing Cell, this is the top level of control. In the larger scale FMS, this level is directly below the shop floor control. It is responsible for the operation of an individual cell or unit in the FMS, such as a storage and retrieval system or a cell of machines dedicated to a particular family of related parts. It is usually implemented with CNC technology, a mini- or super-microcomputer which can control several machines at once, in real time. It stores the process plans for each part made at the cell, as well as instructions for automated inspection if the cell has that capability. It also keeps statistics on the performance of the machines within the cell and maintains control charts based on the results of the automated inspection.

3. **Machine Control**: Each machine within a cell must be individually controlled. In a less automated system, these would be the functions carried out by a programmable logic controller (PLC) or numerical control (NC) unit. In an FMS, these devices may still be used, under the supervision of the cell control computer. Alternately, a personal computer (PC) or other microcomputer may be used to link each machine to the cell controller, or the cell controller may be linked to each machine directly. Either way, it is a separate function, with the tasks of running the part programs and monitoring the machine performance, detecting tool wear and breakage, and handling any sort of error conditions which might arise.

6.3.3 Material Requirements Planning

Material requirements planning, or MRP, is one part of the production system in execution and control function and a very important part. It is the function that ensures the materials needed are known, available, costed, and available on time, as needed, for production [6]. It sounds simple, and in principle it is. However, in a large and complex factory, with dozens or more products, each composed of hundreds or more components, this becomes a gargantuan task. Fortunately, it is all based on simple math and bookkeeping, and so can be explained, if not implemented, with a minimum of difficulty.

An MRP system is based on one simple equation:

$$[\text{what we need}] = [\text{what we want}] - [\text{what we have}]$$

which is to say, we must calculate how many units of raw material and components are necessary to manufacture our planned production volume and subtract from that the number of units that exist, or will exist, in inventory. This leaves the number of units that must be acquired. It sounds simple, but there are complicating circumstances that must be remembered. One is magnitude: there may be a large number of products, composed of a large number of components. Another is timing: components must be ordered early enough for them to arrive on time, and raw material must be ordered early enough for to produce the components needed on time. Finally, there is time-phasing: an entire year's worth of material is not needed all at once but will be spread out over an appropriate period.

6.3.3.1 Master Production Schedule

The MRP system is driven by a master production schedule, which is an output from our strategic manufacturing planning functions. Someone has decided how many units of each of our products we must have, ready to deliver, for each period of time. The methods employed to develop this master production schedule were discussed earlier in this chapter and come from sales and demand forecasts, firm commitments from customers, desired inventory levels, etc.

A master production schedule is a two-dimensional matrix: it expresses the quantity desired for each product during each time period. The size of a time period varies from one company and industry to another but is typically a week, month, or quarter. Daily or yearly time periods are possible but are rarely practical. In the case of a monthly period, the master production schedule would dictate the required production of each of our products for each month. These production periods are also known as "time buckets."

This raises the question of how many time periods or buckets should be included in our schedule. This also varies from company to company but should be as large as each company finds feasible. The schedule can always be updated if demands should change. The length of the master production schedule is called the "planning horizon."

To summarize, the master production schedule is a two-dimensional matrix, with each product listed down the left side and each time period within our planning

PERIOD	JAN	FEB	MAR	APR	MAY
X-11 Computer	300	400	500	500	500
X-15 Computer	800	800	800	1000	1000
X-22 Computer	400	200	200	400	500

FIGURE 6.7   Master production schedule.

horizon listed across the top (see Figure 6.7). These numbers are used to derive the "what is wanted" values in the MRP equation given above.

6.3.3.2   Bill of Materials

Another input to our MRP system is called the Bill-of-Materials (BOM) file. This is used to calculate the details of the "what we want" values in our MRP equation. While the master production schedule tells us what we want to sell, the BOM tells us what we want to produce it from.

There is a separate BOM file for each product in our catalog. In a well-automated factory, a BOM file can be automatically produced from the CAD database for each product. Each file contains a hierarchical list of each subassembly in the product, each sub-subassembly in the subassembly, and so on, down to each individual component. A typical BOM format is called the "indented bill of material" list, as shown in Figure 6.8.

The level of automation and data integration in a factory greatly influences the ease with which the BOM is integrated into the MRP system. In a completely manual system, large amounts of bookkeeping must be performed by hand, by armies of clerks, data entry, and bookkeeping personnel. As a complete CIM system is approached, more and more steps in the information-processing function can be automated, resulting in a hands-off computer-driven MRP system.

6.3.3.3   Inventory Record File

The other input needed for the MRP equation is the "what we have" term. In other words, which components and raw materials are located in inventory. More precisely, how many of each item will be on hand, and not already spoken for, at the time period when it will be needed. Time phasing must be scrupulously followed at this phase of the MRP calculations.

This information is stored, and reported, in the inventory record file. This maintains the information of how many of each item of raw material, component, subassembly, and final product will be on hand for each time period throughout the planning horizon. The format varies from one installation to another, but there are generally be lines for total inventory, projected inventory, inventory already committed, and inventory available for use in future production.

X-11 Computer

System Unit (SU-57)
Chassis (C-17)
Power Supply (PS-12)
Hard Drive (HD-94)
Floppy Drive (FD-52)
Wiring Harness (WH-61)
Mother Board (MB-22)
CPU (CP-12)
Memory Chips (MC-91)
Housing (HA-27)

Monitor (MO-11)
Screen (SC-7)
Wiring Harness (WH-51)
Circuit Board (CB-17)
Housing (HA-28)

Keyboard (KB-77)
Circuit Board (CB-99)
Key Assembly (KA-11)
Housing (HA-29)

FIGURE 6.8 Indented bill of material file.

Again, the inventory records can be completely manual, highly computerized, or anywhere in between. The MRP system can function with any type of inventory system but is most easily performed with a computerized inventory system. The overall relationships between the various segments of the MRP system are shown in Figure 6.9.

The process described above takes the master production schedule and "explodes" it, in accordance with the BOM, into a huge list of raw materials. It also spreads it out over time to accommodate the various lead times involved in ordering and manufacturing each component and subassembly. The result is a highly scattered and fragmented list of items and order times. A certain amount of consolidation is often in order.

One concept to consider is that of "common use items." As previously discussed with Design for Manufacture, it is prudent to design products with common or "preferred" parts whenever possible. Another consideration comes from standard inventory theory concepts such as Economic Order Quantity (EOQ). If the MRP system says to order 9,000 units of some item in a given month, but there is volume discount for ordering 10,000, it may be advantageous to "stock up," despite the slightly increased carrying cost.

A further issue to consider is that of uncertainty. Is manufacturing lead time really a constant? How about ordering lead time? How crucial is each part to the overall

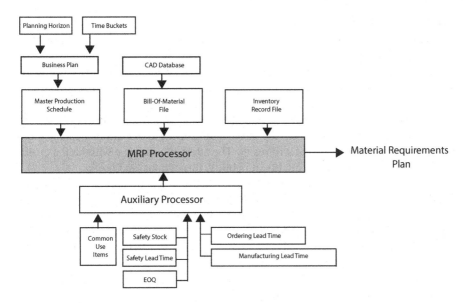

FIGURE 6.9   Elements of an MRP system.

process plan? When there is some doubt about the time required to finish a production plan, it is sometimes advisable to use "safety lead times," a cushion of time to make up for possible slippages in the schedule. A similar concept is "safety stock": excess inventory to account for uncertainty in demand, potential quality problems, or uncertain yield.

6.3.4 Just-In-Time Manufacturing

Addressing manufacturing and inventory effectiveness is the JIT approach to manufacturing planning and control. It is often hailed as revolutionary and the only competitive way to run a factory in the modern economic environment. Few people, though, really understand what a JIT system means or what it includes.

While a JIT system can revolutionize the operation of a factory, it is not based upon any new concepts. JIT must be implemented with all of the traditional techniques that most production people have known for years. What it changes is the philosophy that is used to run the factory, not the tools. The tools are used with a new attitude, which enables them to be used much more efficiently.

JIT is like jumping off a boat. It does not include any new skills that you did not already know nor does it require any. Rather, it puts you in a position where you must use what you know, and in the most efficient possible manner, or disaster will befall. This sounds dangerous and can be. Many would-be JIT practitioners have failed getting their JIT system working properly, but when everything comes together, what exists is an efficient, high-quality manufacturing process.

6.3.4.1 Inventory Exposure

The situation which forces efficient operations in a JIT system is a lack of inventory. The JIT philosophy sees inventory as an unnecessary evil: it artificially buoys up an operation that would otherwise sink and hides the problems that would otherwise be apparent. All efforts, therefore, are focused towards the reduction of inventory, both in terms of finished products and work in process.

This reduction has immediate benefits in the reduction of carrying cost and storage space and in an increase in responsiveness. But these are minor benefits compared to the other changes that will result. The attempt to survive without the crutch of excess inventory improves quality, reliability, throughput, and profits.

How is inventory a crutch? It allows mistakes to go unnoticed. Suppose the requirement is to deliver 100 units of a product at a specific time. If there is only 100 units, they must all be good. If 2% are defective, the order will be two units short for delivery. The process has "drowned" instead of swimming. But, on the positive side, the process detected the defectives and is able to trace them back to their cause, and ultimately cure the problem. If there were an extra 100 units in inventory, the Shipping Department would have "borrowed" the extra two it needed. The defects would not have disrupted the manufacturing operation, but there likely would not have been alert to the need for improvement.

This principle also holds for work-in-process inventory at each step of the manufacturing process. Each department consumes a little more material, and uses a little more manpower, than should be necessary. Finding all of these little inefficiencies is virtually impossible, as they are spread out over the entire operation. No single problem is large enough to stick out, but the cumulative effect is that we are carrying a large amount of waste in the system, in terms of materials, effort, and time, without ever knowing it is there.

6.3.4.2 Pull-Based Production

Implementing JIT requires switching from a "push"-based system to a "pull"-based system [7]. Each step in the manufacturing process must be based on a signal from a downstream department, signifying a need for the upstream department to perform its function. This is a system based upon demand, rather than supply. When department $N+1$ requires 25 units of work from department N, department N will produce it. If each part requires two components, department N will "pull" 50 units of components from department $N-1$ (see Figure 6.10).

How is this an improvement? For one thing, no station is doing more than is necessary. Department N is not allowed to produce a twenty-sixth unit, so it does not. It cannot make a few extra, "just in case" there is a problem. (JIT is often called the "antidote" to the Just-In-Case philosophy.)

Another forced improvement is quality. If department N must deliver 25 units, and is not allowed to produce a twenty-sixth unit, all 25 had better be good! Quality has been forced. This does not mean 100% quality will automatically occur; it just means that if it does not, the repercussions will be felt all the way down the line to shipping, billing, and accounts receivable. The problem will be noticed, and steps will be taken! JIT does not specify what those steps should be; that is up to the individual process engineers and will generally be standard techniques that have been known for years.

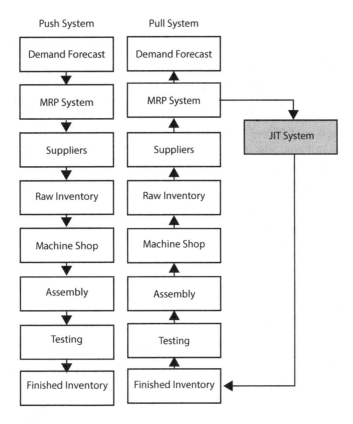

FIGURE 6.10   Information flow in push & pull systems.

6.3.4.3 JIT Implementation

A JIT manufacturing system is implemented by putting this demand-based system in operation. This requires redesign of many of the process planning procedures. All of the changes detailed below lead to a decrease in inventory level of finished goods and work in process and an increase in overall quality. The following techniques must be implemented:

- **Reduction of Lot Size**: The ideal lot size for a JIT system is one. Sometimes this is just not feasible, but often it is. In any event, the lot size should be reduced as much as possible, and policies should be instituted to constantly attempt to reduce it further, until a lot size of one is reached.

  Under no circumstances should a lot size exceed 10% of a day's production. This 10% value is a good starting point for a new JIT system, and it should be constantly lowered. What if a department produces less than ten items per day? Is a lot size of less than one feasible? Certainly, since any large, complex unit will be composed of subassemblies or produced by a number of steps. These should be broken down and implemented as separate lots.

- **Elimination of Discrete Batches**: Production should be thought of as a continuous flow, not batches of a discrete number of units. This is literally true when lot sizes have been reduced to one and is approximated when lots are small enough to complete in less than an hour. This smooth flow of products allows any level of demand to be met without waste, and for smaller and smaller orders to be produced profitably. This also allows for customized orders to become more feasible.
- **Production of Mixed Products Simultaneously**: It should be possible to produce all products at one time, in any possible ratio of individual levels. This, again, is strictly possible only when lot sizes have been reduced to one but is approachable to a reasonable extent when the 10% rule has been obeyed. This mixed mode of production also allows us to meet orders of any size or specialty in a profitable manner, with minimal waste of effort or production.
- **Reduction of Setup Times**: This is the key to many of the above steps. One of the main things which motivates large lot sizes is the overhead of setting up our equipment for a specific production process. If setup time is zero, profitable lot size by definition is one. As long as setup time is sufficiently small, though, sufficiently small batches will result.

  Often, setup time leads to a trade-off in the production engineering phase of process design: is it worth investing in more expensive equipment to reduce the setup time of a specific item? If lot sizes are to be large, the more expensive flexible equipment is rarely justified. If lot sizes must be small, it generally is not worthwhile to spend more for specialized equipment such as multi-purpose jigs and fixtures and programmable machinery. In a JIT system, the primary goal is to reduce lot size. Therefore, get the equipment. The payoff will come in the benefits to be reaped farther down the road.
- **Commitment to "Zero Defects" Goal**: JIT necessitates removal of quality issues. Suppose there are have two options on to purchase of production equipment: one system of machinery is fairly inexpensive but will produce product that is defective 2% of the time. The other brand of machinery costs twice as much but will yield good product 99.99% of the time. Is the better equipment worthwhile? Suppose the cheaper machine costs $5 per day in wasted materials and energy but is cheaper by $100,000. In a short-sighted sense, the waste is justified, but the JIT philosophy says no, it is not. Allowing the waste in the system would destroy the demand-based "pull" system, would create excess work in process, would obscure inefficiencies in the system, and would undermine the entire procedure.

  Quality improvement puts our system into an upward spiral of further improvements. As quality improves, batch sizes decrease, inventory decreases, subtler problems become apparent, and control over the process increases. True JIT is more closely approached, and further quality improvements follow.
- **Increased Worker Understanding**: Workers in the factory must understand more of the production process than just one workstation or department.

There are two reasons for this. First of all, they are working on a pull system: each station will only function when the next downstream station requests a batch. When this has been satisfied, what will the personnel do? In an ideal situation, they move to any other station which has work to be done. It is the workstations and their specific functions which must wait for a pull, not the employees. The more tasks that an employee is able to perform, the more fully utilized that person will be. They will also see the overall operation of the factory that much better.

The second reason is based in the JIT philosophy of continually striving for improvement. The best source of ideas for process improvement are the employees who are doing the work. The more stations that an employee can operate and the broader his view is of the operations at large, and the more useful his insights and suggestions will be. One of the main reasons for reducing inventory was so that problems became more evident. The other side of the coin is to enable the workers to see the problems when they surface.

- **Commitment by Management**: The importance of this item cannot be overstressed. JIT cannot be implemented in one department at a time; it must follow the production process of at least on product line from beginning to end. This will require a fairly high level of management agreement right from the start.

The first thing that happens when a product line is converted to JIT control is that it will fall flat on its face. Production schedules will not be met, deadlines will be blown, deliveries will be late, and customers will be dissatisfied. This cannot be helped! It is the only way that problems and inefficiencies will be found! It is the essence of the JIT process.

The second thing that will happen is that management will want to pull the plug on the whole idea. They will see that the state of the company is going from reasonably well to horrible and will want to give up. This must not be allowed. Management must be made to understand, right from the start, that this is an expected, and indeed necessary, phase of the JIT conversion.

The third thing that will happen is that the process will improve again. As problems are found and corrected, production will begin to catch up with the required schedule. Process efficiency and product quality will vastly exceed the pre-JIT levels. Management will see that its commitment was justified.

6.3.4.4 Pull-Based Startup

How is this "pull" system instituted? Two things must be done upon reviewing customer orders. First, the master production schedule must be adapted to put the initial pull into the production system, and, second, each department or workstation must be given a means of pulling on the department upstream from itself.

Once this customer order production pull is started, how is it propagated through the entire production process? There is no requirement that this be accomplished in a specific way, and each company is free to establish whatever procedure works best for it. Some of the more popular are summarized here:

1. **The Verbal Method**: In a small facility, there may be no need for formal procedures. Merely walking over to the upstream department, or picking up the phone, may be sufficient to get the message across. This is fine for small job shops and low-volume producers but is not recommended for larger operations.

2. **The Floor Outline Method**: A slightly less casual method is to draw an outline of the floor between each two successive departments or workstations. This outline should enclose an area just large enough to hold one lot worth of work in process. The downstream department draws its input from within the outline. The upstream department keeps an eye on the area. When the outlined area is empty, that is the signal to go to work filling it up again. When sufficient output has been produced to fill the outline, the downstream department stops.

   The downstream department must understand that it is not to start work again until the area is empty. Otherwise, there is a permanent inventory of work in process, and the JIT philosophy will have been violated.

3. **The Empty Pallet Method**: A variation on the floor outline method is to use an empty pallet or container. Exactly one container exists for each interdepartmental relationship. That is, there must be one container to shuttle between departments $N-1$ and N, one to shuttle between departments N and $N+1$, and so on.

   The containers are of such a size as to hold exactly one lot of the work in process that travels between the two departments. Department N draws its input materials from the container until it is empty. It then sends it back upstream to department $N-1$. The empty container is the signal for department $N-1$ to begin production again, and fill it up. When full, it is sent on downstream to department N.

   This is basically the same as the floor outline system, except that it can be used for departments that are not near to each other. It is therefore more convenient for large factories with complex process plans.

4. **The Kanban Method**: This is an even more sophisticated technique, but it is still based on the same principles as the methods presented above. Instead of sending empty pallets between departments, authorization cards (i.e., Kanban) are sent. When a downstream department is ready for a lot of input, it sends a Kanban to the previous department upstream, authorizing one batch of work.

   The Kanban system is more flexible than the previous methods because it is based purely on exchanges of information to signal production runs, not physical entities such as pallets. The Kanban can be electronic signals sent over a computer network. This enables extremely large factories with remote and highly interconnected production processes to operate in the JIT mode.

   Some systems use a dual Kanban approach: one type of Kanban is a production signal, telling the upstream department to begin production. The other type is a transportation Kanban, authorizing the motion of either an empty container to an upstream department, or a full container to a downstream department.

Once a JIT system is begun, the initial result will be chaos. As stated earlier, the factory is not able to handle the new requirements, and falls flat on its face, as surely as if a crutch had been removed. That is exactly what will have happened, as the crutch of inventory will have been eliminated.

The final result, however, will be wholesale improvements in efficiency of the overall operation. These include:

1. **Vastly Improved Quality**: Since there is no room for defects in a JIT system, they are necessarily eliminated, or vastly reduced.
2. **Minimized Inventory and Storage Costs**: Since inventory of finished products and work in process are virtually eliminated, no money is tied up in unsold material.
3. **Increased Equipment Utilization**: Since setup times must be reduced in order to operate in JIT mode, the equipment spends more of its time in useful production work.
4. **Minimized Waste**: Since there is little or no room for error in the new system, losses due to wasted material, energy, and time are eliminated as much as possible.
5. **Reduction of Paperwork**: Since there is no longer any significant inventory, there is no need to keep track of it. There is also no need for a central production planning facility to tell each department what it should be doing, since the pull system distributes that information as it is needed.

6.3.5 Technology Leveraging

There are many new technologies that have been developed over the past few decades, and many are still evolving. Manufacturing organizations need to use these technologies when and where they make sense. This section will briefly discuss these evolving technology areas.

6.3.5.1 Mechatronics

Mechatronics is the marriage between mechanical and electronic technologies [8]. As computers became more and more important in all types of technology, especially embedded computer systems, this blending of electronic and mechanical has become ever more important.

Mechatronic applications include robotics, home automation, control systems, and of course manufacturing. Any situation where electrical actuators cause motion, but where complex computer algorithms control and direct the motion, is an application for mechatronics [9]. The Internet of Things (IoT) is heavily dependent on mechatronic technologies. Small, affordable single-board micro-computers like the Arduino and the Raspberry Pi have made mechatronics more accessible to hobbyists, as well as to industrial applications [10,11].

6.3.5.2 Artificial Intelligence

AI includes a broad range of techniques used to create machines (particularly computers) that work in a way that mimics the human brain. Programmers study how humans solve particular problems and attempt to create programs that follow the same processes, in order to provide similar outcomes. Successful application of this methodology allows computers and other machines to handle complex tasks that were once thought impossible.

Examples of AI include speech recognition and facial recognition programs, natural language processing, chat-bots, machine vision, autonomous navigation routines, and some advanced search algorithms. Also, machine learning and expert systems (see below). When your favorite movie streaming app recommends new movies to you based on your past viewing, that can be considered a form of AI.

Interestingly, what is considered AI and what is just accepted as clever programming tend to shift over time. Tasks and programming methods that were considered AI and "cutting edge" 20 years ago are now just considered common programming techniques. Odds are, today's AI will be considered mere programming in another few decades.

6.3.5.3 Machine Learning

Machine learning is one type of AI, based on the concept that machines and computers can increase their store of knowledge while performing their intended functions. That is, they learn by experience. They must start with some beginning set of data and simple models of how to interpret that data. But, more importantly, they are also equipped with algorithms to let them refine their models as more data is acquired. In this way, they can continue to evolve and improve as time goes on, in the same way an intelligent person or animal might. Machine learning is closely related to AI and data analytics.

6.3.5.4 Automated Machining/Processes

Many machining processes and other manufacturing processes are controlled by computers, an architecture referred to as Numerical Control or sometimes Computer Numerical Control. While this is a well-established technology, new techniques are always being developed, especially in the realm of program generation, and it is wise to keep abreast of advances in the field.

6.3.5.5 Additive Manufacturing

Few new technologies have been as revolutionary in recent decades as that of additive manufacturing. While traditional machining starts with a blank that is larger than the finished product and selectively removes material, additive manufacturing (often colloquially termed "3D Printing") gradually adds material to the product until it achieves its desired size and shape.

Additive manufacturing with polymers and resins has been around for decades but has limited use in heavy industry. But recent breakthroughs (as well as recently expired patents) have led to significant advances in the use of additive manufacturing for metal products.

One important form of additive manufacturing is Selective Laser Sintering (SLS) which uses metal powders, along with lasers to sinter (fuse) the powder together to form the product. SLS and related processes such as Selective Laser Melting (SLM) are making great strides, both in the research arena and in industry. These are especially useful in high-price, low-volume products such as medical and aerospace components.

6.3.5.6 Advanced Robotics

Industrial robots have been around for decades and are excellent for manipulating payloads such as workpieces being loaded onto (or unloaded from) a machine tool, a pallet, or a conveyor. They are also used for precise path control tasks such as painting, welding, or grinding. They excel in material handling tasks that are too dangerous for human workers, too repetitive, or require more strength or precision than a person can achieve.

Modern advanced uses of robotics within industrial settings include robots with a variety of evolving capabilities:

* Use of AI and machine learning to make autonomous decisions about the tasks to be performed.
* Ability to travel through previously impossible situations, such as stairways, tunnels, and up and down walls.
* Advanced sensing and situational awareness.
* Micromanipulation of tools and workpieces too small for human workers to handle.

These are tasks within the manufacturing industry. Outside of manufacturing, advanced capabilities for robots include the ability to mimic complex human functions, autonomous navigation, and microbot and swarm robots that will someday enter a human body and conduct medical repairs and procedures at the micro level.

6.3.5.7 Autonomous Technologies

Autonomous technology enables devices to make decisions and act appropriately in unknown situations, without needing to be controlled by a human operator. It is enabled by a combination of sensor technologies, AI, and logic-based analytics. Popular examples of autonomous technology would be self-driving cars and self-navigating drones, as well as self-guided floor sweeping robots. Manufacturing examples would include AGV systems, automated inspection systems, and automated diagnostic and repair systems.

6.3.5.8 Man/Machine Interface

Man–machine interface (MMI) (also known as human–machine interface) is an old concept that is evolving in many new ways. It the simplest sense, a speedometer in a car is part of an MMI, so is the temperature knob on an oven in a kitchen. In an industrial setting, any knob or dial or readout that lets a person adjust a setting, or read a condition, on an industrial machine is also an example.

In the computer age, design of a MMI is part technology and part ergonomics. The classic GUI (graphical user interface) on a Windows-type computer is an MMI consisting of a mouse and pointer, icons, and windows. System designers must have as much knowledge of human psychology as they do of computer programming to make an interface that is both intuitive and powerful.

With more and more information being used in the manufacturing world, and more and more powerful machines to be controlled, the MMI must evolve along with the trends. New and emerging technologies include speech recognition, natural language processing, virtual and augmented reality, and, ultimately, direct neural interface, also known as brain–machine interface.

6.3.5.9   Internet of Things

IoT and Industrial IoT (IIoT) are terms for the way that all manner of devices share information, both locally and over the internet. The terms are primarily applied to the sort of devices that were not connected in the past. Home automation is a prime example: smart thermostats, smart light bulbs, smart doorbells, smart security systems, etc., combine to provide a home that can react to its environment, coordinate various functions, and can be controlled from a central point, or remotely. The rise of, and miniaturization of, embedded computer systems and wireless network interfaces is making IoT both feasible and economical.

Industrially, IIoT can have a tremendous impact on manufacturing. Such information as equipment status, inventory levels, work-in-process location, process scheduling, and maintenance requirements can be shared across the factory, enabling smarter and more timely decisions to be made, which serves to enhance product quality and processing efficiency.

The biggest concern regarding IoT and IIoT is security. Since it mostly uses the internet for communication, it is subject to all the inherent security risks that come along with it. Information can be stolen, and control of operations can be hijacked, if sufficient security precautions are not taken.

It should be pointed out that IIoT is the very soul of Industry 4.0, otherwise known as the Fourth Industrial Revolution. It is also an ongoing and evolving concept, and appears to only be in its infancy at this point, with great strides yet to be made in both the near future and the years beyond that.

6.3.5.10   Big Data and Analytics

In the modern internet age of Industry 4.0, the world is awash in data. There is so much data, and in such varying formats, that analysis by traditional tools (spreadsheets, etc.) has become infeasible. Also, the data are constantly changing, both in value and in format. Extracting useful information from this waving sea of data requires newer and more intelligent tools, often collectively referred to as data mining and analytics.

Techniques for analyzing big data include new computer architectures, new file structures, AI, and heuristic search mechanisms. Examples of big data in the manufacturing arena include inventory data, sales data, market analyses, price volatility, maintenance information, manufacturing equipment operation, material tracking, and production data.

6.3.5.11   Virtual/Augmented Reality

Virtual reality is a system where a user is immersed in a computer-created environment. Typically, it uses a headset that covers the eyes and presents a computer-generated view. As the user turns their head, the scene shifts as it would in real life. There may also be sensor gloves that allow a user to "feel" objects in the virtual world when "grasping" or "touching" them.

Augmented reality is a system that overlays virtual elements into the real world. It typically uses a transparent screen worn over the eyes where virtual elements are shown, which can be perceived along with the real environment, which is seen in the normal way. This allows the computer to supply extra information to the user, based on where he is and what he is doing. Alternatively, a fully immersive headset could be used, which displays both photographic images of the real world, and the extra computer-supplied information.

Both concepts have use in advanced manufacturing technology. Virtual prototyping allows a user to inspect a proposed design of a product without a real item ever being manufactured. It can be held, moved, rotated, and studied in ways that still images cannot. This is particularly useful for ergonomic study, such as layouts of instruments or controls on a dashboard, or accessibility studies for maintenance design. It is also extremely useful in training situations.

6.3.5.12 Supply Chain Visibility

Manufactured products spend much of their time in the process of material handling. This includes both internal transit between workstations and external transit between suppliers and customers and between various plants of the same company. One estimate put the typical time in transit at over 90% of overall manufacturing lead time.

During all of this time, parts can be lost, diverted, or damaged. Therefore, it is important to be able to track them as they travel. This capability is known as supply chain visibility. It applies to raw material, components, subassemblies, and finished products on their way to the end user.

Supply chain visibility is essential to the robustness of the supply chain, that is, the ability to recover from errors. If parts get mislaid or sent to the wrong place, it is essential to know that as soon as possible so that recovery operations can begin. If they are running behind schedule, it is important to re-plan production schedules as soon as possible. If parts are damaged in transit, it is important to have them replaced as soon as possible.

It should be noted that supply chain visibility is all about information and ensuring that the information is both accurate and timely. That makes it a perfect example of Industry 4.0 technology and philosophy.

6.3.5.13 Block-Chaining

A block-chain is essentially an extremely secure digital ledger of transactions. Its most common use is in the processing of crypto-currency transactions, but it has potential to be useful in contracting, supply chain management, and in ensuring the integrity of quality or certification data.

Block-chains are based on the idea of a *hash*. A hash is sort of a summary; a large record of data, of arbitrary size, is converted to a small, fixed-sized code using an algorithm. Anyone who runs the same algorithm on the same original record will get the same hash. However, there is no way to reproduce the original record from the hash; only to verify it. (Passwords are frequently stored this way.)

Each time a new transaction is registered on the digital ledger, it includes a hash of the previous transaction along with the new information about the current transaction. And the next transaction will include the hash of that transaction. In this way, each block (record of a transaction) will contain information about all previous transactions, in hashed form, and can therefore be verified. Any alteration to a previous bock will make the hashes incompatible, and the alteration will be obvious.

Block-chains work because of this cascading system of hashes and because they are stored on a publicly accessible, decentralized, peer-to-peer network. No one entity has access to all of the records, which would be required to make any unauthorized changes.

6.3.5.14 Cyber-Security

Cyber-security includes all steps and precautions taken to prevent malicious and unauthorized access to computer systems, networks, and data. The purpose of the unauthorized access could be to steal information (technical or financial or personal), to disrupt operations, or to take over and hold systems for ransom. Or they could just be malicious mischief that does no good for anyone. As the industrial world becomes more and more dependent on computers and other information systems, cyber-security becomes more and more important.

Many cyber-attacks rely on fooling users into doing something unsafe. *Phishing*, for example, is the process of masquerading as a trusted source to elicit a response from a user which reveals sensitive information. An email with a counterfeit "from" field may fool an employee into thinking he is responding to a colleague inside the company and cause him to reveal sensitive information such as a password.

Other attacks depend on overwhelming a computer or network with unexpected traffic so that its resources are unavailable for legitimate work. This is called a *Denial-of-Service* or DoS attack.

Ransomware is a type of malicious program that takes over control of a computer or network, and either freezes all operations, or seizes sensitive information which it then threatens to expose to the public. The owner of the attacked system is held for ransom; control is not returned until the owner has paid the ransom demanded by the attackers.

Keylogging is the act of surreptitiously recording the keys pressed on a computer keyboard by a user. Both software and hardware keyloggers exist. While these have legitimate uses, they are often used nefariously in order to steal sensitive information such as passwords or account numbers.

This is just a small sampling of the types of cyber-attacks that exist. A large and growing number of types of attacks are in common usage. Many cyber-security techniques have been developed to attempt to prevent these and other types of cyber-attacks from being successful.

A *firewall* is a hardware or software implementation that examines all information coming to a computer (or a network) from outside sources, such as the internet. The firewall watches for known threats, or likely threats, and attempts to block them before they can enter the system.

Antivirus software is similar but runs on a specific computer. When it detects malicious software, it attempts to isolate it or shut it down. Of course, both firewalls and antivirus software require that the threats are known and previously encountered in order to recognize them.

Encryption involves using a private code to translate or encode a message into an unusable or unreadable format before sending it over the internet. Only the intended recipient has the key to decoding the message, so that the data can be sent safely, even if it falls into the wrong hands.

User education is possibly the most important defense against cyber-attacks. Since many threats to cyber-security rely on users being fooled into unsafe behavior, this can have a large impact on establishing the security of the system.

These have just been a few examples in a large and constantly changing environment, and the interested reader is urged to consult many of the existing resources available on the topic.

6.4   EXPERT SYSTEMS

One of the most powerful tools available for management control of manufacturing is the expert system. Once thought of as a part of AI, expert systems have evolved to the point where they are more commonly thought of as just another programming technique. But they remain powerful and interesting user-friendly tools [12,13].

6.4.1   EXPERT SYSTEM DEFINITION

An obvious question is "what is an expert system?" Many definitions exist, and not all are completely compatible. For TMA purposes, an expert system is defined as a computer program that relies on (1) knowledge and (2) reasoning to heuristically arrive at a decision or perform a task that is usually accomplished by a human expert.

It is important to recognize that this definition sets expert systems apart from the myriads of other types of computer programs. Several concepts are implicit in our definition, each of which leads to a specific strength of the expert system approach.

- Knowledge is used rather than data. Data is raw information, not directly usable. As a consequence, it is not directly obvious which pieces of data are appropriate to a specific application. Knowledge tends to be verbal rather than numeric; a piece of knowledge may be: "The machine is broken," whereas a piece of data would be: "$X = 5.25$." The use of knowledge rather than data makes the expert system easier to build, understand, update, adapt, and maintain.
- Knowledge and reasoning abilities are independent functions of the system. In a standard computer program, the procedures to be followed, the rules and case statements and control algorithms, and the information concerning the task at hand are all bound together in one large system of programs and sub-programs. To update the system, one needs to search through piles of control code to find the specific line of code which contains the information sought. In an expert system, the knowledge is stored in individual chunks (called rules and facts), without regard for the order in which they are to be used, or the purposes they serve. The control and use of this knowledge

is governed by an inference engine which has the ability to reason on any problem, as long as the necessary knowledge is made available.

- Decisions are made based on heuristics rather than algorithms. A heuristic can be described as an imprecise procedure for solving a problem. It is in this very imprecision that the power of the heuristic approach lies: since it is loosely defined, it has a broader range of application, and is a more robust approach. An algorithm will probably give a more correct answer to a problem than a heuristic, but the heuristic will give a good enough answer far more often and will succeed acceptably in situations where the algorithm would break down and give no solution at all.

- The expert system emulates the thought process of a human expert. Humans tend to use heuristic approaches rather than algorithmic approaches, and their heuristics evolve over time. This is why a human expert is better than a novice at many tasks: he has built up a good system of heuristics to solve a variety of problems in a specific domain. Similarly, expert systems are most effective at tasks which require specific knowledge and procedures, rather than general purpose reasoning.

From these concepts, one can begin to see some of the advantages of the expert system approach. First, expert systems are better than ordinary computer programs.

- Since they use heuristics, they are more robust and can find workable solutions to problems where an algorithmic program might have failed to find a solution at all.
- They are more easily updated, since knowledge is maintained separately from the reasoning mechanism; a new fact or rule can merely be inserted into the knowledge base, and the reasoning mechanism will use it when appropriate.
- They explain their solutions. As we shall see, expert systems can easily be endowed with a "justification" feature, so that they not only provide solutions to problems but can also explain their reasoning if asked.

Second, expert systems are better than human experts.

- They don't get sick, go on strike, or retire.
- They are easily duplicated. A human expert requires years of expensive training. To clone an expert system, the only cost is for an additional computer and a copy of the software.
- Again, an expert system can explain its reasoning, whereas a human expert often has difficulty justifying decisions that he knows "just feel right."

6.4.2 EXPERT SYSTEMS VERSUS ARTIFICIAL INTELLIGENCE

The concept of expert systems arose out of the AI research community. It began as a tool to attempt to understand the mechanisms of human cognition and to emulate human thought processes [14].

With the development of expert systems that had the ability to perform certain types of reasoning, it became clear that they could be used to perform useful tasks. A new segment of the AI community began to concentrate on developing expert systems to be used for their own ability, rather than as a research tool. New programming techniques were developed, and the expert system moved out of the realm of AI and into the world of computer science and programming technology, creating a new niche sometimes known as "knowledge engineering."

As is true with many AI concepts, as soon as expert systems became useful tools with profitable applications, they ceased to be regarded as AI. Certainly, the development of specific expert systems for specific tasks is more knowledge engineering than AI work. But fundamentally, they still conform to the goals and concepts of the AI community.

6.4.3 EXPERT SYSTEM STRUCTURE

The structure of any expert system is based on four key elements, each of which has an analog in the human expert. The elements are the knowledge base, inference engine, user interface, and explanation facility (see Figure 6.11).

The *knowledge base* is where the information resides in the expert system and varies from one application to the next. Two expert systems, for two different purposes, may be identical in all respects except for the knowledge base.

The knowledge base can be thought of as the "education" of the expert system. Its analog in a human expert is the years of training, experience, and practice that a person accumulates on the way to becoming an expert. Like the expertise of a human expert, the knowledge base is the most difficult element to acquire.

The knowledge base itself can be divided into two parts: the rule base and the fact base. The rule base consists of a collection of rules, often of the form "IF <condition> THEN <conclusion>." These rules are completely general, often including

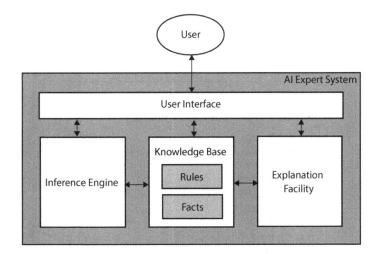

FIGURE 6.11   Expert system elements.

variables, and form the basis for decision-making in all possible scenarios of the domain in which the system is an expert.

The fact base, by contrast, contains simple statements about the condition of the world for the current problem at hand. Generally, the fact base will grow as the expert system is reasoning about a problem and new facts are discovered, whereas the rule base will only change if the knowledge engineer in charge decides that it should be updated.

Rules and facts together lead to the new facts, in much the way a syllogism functions in formal logic. For example, the rule "IF X is a man THEN X is mortal" combined with the fact "Socrates is a man" leads to the new fact "Socrates is mortal," which can then be added to the fact base.

The *inference engine* is the part of the expert system that performs the reasoning. It can be compared to the raw intelligence of a human expert. As with humans, different types of inference engines exhibit different levels of intelligence, but the same inference engine could be combined with many different knowledge bases to reason about many different domains of expertise.

Many different forms of inference engines exist, but all are designed to perform the same task: examine current facts, and use available rules to generate new facts. One type of inference is called "discovery." In this mode, all rules and all facts are used to generate all possible new facts, without regard to their possible relevance. In another type of inference, called "determination," the use of the knowledge is more goal directed. A specific "hypothetical" fact is suggested, and only those rules and facts are used that are necessary to determine if the hypothetical fact is true or false.

The technique used for this reasoning is called "chaining," and both "forward" and "backward" chaining exist [15]. Forward chaining is used for discovery (see Table 6.2). Each fact in the knowledge base is compared against the "if" part of each rule. If the conditions in the "if" section are satisfied, the rule is "fired," which means that the inference engine will execute the "then" part, by creating the facts it specifies. These new facts are also compared to the "if" portion of each rule, and so on, until no new firings are possible. At this point, all possible facts have been "discovered."

TABLE 6.2

"Discover" Using Forward Chaining

Current Rules	Current Facts
1. IF A and B THEN C	1. A
2. IF U or V THEN D	2. B
3. IF C or D THEN E	3. Y
4. IF X or Y THEN F	4. V
5. IF E and F THEN G	
6. IF G THEN H	
7. IF G THEN I	
8. IF D and I THEN J	

TABLE 6.3

"Determination" Using Backward Chaining

Current Rules: same eight rules as in Table 6.2
Current Facts: same original four facts as in Table 6.2

Task: Determine if J is a true fact.
Inference Engine Steps:
Goal is J
Rule 8 can prove J, so D and I are new goals
Rule 2 can prove D, so (U or V) is a new goal
V is a fact (Fact 4), so D is proven
I is only goal left to determine if J is true
Rule 7 can prove I, so G is a new goal
Rule 5 can prove G, so E and F are new goals
Rule 3 can prove E, so (C or D) is a new goal
D has already been proven, so E is proven
F is only goal left to determine if J is true
Rule 4 can prove F, so (X or Y) is a new goal
Y is a fact (Fact 3), so F is proven

Backward chaining (see Table 6.3) reverses this process, and it is used in determination. The expert system is asked to determine whether a candidate "fact" is true. The inference engine compares this fact to the "then" portion of each rule. If a rule is found which can generate the candidate fact, this rule's "if" portion becomes the new candidate. The inference engine then proceeds to attempt to prove the validity of the facts in the rule's "if" portion. Eventually, the hypothesis is either proven or the expert system runs out of facts and rules, thus determining that the original candidate fact cannot be said to be true.

Inference Engine Steps
Rule 1 uses Fact 1 and Fact 2 to get new fact: C (Fact 5)
Rule 2 uses Fact 4 to get new fact: D (Fact 6)
Rule 3 uses Fact 5 to get new fact: E (Fact 7)
Rule 4 uses Fact 3 to get new fact: F (Fact 8)
Rule 5 uses Fact 7 and Fact 8 to get new fact: G (Fact 9)
Rule 6 uses Fact 9 to get new fact: H (Fact 10)
Rule 7 uses Fact 9 to get new fact: I (Fact 11)
Rule 8 uses Fact 6 and Fact 11 to get new fact: J (Fact 12)

All intermediate goals have been traced back to known facts, therefore, J is true.

The *user interface* enables the expert system to communicate with a user. The form of the interface used depends on the intended audience of the expert system, and many systems contained a variety of user interfaces. The user may be well versed in the area of expertise in question and able to respond to queries with little prompting, or he may be a complete novice in need of constant hand-holding. The user may be a computer expert who is comfortable with a crude interface, or he may be a neophyte who would appreciate complete user-friendliness with graphics, menus, and copious online help.

Whatever the form, the user interface must allow for the following communications between expert system and user:

- The user must tell the expert system what the task is.
- The expert system must tell the user what facts it requires.
- The user must tell the expert system the current state of the world as it relates to the situation in question.
- The expert system must convey its conclusions to the user.
- If asked, the expert system must be able to explain its decisions to the user.

The explanation *facility* is the element that allows the expert system to justify its conclusions. It is an important facility for several reasons:

- Users may not trust the expert system and wish to be reassured that the conclusions are valid.
- Users may wish to learn from the expertise that the expert system possesses, rather than follow its suggestions blindly.
- The developer of the expert system may want to ensure that the knowledge base is properly codified.

The development of the explanation facility parallels the development of the rule base. Frequently, each rule in the knowledge base will include, in addition to an IF and a THEN clause, a WHY clause. This clause is used when the user requests an explanation of a conclusion.

6.4.4 EXPERT SYSTEM DEVELOPMENT

Now that a basic definition of an expert system has been established, the "hows and whys" of applying expert system technology to aid in achieving TMA will be discussed. Suppose that it has been decided to develop an expert system to help make decisions in our inventory control area. The development team is now tasked to efficiently construct a working expert system.

6.4.4.1 Expert System Domain

The first task is to properly define the *domain*. In general, the domain for this task is inventory control, but it is important to be as specific as possible. Will it include keeping track of current inventory, or just making decisions? Will it calculate optimal order quantities? Will it allow for backorders? Will it provide answers to specific questions only, or will it volunteer information that the user may not think to request? These, and other questions, must be answered before starting development.

What defines an appropriate domain for an expert system? The domain must not be too broad. Handling the accounting for an entire corporation would be a Herculean task. A good rule of thumb is: can one or two well-experienced people perform the task satisfactorily? If not, it may be necessary to downsize the domain purview.

The domain must not be too general, either. Tasks that require "common sense" are rarely appropriate for expert system implementation. One reason for this is that even a human expert will have trouble explaining his reasons for making a decision. If something "just makes sense," we have very little basis for coding a rule in our knowledge base. A rule of thumb to keep in mind is that the task should be solvable with expertise that has been learned over the course of time. In other words, *experience*. If a novice is just as good at completing a task as an experienced expert, then it probably doesn't' make sense to develop an expert system to perform the task.

6.4.4.2 Domain Expert

Once our domain has been specified and scoped appropriately, *domain expert* is required. This is the source of the knowledge that will go into the knowledge base. Frequently, the domain expert is an actual person, such as the person who has been performing the task in the past. It is noteworthy that a domain expert rarely becomes expendable after being immortalized in an expert system. For one thing, software requires as much maintenance as hardware; as company policy changes, the knowledge base must be updated. Furthermore, any large software project will require years of debugging and optimizing. In additions to this, the system will probably be implemented in several phases of gradually increasing complexity, and many years will pass before all tasks are automated.

As a final incentive to the balking human expert, there is the "mental wrench" effect. Often, the human domain expert who was used to generate the knowledge base becomes the user of the system in the future. The human expert finds that the expert system helps to magnify their own abilities within the domain. It frees the expert from the tedious parts of the development task, allowing greater focus on the more creative aspects. The expert can set up complex "what if" scenarios to test out solutions to problems and can make decisions in a fraction of the time formerly required... and may even learn more about the domain through the experience of coding their knowledge in a formal system.

6.4.4.3 Knowledge Engineer

In the expert system development project, there must be someone who is well versed in the theory and mechanics of expert system technology. This person is referred to as a *knowledge engineer*.

The knowledge engineer is the bridge between the human expert and the expert system. While there is generally not a need for prior knowledge of the domain area, the engineer should be able to grasp and understand the principal concepts quickly. The knowledge engineer must be enough of a psychologist to help the domain expert formalize their own knowledge and be enough of a computer scientist to code that knowledge into the form needed by the inference engine.

If knowledge engineering sounds like a daunting task, there is hope. With the development of newer and better shells (see below), the knowledge engineer's task is becoming easier. For many small-scale expert systems, the knowledge engineer and the domain expert are the same person. This is particularly true in technological domains, where the domain expert is already familiar with computing and programming techniques.

6.4.4.4   Expert System Shell

In the early days of AI programming, the only expert system tools available were string-processing computer languages such as LISP and Prolog. These required large amounts of programming to accomplish even relatively minor tasks and ran on expensive mainframe computers.

Nowadays, a multitude of expert system "shells" exist on the market, with prices ranging from hundreds of thousands of dollars down to several hundred dollars. A shell takes care of the busy work that must be done to keep track of rules, facts, variables, etc. Shells are usually written in a standard computer language such as C or LISP, and give the user a higher-level "knowledge language" that is much easier to use.

The shell provides all elements of the expert system except the knowledge base. It includes an inference engine which can perform forward or backward chaining, or both; it includes some sort of ready-made user interface, often with a built-in menu structure and graphics interface; it includes an explanation facility which can communicate the "why" fields of rules to the user; and often the shell will include a "development environment," which provides basic tools for the knowledge engineer to aid in construction of the knowledge base. This could include a text editor, an interactive debugging facility, and sometimes an "intuition facility," which can extract rules from decision tables.

6.4.4.5   The Development Process

The very first step in the expert system development process, which must be completed before actual work begins, is the task of assembling the essential elements listed above. The domain of the system must be properly scoped out, with a task that is well defined, sufficiently focused, and capable of solution by the use of expertise. A domain expert, or experts, must be found; either a person who can solve the task, a collection of well-documented case studies, or perhaps a good textbook. A knowledge engineer, or someone willing to learn to become one, must be hired or appointed to the task. Finally, the appropriate development platform must be acquired, probably an expert system shell with some sort of development environment.

Step 1: Structural Design

After these preliminaries are taken care of, the first step in the process is to design the structure of the expert system. The issues to address are such questions as: what will be the system inputs?...what will be the outputs?... what data is available?...what kinds of rules will be used? ...what will the principle variables be? ...what will the intermediate variables be? ...will there be different classes of rules?...and what will the classes be?

Answering these questions correctly is not essential. There will be opportunities to revise them. Most expert systems are developed "organically," in that they grow gradually from a small prototype and evolve as they grow. It is best to make reasonable decisions on the structure of the system and then see how well they work before revising them, if necessary.

It is also wise to focus on a subset of the domain at this point and to code some rules in the chosen structure. If these seem to make sense, the structure can be considered useable.

Step 2: Knowledge Acquisition

At this point, a well-defined structure for the expert system knowledge has been developed. It has been decided what sort of rules are needed, what they should do, and how they should interact. The next step to take is to extract the knowledge for the complete domain from the domain expert.

Attempting this step before Step 1 would be a mistake. The insights gained from the structural design phase help guide the knowledge acquisition process. This is especially important if the domain expert has trouble explaining his methodology. When presented with rule structures, he can usually "fill in the blanks" with less trouble.

Historically, knowledge acquisition is considered the "bottleneck" in the development process. It can be slow and tedious and requires many iterations and re-interviews with the domain expert. But care should be taken that the step is performed completely and accurately. The quality of the finished expert system depends almost completely on the accuracy, precision, and completeness of the knowledge acquired.

Step 3: Coding

The next step is to convert the knowledge collected in Step 2 into the structure developed in Step 1. This should be a fairly straightforward step if the previous steps were conscientiously handled.

This is also the step in which the expert system shell's development environment comes in most handy. The task difficulty can vary by orders of magnitude between a well-organized development system and a poor one.

Step 4: Validation

After the knowledge is coded, it is imperative that it be tested. Use the expert system from the perspective of the end user, assuming the level of understanding that they will have of the domain.

Case simulations should be run, and the solutions offered should be compared with those of the human expert. Initially, try running the examples that were used to develop the rules in the first place. If these do not yield the expected solutions, there is a coding error. Once these examples have been validated, try making up some new examples that the system has not seen before. Compare its answers to what the domain expert would do in the same situation. In this way, the scope of the system can be determined. Ideally, it should be able to solve as many types of problems as are likely to occur within the realm of the system's domain.

Step 5: Growth

As was stated earlier, the best of expert systems will grow organically, getting larger and more complete with time. Try following the first four steps on some significant subset of the knowledge domain first, and see how well the system works. If it performs adequately, add more and more capability. If not, do not be afraid to throw it away and start from scratch with a new structural design for the knowledge. The insight gained from the first attempt will make the second one all that much better.

Also, be prepared to never put the project completely to bed. As time passes, new knowledge is created. The domain will evolve as product lines

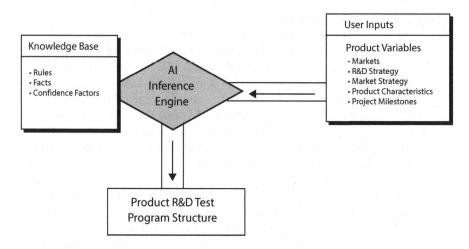

FIGURE 6.12   Conceptual expert system.

are updated, company policies change, and organizational goals are reevaluated. Occasionally go through the knowledge base and weed out old rules that never fire anymore (some shells provide a mechanism for tracking rule usage).

6.4.5   EXPERT SYSTEM TOPICAL APPLICATION AREAS

Following are a variety of examples of expert system applications which have proven useful in various facets of the manufacturing industry [16]. This is by no means a complete list but merely some interesting topical applications. Figure 6.12 provides an example of how the expert system provides the AI for a product research and development test program structure.

Design

- Prediction of stress cracking
- Design of heat fins
- Evaluation of design for manufacturability
- Design of v-belt drive systems
- Design of wire ropes
- Design of reciprocating pumps
- Design of molds
- Cost estimation

Production

- Machining condition (speed, feed, etc.) selection
- Process planning for machining or assembly

- Job scheduling
- Inventory order control
- AGV path planning
- Manpower resource planning
- MRP and MRP-II high-level control
- Robot selection

Control

- Fault diagnosis
- Alarm monitoring
- In-process inspection
- High-level control of PLCs
- Online tool selection
- Online tool monitoring
- Online component failure prediction

Maintenance

- Troubleshooting diesel-electric locomotives
- Troubleshooting electrical systems
- Troubleshooting HVAC systems
- Troubleshooting industrial robots
- Diagnosis of vibration problems in turbomachinery

TMA Case Study: AI-Based Tool for New Technology R&D Test Program Definition

OVERVIEW

There are a large number of tests and test sequences that can be applied during a new product research and development (R&D) project, each specific test program must be properly designed and optimized relative to the project and product strategic R&D objectives and characteristics at hand. This involves considering the marketplace and hardware technology, complexity, and end-item application characteristics, as well as the anticipated volume (or exposure) and cost constraints of the particular product being developed.

ISSUE

Defining the proper test program is an iterative process that involves trade-off analysis to maximize cost-effectiveness. It is desirable to use company experience to guide engineering management in developing an AI-based tool which leads the user through test program planning in a disciplined and efficient manner.

STRATEGIC OBJECTIVE

The strategic objective is to use an automated, AI-based system to aid engineering in defining and implementing an optimized new technology R&D testing program. This will use, in part, historical company experience knowledge to expedite development while optimizing test effectiveness, time, and cost.

CASE BACKGROUND

A decision logic process was developed to facilitate the development of AI-based expert system to define the most effective test program plan for a given project. As the process is carried out, the resultant sequence of "yes" and "no" answers, together with the rationale for each decision, is used to develop an applicable product development test plan.

The logic process requires the user to define the total development project strategic test plan, the product characteristics, and R&D objectives, and then asks if each level of testing (i.e., experimental, Design Verification Test, Design Approval Test, reliability, and field) is necessary. If the answer is "yes" for a particular test, then the test parameters must be defined. If the answer is "no," then that particular test is deemed to be not advantageous to the scheme of the product R&D effort and the logic user advances to the next question. Once the applicable test sequence is defined, the optimal effectiveness of the candidate test program plan is determined using a constrained optimization model. The model optimizes effectiveness relative to the estimated cost and time of the program plan with the known cost and time constraints.

If the plan satisfies the defined R&D strategic testing objectives and is evaluated to represent an acceptable effectiveness level (based on available data and engineering intuition), then it is formalized. If the plan does not meet acceptability requirements, then the test sequence, rigor, cost, time duration, etc., must be revised. This requires recycling the decision logic process until a test program plan has been defined which is acceptable.

The decision logic questions, along with some guidance to help in making the correct decisions, are presented in the text which follows.

Question 1: Is the product strategic R&D plan defined?
This question addresses all of the key strategic issues of the product (and project). Such a plan is derived based on the performance of both external and internal analyses.

A "no" answer means that a strategic plan must be defined at this time. The strategic plan is defined based on:

1. **Customer Analysis**: Segments, Motivations, and Unmet Needs
2. **Competitive Analysis**: Identity, Strategic Groups, Performance, Objectives, Tactical Strategies, Culture, Cost Structure, and Strengths Weaknesses Opportunities Threats (SWOT Analysis)
3. **Industry Analysis**: Size, Projected Growth, Industry Structure, Entry Barriers, Cost Structure, Distribution System, Trends, and Key Success Factors

4. **Environmental Analysis**: Technological, Governmental, Economic, Demographic, Cultural, Scenarios, and Information Need Areas
5. **Performance Analysis**: Return on Assets, Market Share, Product Value and Performance, Relative Cost, New Product Activity, Team Member Development and Performance, Employee Attitude and Performance, and Product Portfolio Analysis

If the answer is "yes" or if a "no" answer has been satisfactorily addressed, then the product characteristics are defined and all pertinent supporting information is noted on the worksheet. The logic process then proceeds to Question 2.

Question 2: Are the product characteristics defined?

This question addresses all of the key characteristics of the product. These are derived based on the product's intended operational profile and commercial application.

A "no" answer means that these characteristics must be defined at this time. The characteristics are defined based on the technology to be used, product design complexity, end application, potential market volume (or exposure), and availability (i.e., reliability and maintainability) requirements.

If the answer is "yes" or if a "no" answer has been satisfactorily addressed, then the product characteristics are defined and all pertinent supporting information is noted on the worksheet. The logic process then proceeds to Question 3.

The information documented as part of Questions 1 and 2 is used to establish the R&D strategic objectives for the product. These objectives dictate the R&D steps and activities necessary to "prepare" a product for manufacturing and commercialization.

The following questions are answered based on the strategic objectives defined.

Question 3: Is laboratory testing necessary?

This question addresses the need for the performance of formal laboratory testing, either indoors or outdoors. A "no" answer means that laboratory testing is not necessary as part of the R&D effort and the logic process moves to Question 8.

If the answer is "yes," then the appropriate laboratory tests and their parameters are defined by answering Questions 4 through 7.

Question 4: Is experimental testing appropriate?

This question addresses whether or not experimental testing should be conducted as an integral part of the product R&D process.

A "yes" answer means that the experimental testing is to be conducted and the test parameters must be defined at this time. The test parameters and typical values consist of the following:

1. Test Item Configuration: Prototype
2. Number of Test Items: 1
3. Test Duration: As Required
4. Measurement: Performance and Visual
5. Test Conditions: Laboratory (Ambient)
6. Acceptance Criteria: Meets Product Specification

7. FRACA: Not Applicable
8. Test Monitoring, Data Recording, Loading: Manual, Simulated Load

If the answer is "no" or if a "yes" answer has been satisfactorily addressed, then the test parameters and their values are defined and noted with all pertinent supporting data. The logic process then proceeds to Question 5.

Question 5: Is design verification testing appropriate?

This question addresses whether or not design verification testing should be conducted as an integral part of the product R&D process.

A "yes" answer means that the limited laboratory testing is to be conducted and the test parameters must be defined at this time. The test parameters and typical values consist of the following:

1. Test Item Configuration: Early ETU (focus on high-risk components)
2. Number of Test Items: 1–3
3. Test Duration: As Required
4. Measurement: Performance and Visual
5. Test Conditions: Selected Field Environments
6. Acceptance Criteria: Meets Product Specification
7. FRACA: Fully applicable
8. Test Monitoring, Data Recording, Loading: Manual, Simulated Load

If the answer is "no" or if a "yes" answer has been satisfactorily addressed, then the test parameters and their values are defined and noted on the worksheet along with all pertinent supporting data. The logic process then proceeds to Question 6.

Question 6: Is design approval testing appropriate?

This question addresses whether or not design approval testing should be conducted as an integral part of the product R&D process.

A "yes" answer means that the full laboratory testing is to be conducted and the test parameters must be defined at this time. The test parameters and typical values consist of the following:

1. Test Item Configuration: Advanced ETU
2. Number of Test Items: 1–3
3. Test Duration: As Required
4. Measurement: Performance and Visual
5. Test Conditions: Full Environmental Test Sequence
6. Acceptance Criteria: Meets Product Specification
7. FRACA: Fully Applicable
8. Test Monitoring, Data Recording, Loading: Manual, Simulated Load

If the answer is "no" or if a "yes" answer has been satisfactorily addressed, then the test parameters and their values are defined and noted on the worksheet along with all pertinent supporting data. The logic process then proceeds to Question 7.

Question 7: Is reliability testing appropriate?

This question addresses whether or not reliability testing should be conducted as an integral part of the product R&D process.

A "yes" answer means that the reliability testing is to be conducted and the test parameters must be defined at this time. The test parameters and typical values consist of the following:

1. Test Item Configuration: Laboratory Qualified or Advanced ETU
2. Number of Test Items: 2–4
3. Test Duration: 3,000–10,000 Cumulative Hours
4. Measurement: Performance and Visual
5. Test Conditions: Accelerated Environmental Test Cycles
6. Acceptance Criteria: R-achieved meeting R-specified
7. FRACA: Fully Applicable
8. Test Monitoring, Data Recording, Loading: Manual, Stimulated Load

If the answer is "no" or if a "yes" answer has been satisfactorily addressed, then the test parameters and their values are defined and noted on the worksheet along with all pertinent supporting data.

The logic process then proceeds to Question 8.

Question 8: Is field testing necessary?

This question addresses whether or not field testing should be conducted as an integral part of the product R&D process.

A "yes" answer means that the field test is to be conducted and the test parameters and their values must be defined at this time. The test parameters and typical values consist of the following:

1. Test Item Configuration: Pilot Unit
2. Number of Test Items: 3–20
3. Test Duration: 1–2 Years
4. Measurement: Operational, Periodic Performance and Visual
5. Test Conditions: Set of Representative Operational Locations
6. Acceptance Criteria: Meets Operational Goals
7. FRACA: Fully Applicable
8. Test Monitoring, Data Recording, Loading: Automated, Actual Load

In defining the sizing parameters for the field test, the expert system determines values for the number of test units, the length of the test, and the frequency of measurement and inspection of the test units.

If the answer is "no" or if a "yes" answer has been satisfactorily addressed, then the test parameters and their values are defined and noted on the worksheet along with all pertinent supporting data.

Once the basic test plan is defined, its optimal effectiveness is then determined with respect to cost and time considerations. The logic process then proceeds to Question 9.

Question 9: Is the test plan acceptable?

This question addresses whether or not the product R&D test program plan defined will adequately support technology transfer. Answering of this question is dependent on the responses made for Questions 3–8, and engineering management intuition assessment in comparing the test program plan defined with the basic strategic issues impacting the product and overall R&D project.

A "yes" answer means that product R&D test is acceptable and the applicable test(s) are integrated into a formal product R&D test program plan.

A "no" answer means that the test plan must be recycled through the decision logic to Question 3 in order to adjust the test plan relative to (1) content, (2) scope, (3) rigor, (4) effectiveness level, (5) time, and/or (6) cost. Recycling is performed until a "yes" answer is achieved for Question 9.

RESULT/CONCLUSION

The AI-based expert system recommends a basic test plan, which upon acceptance, leads to the development of a formal product R/D test program plan. This test plan establishes exactly what must be done, who must do it, and when it must be completed. It describes in detail the product to be tested, the environments that it will be subjected to, and the test schedule. The formal plan outlines, in detail, the engineering management responsibilities in overseeing the test(s) implementation, as well as specific supporting personnel responsibilities.

An effectiveness-balanced weighting factor is redefined for each applicable test effectiveness factor based on the following relationships:

1. *Risk Reduction Factor, F1*

   For each applicable test, the derived effectiveness weight provides a means for subjectively scaling the rigor and completeness with which the test will be implemented.

2. *Operational Suitability Factor, F2*

   This factor only applies to field testing. The derived effectiveness weight provides a means for subjectively scaling the rigor and completeness with which the test will be implemented.

3. *User Acceptance, F3*

   This factor only applies to reliability and field testing.

For "reliability" tests, rigor/completeness allocations depend on dedication to the elimination of failure modes as top priority, if accelerated (overstress) tests will be used, and if immediate analysis and effective corrective action for all failures will take place. For "field" tests, rigor/completeness allocations depend on how broad the field test will be and if it will completely assess the operational effectiveness and user suitability of the new product. It will be related to the number of pilot units to be employed, as well as the intended quality of construction. Once the new/adjusted F1, F2, and F3 values are determined for the test(s) defined to be necessary as part of the product R&D test program, the specific activities (for each test) corresponding to each of the effectiveness factors are identified. Each effectiveness weight is then allocated to each activity based on the activities' importance, as well as cost and time impact to the overall product R&D test program.

Case Study Question: *How much influence should AI have in dictating R&D testing programs for new technologies advancing towards commercialization? Is optimization possible if left in the hands of engineering management alone?*

6.5   SUMMARY

This chapter addressed the management control required as part of TMA. Control is a fundamental element of strategically directing an efficient manufacturing enterprise. This is critical for implementing the tactical strategies that are part of the overall organizational strategic plan.

The entire organization must realize that together they serve a common purpose and goal for business profitability and long-term sustainability. It is imperative that all walls between departments be eliminated and a broad focus on the manufacturing activity be established by each. A free flow of information must exist throughout the corporation. Obviously, the organizational structure and mindset must facilitate this to occur.

Additionally, structure and discipline must be applied to facilitate effective and consistent decision-making. There are numerous qualitative and quantitative tools available. However, perhaps the most effective tool for every complex activity (and simplistic activities, as well), it is imperative that a project management plan be developed. This involves defining and documenting key project milestones and time frames. It is also productive to enhance a plan by defining a project management network or looking for the "optimal" solutions to decisions via mathematical routines.

Ideally, one wants to always make accurate, timely, and consistent decisions for new and especially repeat, situations. This is possible with the advent of expert systems. Expert system technology is widely available and enables the capture of human expert knowledge for continual re-application simultaneously throughout the business and manufacturing environments of an organization.

QUESTIONS

1. Develop a management network diagram for a current work or school group project. Identify the critical path and the amount of slack.
2. Discuss the trade-off between constrained optimization model complexity and accuracy.
3. What does "feasible region" mean?
4. Discuss the difference between "knowledge" and "data." Give three examples of each.
5. Discuss the difference between an "algorithm" and a "heuristic." Give two examples of each.
6. Select some simple item which you can disassemble (pencil sharpener, stapler, etc.). Suggest ways to redesign the item in keeping with the guidelines for DFM.

7. List three advantages of CAD that your company would benefit from.
8. Discuss the difference between an FMS and an FMC.
9. Describe two benefits to be gained by increasing employee understanding of many parts of a company's production processes.
8. Describe the relationship between time buckets and planning horizons.
9. From where do the inputs to the basic MRP equation come?
10. Is a lot size of less than one at all similar to the simultaneous production of mixed product lines? Why or why not?
11. Discuss several ways to implement a pull system for a production line in your factory.

REFERENCES

1. Kelly, A. (1987). *Maintenance Planning & Control*. London: Butterworths.
2. Moore, F. (1969). *Manufacturing Management*. Homewood, IL: Richard D. Irwin, Inc.
3. Eppen, G. & Gould, F. (1985). *Quantitative Concepts for Management: Decision Making Without Algorithms*. Englewood Cliffs, NJ, Prentice-Hall, Inc.
4. Boothroyd, G., Poli, C. & Murch, L. (1982). *Automatic Assembly*. New York: Marcel Dekker.
5. Groover, M. (1987). *Automation, Production Systems, and Computer-Integrated Manufacturing*. Englewood Cliffs, NJ: Prentice-Hall.
6. Vollman, T., Berry, W. & Whybark, D. (1988). *Manufacturing Planning and Control Systems*. Homewood, IL, Richard D. Irwin, Inc.
7. Hernandez, A. (1989). *Just-In-Time Manufacturing: A Practical Approach*. New York: Prentice-Hall.
8. Bishop, R. (2006). *Mechatronics: An Introduction*. Boca Raton, FL: CRC Press.
9. Richardson, M. & Wallace, S. (2013). *Getting Started with Raspberry Pi*. Sebastopol, CA: MakerMedia. ISBN: 978-1449344214.
10. Blum, J. (2019). *Exploring Arduino: Tools and Techniques for Engineering Wizardry* (2nd ed.). New York: Wiley. ISBN: 978-1119405375.
11. Cetin, S. (2007). *Mechatronics with Experiments* (2nd ed.). New York: Wiley. ISBN: 978-1-118-80246-5.
12. Parsaye, K. & Chignell, M. (1988). *Expert Systems for Experts*. New York: John Wiley and Sons.
13. Waterman, D. (1986). *A Guide to Expert Systems*. Reading, MA: Addison-Wesley Publishing Company.
14. Patterson, D. (1990). *Introduction to Artificial Intelligence and Expert Systems*. Englewood Cliffs, NJ: Prentice-Hall.
15. Giarratano, J. & Riley, G. (2005). *Expert Systems: Principles and Programming* (4th ed.). Chicago: Course Technology. ISBN-13: 978-0534384470.
16. Kusiak, A. (1990). *Intelligent Manufacturing Systems*. Englewood Cliffs, NJ: Prentice-Hall.

7 Organizational Leadership

Management control and leadership are critical factors in the success of an organization regardless if it is public, private, for profit, or nonprofit. Accepting this, it is easily recognized that organizational leadership plays an integral role in implementing and attaining TMA. To move the entire organization strategically forward, organizational leadership needs to be embraced to ensure that TMA does not end up on the "industry trend" trash pile. It ultimately is the organization's strategic commitment to continual organizational leadership improvement and evolution that will enable TMA to position the organization as a global benchmark.

Effective organizational leadership is demonstrated by maintaining a direct and expedient course towards organizational strategic goals and objectives. Leadership is demonstrated through the desire to ensure that products are insightfully and optimally designed upfront and subsequently manufactured and provided to the customer without any loss of intended design integrity.

This chapter addresses organizational leadership topics and tools that align with TMA activities. These concepts and tools are not presented as new and innovative techniques. They are presented as underused and very effective and efficient means to maintain management control.

7.1 MOTIVATION

A fundamental problem in organizations is complacency. This results when people are no longer motivated to advance the well-being of the company or themselves. Complacency makes it very hard to maintain a progressive and aggressive organizational attitude. Complacency is an infectious disease, which, if not timely addressed, can have catastrophic effects on overall organizational effectiveness, growth, and sustainability. The key to combating complacency is to motivate and empower people. If this is not possible, then complacent people must be removed from the organization. The bottom line is that top management is responsible for providing the leadership necessary to motivate workers to function in a state of high creativity, entrepreneurialism, and productivity. An objective of effective leadership must be to motivate people to not just be happy with their job but to continually look for ways to improve it and make it more rewarding both personally and for the company.

Ideally, every person should intrinsically enjoy making an effective contribution towards organizational growth and profitability. However, although all workers are capable of self-motivation, management generally must provide the motivational spark via the proper leadership style. The leadership style by which motivation is achieved must remain flexible to accommodate change, as necessary, in order for the

DOI: 10.1201/9781003208051-9

organization as a whole to avoid the complacency trap. Particularly, in light of the broad purview of personalities, educational backgrounds, and organizational levels contained within a company.

7.1.1 LEADERSHIP STYLE

Four factors are identified as being associated with effective leadership [1]. The first factor is *support*, which involves making people feel worthwhile and aware of management consideration. The second factor is *interaction facilitation*, which involves fostering the development and effective use of people and establishing a good communication network. The third factor is *goal emphasis*, which involves making people aware of the goals of the organization and instilling a sense of enthusiasm for achieving these goals. And, the fourth factor is *work facilitation*, which involves helping people achieve the corporate goals by effectively administering, planning, and coordinating resources.

These four factors are certainly important and should be a fundamentally employed at every level of organizational leadership. However, they must be applied within the perspective of the target worker. For example, it likely that the exact same leadership style will not be effective for workers on the production floor and workers in the corporate office. These two sets of people have different personalities and divergent organizational expectations.

It is clear that in attaining all strategic goals, including those that are TMA specific, it is important to determine the appropriate leadership style required. This must be done at each successive organizational management level.

In general, three variables are considered in determining the appropriate leadership style: (1) forces in the projected leader, forces in the other organizational members, and forces in the situation [2]. The best leadership style depends on management values, personality characteristics, and feelings in various situations. Further, it depends on whether people are ready to accept responsibility, interested in company problems, and in tune with the organization's goals.

These three variables can be considered together using Vroom's prescriptive model of appropriate leadership style [3]. The leadership style serves as the cornerstone of the motivational strategy. The model is exercised using a decision logic which defines five alternative leadership styles as identified below and keyed to Figure 7.1.

1. Decisions are made solely by the manager based on whatever information possessed.
2. Decisions are made solely by the manager, but additional information is gathered from workers.
3. Decisions are made by the manager, but significant input is derived from workers on an individual basis.
4. Decisions are made by the manager, but significant input is derived from workers on a group basis.
5. Decisions are made by the manager *and* the workers as a group.

By exercising the decision logic, the optimal management leadership style is defined.

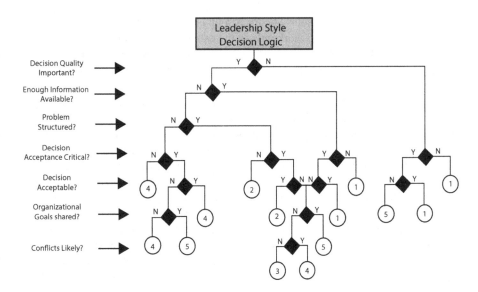

FIGURE 7.1   Leadership style decision logic.

7.1.1.1   Organizational Psychology

Organizational psychology is a broad and very interesting field. Much research has been performed and the field itself continues to be an attractive research area. Several prominent organizational behavior models and motivational theories have been defined. These provide a foundation for establishing corporate, divisional, and departmental motivational strategies.

There are four important organizational behavior models: (1) autocratic, (2) custodial, (3) supportive, and (4) collegial [4]. The autocratic model is founded on the power base of the manager and strict compliance of the worker. It was most prevalent over 90 years ago. In the 1920s and 1930s, it yielded to the custodial model, which proved more successful. This model is founded on the economic resources made available to the worker and his/her satisfaction. The supportive model is widely accepted and predominates in many organizations. This model relies on leadership and the subsequent motivation of the worker. Currently, many progressive organizations use the collegial model. The collegial model requires mutual contribution from both management and the workforce. The result is a firm commitment to the job at hand and to the organization.

Some of the most widely known motivational theories include those presented by Maslow, Vroom, McGregor, and Herzberg. Maslow's Hierarchy Theory states that people progress through a hierarchy of needs. These consist of physiological needs, safety needs, social needs, esteem needs, and self-actualization needs. Each level of needs provides a source of motivation. Once a level is satisfied, the person motivated to achieve the next level.

Vroom defines the Expectancy, or Path-Goal, Theory. He states that motivation depends on what a person expects to receive for his/her efforts. This expectation is not limited to first-order needs such as successful task completion but includes second-order needs to which the first-order needs might lead (e.g., a pay raise).

McGregor defines the famous Theory X and Theory Y notions. Theory X states that workers are lazy and will work only to the extent that their work is instrumental to getting rewards. Theory Y on the other hand states that workers are not lazy but are self-motivated to set and achieve their own goals which are generally consistent with the goals of the corporation.

Herzberg defines the Motivation-Hygiene Theory. This theory states that people have two sets of needs. The first set are those related to the job context, that is, salary, working conditions, supervisory policies, etc. People are never satisfied by these factors but merely neutral and their absence leaves them dissatisfied. The second set make up the content of the job, that is, challenge, the need to use one's full capacities, the need to be resourceful, etc. The absence of these two sets of needs leaves people neutral, and their presence makes them satisfied.

There are many more motivational and organizational behavior theories. Those identified here are only a sample of some of the more widely known. The reader is encouraged to explore this area of psychological research. One point should be obvious: no one theory provides all the solutions for all possible situations. Corporate cultures are complex phenomena and require a motivational strategy based on a hybrid of the numerous theoretical approaches defined.

7.1.1.2 Strategic Motivation Model

The nature of most organizational activities lends them to be depicted by a work path to achieve an end goal which is characteristic of Vroom's Expectancy Theory [5]. From this baseline framework, other theoretical ideas easily rally around Vroom's notion (see Figure 7.2), which together formulate a strategic motivation model.

As shown in the figure, the basic project path and ultimate goal is a function of what is expected to be accomplished and the value of the accomplishment. Once an organizational activity is initiated, individual responses to attain completion are based on one's ability, effort, and role perceptions. An intrinsic need to successfully

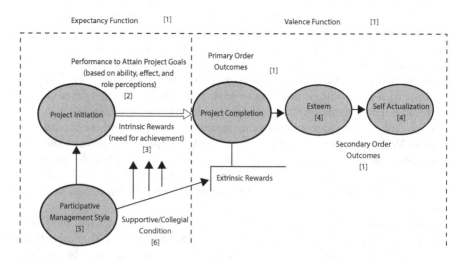

FIGURE 7.2 Strategic motivation model.

complete the activity exists within the organizational team members. To foster this need throughout the activity's duration, a supportive/collegial condition is defined in order to provide the greatest opportunity for creativity in attaining the end goal. Of course, the primary goal is successful completion. Derived from attainment of this is monetary compensation, as well as self-esteem and, ultimately, self-actualization.

This strategic motivation model is intended to provide the most suited environment for the organizational members it impacts. It is designed to emphasize both individual and group creativity to facilitate a modern, innovative, and progressive corporate attitude.

7.2 TEAMING

Management and academic publications have increasingly emphasized the importance of teams, at all organizational levels, for overall organizational success. Teams consist of individuals brought together to establish organizational strategic direction and manage organizational performance. Teaming has expanded in response to a dynamic and complex global operational environment. Teams in general, and particularly strategically established executive leadership teams (ELTs), helps organizations achieve competitive advantage by applying collective expertise, integrating disparate efforts, and sharing responsibility for the success of the organization [6–11].

Teams and their effectiveness play an important role at all levels within an organization. For example, the success of an organization's top leadership is often dependent on the effectiveness of teams. It is broadly recognized that there are four types of teams: (1) work, (2) parallel, (3) project, and (4) executive leadership (or top management) [12].

Work teams are the type of team most people think about when discussing teams. Work teams are continuing work units responsible for repetitive, ongoing activities in an organization such as producing goods or providing services. Their membership is typically stable, usually full time, and well-defined parallel teams pull together people from different work units or jobs to perform functions that the regular organization is not equipped to perform well. They are used for problem-solving and improvement-oriented activities. Project teams are time-limited and focus on tasks that are non-repetitive in nature and involve considerable application of knowledge, judgment, and expertise, which produce one-time outputs (e.g., a new product or service)

The fourth team type, which is a focal interest for attaining TMA, is the executive leadership team (ELT). In general, ELTs coordinate and provide strategic direction to the overall organization. The use of ELTs has expanded in response to the turbulence and complexity of the global business environment. The ELT helps an organization achieve competitive advantage by applying collective expertise, integrating disparate efforts, and sharing responsibility for the success of the organization. A key strength of ELTs is in their use of "objective measures of organizational performance" to identify their effectiveness, which highlights the need for some level of effectiveness accountability.

The management literature on team effectiveness has typically focused on demography and team performance [13] and the industrial-organizational literature has begun to focus on less-traditional aspects of team composition, such as the mix of cognitive ability, personality, and effectiveness [14].

7.2.1 BASIC ASSUMPTION THEORY

Basic Assumption Theory (BAT) provides the theoretical framework for the investigation of individual predispositions for team behavior. It was conceptualized by Wilfred Bion in the late 1940s and led to his recognition as an important theorist regarding psychodynamic perspectives on groups. Most notably, BAT is the cornerstone of the Tavistock Method as it serves as a framework for the group-as-a-whole approach to understanding group, or team, behavior [15].

As a therapist, Bion worked in a unique way with small groups of psychotherapy patients in that he provided the group with no direction or structure. He emphasized interpretation of group phenomena rather than individual phenomena [16]. As Bion observed his groups of patients, he noticed certain significant emotional reactions in the group. For example, at times, the group appeared to be unanimously expressing a need to run away from the current group situation or to demand that he, the therapist, provide more direction. From these observations, the idea developed that a group could be thought of in terms of dynamically moving through a series of emotional states, or basic assumption cultures, in which some group affective need was inextricably associated with the work the group was trying to do. That is, (1) individual members as contributed towards, acquiesced in, or reacted against these emotional states and (2) group members formed relationships with one another on the basis of their predispositions for the various emotional states.

The BAT presented the idea of a group-as-a-whole instantaneously exhibiting one of two group cultures. The first characterized the group engaged in sophisticated, rational work (i.e., a work group culture) and the second characterized the group acting as if it were assembled for some reason other than the defined work task (i.e., a basic assumption, emotionality, or non-work culture). Bion believed that work and emotional components of group life were so interrelated that one never occurred without the other, and that an understanding of group experience can come about only when both are studied in their dynamic and changing relationships to each other.

The work aspects of group operation are defined as the consciously determined, deliberative, reality-bound, goal-seeking aspects of the group's activities. While work activity can always be perceived in the group, there are times when an analysis of this kind of activity alone cannot explain what is happening. For example, although the group may say it is interested in solving some problem, its behavior seems to lead it away from coming to grips with it. In contrast to work, the emotional preoccupations of the group are nonpurposive, "instinctual," and not under conscious control [17]. The conclusion is that the work activity of a group, or team, is always influenced to some extent by emotional states, or emotionality.

The BAT suggests that at any given time the group-as-a-whole acts as if it is operating in a single emotional state, or basic assumption culture. That is, it will maintain itself (1) through developing intimacy (pairing), (2) through reliance on external authority (dependency), (3) through fighting with or resisting with others (fight), or (4) through fleeing from stress (flight). Team members can be seen either as accepting and expressing the basic assumption or as reacting in some other way to its existence in the culture of the group.

The BAT basis is that a team is either in a nonworking state (driven by a basic assumption or emotionality variable) or in working state. As part of his group, or teaming, seminal work, Herbert Thelen [18] deviated from this by positing that with emotionality, there is always some level of work ongoing, which forms the basis for adaptation of BAT to describing the role of EW (emotionality–work) predispositions in the teaming situation.

Thelen posited that when the team is operating in a dependency-work culture, it is acting "as if" (i.e., on the basic assumption that) the team exists in order to find support and direction from something outside itself – the leader, external standards, or its own history. When it is operating in pairing-work culture, it acts "as if" its function is to find strength from within its own peer group. When it is operating in fight-work or flight-work cultures, it is operating "as if" its purpose is to avoid something by fighting or running away from it, respectively.

Other research has emphasized that the "as if" was important in these definitions since it also indicated what was not meant when a team was operating in a certain EW culture [19]. For example, when a group, or team, is operating in a dependency-work culture, it does not mean that nothing but dependency is expressed, or that the group is consciously aware of its preoccupation with dependency, or that the dependency can necessarily be directly observed. What it does mean is that if the group becomes ineffective, it is relying on an emotionality culture that does not positively associate with work.

Thelen accepted the notion of BAT to describe groups as exhibiting an EW cultural preference, which reflected the collective EW predispositions of the individual team members. However, he went on to speculate that individual reactions to the various EW cultures exhibited in the group might reveal basic tendencies in individual personality [20]. This led Thelen to investigate whether EW, which was meaningful for studying the group-as-a-whole, might also be applicable to individual expressions and behavioral predispositions. In advancing BAT, it has been applied conceptually to mainstream teams. Thelen's research contribution included development of the RGST (see Appendix A), which provided a means to identify individual EW predispositions for team behavior prior to the individual engaging in team activity and impacting team effectiveness.

7.2.2 TEAM EFFECTIVENESS

As the use of teams has increased, so has research attention focusing on team effectiveness Researchers have tended to conceptualize team effectiveness as a function of each team member's individual input relative to the process losses associated with working with others. Original ideas of team effectiveness focused explicitly on teamwork, which included the productivity gains due to team coordination. Through the study of team performance and team processes, a number of models of team effectiveness have been developed [21–23]. Understanding what makes teams effective is important as organizations look to teams as a solution to enhancing overall operational effectiveness. This is especially true of ELTs, whose members often have diverse backgrounds and competencies.

BAT becomes a basis for establishing team membership in order to maximize team performance and team effectiveness. Team performance accounts for the outcomes of the team's actions regardless of how the team may have accomplished the task. Conversely, team effectiveness takes a more holistic perspective in considering not only whether the team performed by completing a task, but also how the team interacted via team processes to achieve the team outcome. This holistic perspective is consistent with BAT [24].

7.2.2.1 Personality

It has been reasonable to accept that the behavior of individual team members can enhance or impede team effectiveness. Interpersonal conflicts, poor communication, lack of team cohesiveness and disagreement over goals, etc., have been recognized as some of the behavioral consequences of dysfunctional team member behavior. From this, it seems self-evident that the behavioral predispositions of team members can play a major role in the success or effectiveness of any team.

There has been much research accumulated supporting the premise that demographic variables (e.g., age, race, gender, seniority), abilities, and personality variables are examples of team member interpersonal characteristics that are related to effective, or ineffective, team behavior. The thought was that this should be especially true of personality variables, which are a "mixture of values, temperament, coping strategies, traits, character, and motivation." In other words, it was research involving personality variables as sought to describe and predict the typical behavior of individuals in team situations.

Several events have led to a resurgence of interest in the personality composition of teams. First, personality has increasingly been found to be a valid predictor of performance. Second, research on groups has increased. And, third, business and industry, as well as other types of organizations such as higher education, have demonstrated a sustained increase in the use of teams, as well as a need for strategies designed to select group members.

As discussed previously, an individual's "group-relevant aspects of personality" affected team effectiveness. Individual personality has continued to be viewed as having an impact on team effectiveness. Supporting the concept of EW individual predispositions, or group-relevant personality traits, defined personality as "an individual's characteristic patterns of thought, emotions, and behavior."

Despite the scarcity of research specifically related to team membership, research in group dynamics has provided a basis for making predictions of how personality likely contributes to team effectiveness. It is often suggested that the individual characteristics of group members, as well as the diversity of skills and traits within a group, were important variables related to team effectiveness.

There have been a multitude of studies that have explored the relationship between personality traits and team effectiveness. However, perhaps the most commonly referenced has been the Five-Factor Model [25] and the Big 5 model [26]. These models relate individual personality and team effectiveness and provided structure and organization for a vast array of research-defined personality traits.

The subject five personality traits defined by these two models include: (1) extraversion, (2) agreeableness, (3) conscientiousness, (4) emotional stability, and (5) openness to experience [27]. Each of these five personality traits was defined to consist of multiple facets or subtraits, which can be assessed independently of the trait that they belong to. The Big 5 framework has become one of the most widely used and researched models of personality.

Each of these five personality traits describes, relative to other people, the frequency or intensity of a team member's feelings, thoughts, or behaviors. It has been broadly posited that everyone possesses all of the five traits to a greater or lesser degree. For example, two team members could be described as agreeable, which by definition states that agreeable people value getting along with others. But, there could be significant variation in the degree to which they were both agreeable. In other words, all of the Big 5 personality traits exist on a continuum rather than as attributes that a person does or does not have.

The Big 5 model offers an integrative framework for personality psychology by focusing on a core set of behavioral traits that were proposed to enable understanding people by knowing how much they display each of these five traits in their lives. However, personality traits continue to be looked at as a way for organizations to maximize the effectiveness of teams.

The Big 5 model presents the personality traits as a general explanatory framework for interpersonal behavior: (1) openness describes team members that are imaginative, sensitive, intellectual, polished versus down to earth, insensitive, narrow, crude, simple; (2) stability describes team members that are calm, enthusiastic, poised, and secure versus depressed, angry, emotional, and insecure; (3) agreeableness describes team members that are good-natured, gentle, cooperative, forgiving, hopeful versus irritable, ruthless, suspicious, uncooperative, inflexible; (4) conscientiousness describes team members that are careful, thorough, achievement-oriented, responsible, organized, self-disciplined, scrupulous versus irresponsible, disorganized, undisciplined, unscrupulous; and (5) extraversion–introversion describes team members that are sociable, talkative, assertive, active versus retiring, sober, reserved, cautious.

Research findings suggest that while team member personality is related to team performance and other variables important for the success of teams, different Big 5 personality variables predict performance in different types of teams and the situational demands of the team task. Regarding openness, it may be important for creative and imaginative team tasks but less important, or even detrimental, when the task is of a more routine nature. Regarding stability, the level of emotional stability has been positively related to team performance for a wide range of team tasks. Regarding agreeableness, it may be important for performance in long-term teams with tasks that involve persuasion or other socially related dimensions. When tasks do not require a high degree of social interaction, the evidence suggests that agreeableness may actually inhibit performance in teams. Regarding extraversion, it has been related to team performance when tasks involve imaginative or creative activity but may inhibit performance when tasks call for precise, sequential, and logical behavior.

Of the Big 5 personality variables, conscientiousness has been found to have the strongest and most reliable correlation with individual performance across job settings. This suggests that conscientiousness should be positively related to team performance across a wide variety of tasks and settings [27].

7.2.2.2 Team Composition

Another key driver of team effectiveness has been team composition relative to personality and demographics. Although there has been developed limited evidence on the role of team composition-based personality on team effectiveness, there is evidence that diversity on other variables (e.g., demographics) impacts team effectiveness.

Regarding personality variables, several research studies indicated positive relationships between team heterogeneity and creativity and decision-making effectiveness. These studies suggest that diversity in membership might be desirable for increasing the quantity of solutions offered and the quantity of alternatives offered. In general, research has suggested that teams composed of members having heterogeneous personalities should produce higher quality decisions and generate more creative ideas than homogeneous teams. Homogeneous personality composed teams should be more cohesive, have less conflict and turnover, and perform well when decision-making and new ideas are less important.

Team composition relative to demographics refers to the configuration of a team based on team member attributes. The ELT literature has attempted to link attributes of organizational leaders with strategic choices and organizational outcomes. Using demographics, researchers have related team composition demographic variables such as age, tenure, education, and functional background to organizational effectiveness outcomes.

However, although many studies have found a relationship between demographic variables and outcomes, they have produced conflicting results. Faced with these inconsistencies, identified are several limitations regarding team composition research. First, demographic characteristics are, at best, imperfect representatives for psychological constructs. Second, ELT research has not definitively linked demographic attributes to team effectiveness. And, third, most demographic studies have failed to account for the impact that situation-specific factors have on team process and performance. From these limitations, ELT effectiveness may vary greatly from one team situation to another, which indicates that demographic analysis, used in isolation, may provide an incomplete explanation of variation in a team's effectiveness over time. This suggests a need for additional research to explain how situational factors and team attributes work together to shape ELT effectiveness.

The EW concept has proved useful in studying team composition, particularly, regarding team functioning and the shifts between team member-shared emotionalities. It has been found that there exists a relationship between team composition, team development, and team effectiveness in that team members would combine with one another to "establish, maintain, or dissolve specific EW cultures" existing in the group-as-a-whole in response to group situational factors [28].

Regarding team composition, the focus has been on individual group-relevant personality traits defined by EW predispositions. However, it has been asserted that the demographic characteristics of organizations, a key factor affecting team composition, also shape behavior patterns (e.g., team communication) and, ultimately, effectiveness. Team composition has remained a rich research area regarding demographics and has long been considered a powerful and effective means of increasing team effectiveness.

7.2.2.3 Reactions to Group Situations Test

Herbert Thelen posited that the simplest possible "prediction of individual group behavior" could be based on the quantitative magnitude of the EW factors [20]. That led to the development and demonstration that the RGST (see Appendix) could accurately predict the EW behavioral pattern in group interaction to be exhibited by individuals.

The RGST provides an EW measurement instrument, that is, the RGST as a means for measurement was essential to understanding the behavior of individuals and the team as a whole. Numerous instruments have been developed for analyzing group therapy processes, group climate and therapeutic dimensions, individual personality, and interactions among group members. However, measurement instruments have typically focused either on the individual or on the group, but not both simultaneously.

The RGST has successfully identified individual EW predispositions for team behavior in advance of actual team activity. In its current form, the RGST has evolved to be a stand-alone instrument removing the need for qualitative analysis of verbal or written statements by respondents. This makes it unique in that it enables variation from research practices that continue to include observational analysis for group study.

As stated previously, the RGST has its origins in BAT and incorporates the four emotionality (E) variables: fight, flight, dependency, and pairing, and the work (W) variable. It was designed as a 50-item forced choice test, which provides the respondent with incomplete sentences that describe common occurrences in team situations. The respondent completes the test sentences by selecting one of two alternatives. These alternative completions are designed to reveal a predisposition for one of the four emotionality variables and the work variable. The alternative completions are arranged in a balanced design so that each emotionality variable and the work variable are paired five times with each other variable item.

The RGST design reflects two key features regarding the team situations and the response selections presented. First, each incomplete sentence (or team situation) includes two components: (1) a stimulus situation that is based on a specific EW variable and (2) a person to identify with (i.e., self, other, or group). Second, each alternative sentence completion (or team situation response) includes two components: (1) a response that is based on a specific EW variable and (2) a person to identify with (i.e., self, other, or group). These features reflect the RGST's design intent to represent "meaningful group situations" and to facilitate the respondent's "free reaction to the response."

These RGST design features provide another dimension to assessing individual EW predispositions for team behavior relative the EW variable-based stimuli and EW variable-based responses. A respondent's selection of an EW variable-based response that is the same as the stimulus presented in the incomplete sentence indicates a preference to continue to operate within, or maintain, the EW behavioral mode presented in the stimulus. Likewise, a respondent's selection of an EW variable-based response that is different than the stimulus presented indicates that the respondent rejects, blocks, or prefers to operate in another EW behavioral mode. This information provides additional insight to the respondent's direct, general response to the stimulus. The Appendix provides the RGST incomplete sentence allocation in terms of EW variable stimulus and alternative response modalities.

The responses to the RGST, taken together, sketch a picture of the participant's predisposition for EW behavior in a team situation. There are fundamentally two ways of using the data from this instrument. First, use the EW variable scores to indicate an individual's preference for, or strength of, group-relevant behavioral traits. Second, use the EW pattern of responses to stimuli to indicate an individual's acceptance or rejection of a group EW operating mode. In general, the RGST is a useful tool to "sensitize participants to the EW dimensions of group relations" and it "offers the chance to manipulate team composition or function as a diagnostic device."

7.3   STRATEGIC HUMAN RESOURCE PLANNING

It is easy for the value of organizational human capital to be under appreciated. Every organization will say publicly that "people" are the organization's greatest assets. The truth in such a public statement is ultimately demonstrated by the organization's sincere attention to its human resources. That attention can take many forms, such as job security, payroll, benefits, culture, environmental safety, team building, education, and participation in organizational strategic decision-making. In any case, people will know if the organization leadership is sincere or not regarding engaging everyone in a dynamic, progressive, and strategic teaming environment and culture. This awareness is what leads people to leave or stay with an organization.

There are four major steps in strategic human resource planning.
Step 1: Analyze Organizational Human Resource Situation
- A.   Assess Team Member Inventory and Capabilities – The inventory assessment comes from the organizational chart included in the strategic business plan. The capabilities assessment comes from advancement potential profiles (i.e., succession planning database) and the performance appraisal.
    Questions answered:
    - How good are our people?
    - What knowledge holes exist?
    - What team members need to be taken care of?
    - What team members need to be removed?
- B.   Define Organizational Changes – This comes from the strategic business plan and projected changes to the organizational chart for successive years (e.g., 2–5 years).
    Questions answered:
    - What are the organizational functional personnel allocations?
    - Is the organization designed for success and sustainability?
- C.   Evaluate Occupational Trends – This comes from analysis of basic wage groups and basic occupational groups.
    Questions answered:
    - What demographic changes are occurring?
    - What is competitor impact on team member retention?
    - Is equal opportunity and equality embraced in organizational growth?

D. Evaluate Productivity Increases – This comes from analysis of cost per
    team member over time
    Questions answered:
    – Have team member investments been beneficial and value-added?
    – Is more getting accomplished in support of strategic goals and
       objectives?
    – Is organizational human resource return on investment and benefit-
       cost increasing?
    – Are more people involved in doing the same, or less, amount of
       work?

Step 2: Forecast Human Resource Needs – This comes from organizational
change definition, occupational trend evaluation, productivity increase evaluation, and succession plans.
    Questions answered:
    – How many team members?
    – How soon/timing?
    – Do major organizational holes exist that will constrain growth and
       advancement?

Step 3: Reconcile the company budget – This comes from comparing human
resource need projections with future financial plans. (Should be expressed
in terms of monetary value compatible with profit objectives.)
    Questions answered:
    – How are human resource investments going to be paid for?
    – Do new team members make business to improve business profit-
       ability, sustainability, and growth?

Step 4: Reconcile human resource need forecast with team member inventory
to identify problems in recruiting, management development, promotions
(or transfers), and team member utilization.

7.3.1 Performance Appraisal

In all organizations, there is a process for employee performance appraisal. This may
be on the ends of a broad spectrum from an informal process to an overly burden-
some process. Ideally, a performance appraisal process and tool is used to motivate
employees and facilitate employees to be valuable contributors to organizational success and sustainability. The process enacted is typically reflective of organizational
exposure and liability to employee claims of discrimination, hostile work environment, unsafe work environment, etc., as well as the value that top management strategically puts on human capital as organizational team members. The bottom line
is that every organization needs to have in a place a process that achieves several
organizational employee behavioral objectives: (1) reward performance, (2) correct
performance, or (3) terminate the employee for lack of performance.

Performance is appraised based on what is expected of the subject position and
career level. Expected performance is defined as that performance requirement cited
in the position description or occupational standard. Performance expectation is not
based on time in a specific position or longevity with the organization but rather to

defined performance characteristics, or key job performance requirements, for the specific position being evaluated. Performance characteristics are compiled under the categories of quality, creativity/innovation, productivity/organization/planning, relationships/communication, and attendance/safety. For each performance characteristic's defined actions, typically, performance is assessed as "above expectation (AE)," "meets expectation (ME)," or "needs development (ND)." It goes without saying, but it needs to be said, that "the best performance evaluation is the honest, unbiased performance evaluation."

There are several functional elements of the performance appraisal/evaluation for each organization team member:

1. General Team Member (i.e., employee) Information
2. Performance of Major Activities
   a. The evaluator lists the team member's major activities (i.e., projects or tasks) completed or ongoing over the past appraisal period. (Emphasis is on "major" activities.)
   b. The team member adds to the list, if desired, and provides "comments" regarding their activity performance.
   c. The evaluator provides a performance assessment for each activity and "comments" upon receipt back from the team member.
3. Performance to Appraisal Period's Top Three Development Objectives
   a. The evaluator lists the appraisal period's top development objectives defined for the team member during the prior year's performance appraisal.
   b. The team member provides "comments" regarding meeting development objectives.
   c. The evaluator provides a performance assessment for each objective and "comments" upon receipt back from the team member.
4. Performance Characteristics
   a. The team member provides "comments" regarding each subsection topic (i.e., quality, creativity/innovation, productivity/organization/planning, relationships/communications, and attendance/safety).
   b. The evaluator provides a performance assessment (i.e., AE, ME, or ND) for each checklist item (using the criteria provided) and "comments" upon receipt back from the team member.
5. Top Three Development Objectives and Measures for the Next Appraisal Period
   a. The evaluator and team member together identify the top three improvement objectives for the upcoming appraisal period. Development objectives address the root cause of performance below expectation and/or other identified performance improvement opportunities. Define meaningful objectives reflecting the team member's job duties and responsibilities, and that will provide a return on investment to the team member and the organization. For each objective, define the measure for monitoring performance, which will be used to monitor development progress.

6. Review/Acknowledgments
   a. The completed appraisal is signed by the team member, evaluator, and the evaluator's supervisor. Note on the form if the team member chooses to not sign the appraisal document.
   b. The evaluator and team member together define the team member's educational plan for the next year. The team member also indicates if she/he has a documented career plan. If not, the team member should be encouraged to create a career plan in collaboration with applicable human resource and management personnel.

For each organizational team member, a performance appraisal score should be derived from the cumulative assessment of the appraisal. Through normalization of scoring for all levels of the organization, it is possible to develop a distribution of scores for the organization in order to facilitate establishing team member reward for the appraisal period's activity. From the distribution of scores, team member performance may be segregated into reward categories, such as (1) above-average monetary increase plus bonus, (2) average monetary increase, (3) no monetary increase, or (4) termination.

7.3.2 ORGANIZATIONAL DEVELOPMENT/TRAINING

As education is a lifelong experience, there are multiple curriculum phases in a person's academic and workforce career. These phases are higher education curriculum and the industry professional development curriculum. These two phases are inherently connected, although typically disconnected in their transition to one another.

Changes in technology, company and organizational strategy, and market direction all contribute to a never-ending cycle of learning. To function in the "real world," the worker experiences two phases of curriculum hardening: (1) *Academic Hardening* and (2) *Industry Hardening* [29]. Academic Hardening gives the theoretical and quasi-practical knowledge needed to obtain a job, with career programs focusing almost entirely on the practical knowledge needs. Most importantly, Academic Hardening teaches students "how-to-learn." Industry Hardening requires using that "how-to-learn" ability to maintain and expand the knowledge needed to keep a job and, ultimately, move up the corporate ladder. Of the two phases, the latter is the most important for personal career satisfaction and longevity, as well as the employing company's profitability and sustainability.

The key to successful workforce development is the successful transition of the student out of the academic curriculum phase (where Academic Hardening occurs) and into the industry curriculum phase (where Industry Hardening occurs). Thus, success relies on the identification and integration of necessary components of each curriculum as they uniquely align with the organization's overall strategic direction and objectives to maximize productivity.

Academic Hardening describes the typical student that progresses through a pretty much pre-planned, required university or college curriculum to get a degree. The curriculums are for the most part based on traditional topics that are fundamental to an engineering field (e.g., mathematics, physics, and design analysis). To provide some direction ownership, students integrate electives into their study to tailor their degree curriculum to a design specialty and personal interests. It is this structured curriculum linked together with having the discipline to make it through the curriculum that provides Academic Hardening.

So, Academic Hardening is more than just attending a university. It is the process of learning "how-to-learn." It is instilling the discipline needed by the student to ultimately fulfill all career ambitions. Without mastering this discipline, the student ends up with a very constrained career, regardless of having or not having a degree.

Industry Hardening is an essential element of every person's career; particularly with the advent of a globally competitive business environment. The onslaught of market penetration throughout the world has forced a new era of company awareness for efficiency. This emphasizes the need to maximize productivity as a key business strategic objective. Often, the productivity improvement approach of choice is to downsize the staff to minimize overhead. The better approach is to develop team members to increase productivity through Industrial Hardening.

Industrial Hardening requires the team member to use the "how-to-learn" ability to research, understand, and apply the latest concepts and technologies. It also requires the company to foster an environment that facilitates applying this "how-to-learn" ability. Both the team member and company benefit from the fact that they are making sound investments in their futures.

However, the investment is only worthwhile if it is strategically focused. There must be emphasis on true business "needs" development and not just education. This is where most companies stumble, they provide educational/training events but the development focus is missing or off-target. Needed is a structured professional development program.

But successful team member development is not just the responsibility of the company. Ultimately, the team member must actively and objectively take the responsibility for development; whether or not the company provides, for example, an engineering development program. Ideally, both the engineer and company equally take part in engineering development, with development emphasized over just education. The result is that both the engineer and the company win by maximizing technical effectiveness and productivity (a key business strategic objective).

7.3.2.1   Stages of Development

It is good management to have a clear objective(s) when implementing, for example, an engineering development program. This gives a benchmark at any point in program implementation to compare what is being implementing with what was desired to be implemented.

A candidate objective is to develop engineering personnel, via a company-focused education program, in practices and technologies (addressing management, engineering, and quality) that are strategically important to long-term profitability and survival. Keys to meeting this objective are (1) teaching the practical use of "state-of-the-art," "leading-edge," and "existing" technologies and practices and (2) teaching company-specific adaptation and application of technologies and practices. The associated goal for this objective should reflect some reasonable time frame to meet near-term "situational" needs, and to meet long-term career development needs.

Keep in mind that in order for meaningful development to occur, it must be directly related to some desired output. From a company perspective, this means developing team members regarding technologies and management techniques that are embraced by the company. From team member perspective, this means developing around technologies and management techniques that are embraced by the company *and* personal career objectives.

Looking at engineer development from a broader viewpoint, it is clear that Industry Hardening results from a multi-stage process. The stages include (1) company/department orientation to get the engineer familiar with policies, rules, facilities, etc.; (2) near-term engineering core to get the engineer functional with company engineering practices as quickly as possible; (3) near/long-term engineering technology core to get the engineer proficient with all company applied/ future technologies; and (4) long-term career development to continually move the engineer towards career goals. These stages form the foundation of a strategically focused development program and forge the long-term relationship between team members and the organization.

7.3.2.2 Technical Development Program

Maximizing the return-on-investment for developing the organization's technical workforce requires defining and implementing a structure program. Four stages of workforce development are defined:

STAGE 1: COMPANY /DEPARTMENT ORIENTATION

STAGE 2: DISCIPLINE DEVELOPMENT CORE: *Topics such as:* Project Management, Universal Computer Applications, Product Development Process, CAD/CAM

STAGE 3: DISCIPLINE DEVELOPMENT SITUATIONAL CORE: Core targets key manufacturing and design technologies

STAGE 4: CAREER DEVELOPMENT: Pool of classes addresses long-term career development

Management

From a management perspective, a specific discipline workforce development program should be a part of the broader organizational development program (if one exists). All of the logistical and administrative issues, as well as class design, development, and instruction should be coordinated by a single focal point in the organization (e.g., Human Resources). From this focal point comes the "what," "who," "when," and "how "question answers for each development stage.

Ensure that each class has an advisor(s) to guide class definition (i.e., outline), content (including example company-specific applications), and instructional design (i.e., presentation materials and methods). Advisors should attend their subject class to support the instructor, as appropriate, as they are selected based on their knowledge and expertise.

Pilot classes (as needed) for new topics and/or instructors. This is a good way to "work the bugs out" of each class before getting in front of a real audience.

Individual Training Curriculum Definition
Define an education curriculum for each discipline area. It is often convenient to have each curriculum comprised of "required" and "elective" courses. "Required" classes are used as a gate in moving to the next career level. The "electives" provide the flexibility to have discipline specific, relevant development. Also, to ensure that the development remains meaningful, development planning should be performed as part of the annual performance appraisal process to create individualized education plans.

Education Contact Hours
Any time an hourly goal is defined for development time, the development objective has probably been lost. Make sure that emphasis is on team member development, not contact hours.

Education Partnering
Partner with other organizations (e.g., consultants, product vendors, and universities). But, establish a partnership only as it is justified to provide the greatest benefit-cost regarding class development time, cost, and practical subject expertise.

Other things to consider include: (1) *Education Cycle*, (2) *Instruction Location*, (3) *Schedule*, (4) *Education Tracking*, (5) *Testing*, (6) *Proficiencing*, and (7) *Class Make-ups*.

Engineers, just like everyone else, continue to learn throughout their careers. The breakthrough learning, and development, occurs when the engineer and company team up together. There are many educational and development opportunities available. These include seminars, university tuition reimbursement, and in-house classes. Whichever medium is used, facilitating team member development is a sound investment to support company profitability and long-term sustainability.

7.3.3 Succession Planning

Succession planning is an important part of moving the organization forward but, unfortunately, is often not given any attention. It aligns with leveraging the expected return on investment from the team members in the near and long term. Fundamentally, it involves systematically assessing team members regarding technical expertise, innovation traits, leadership strengths and weaknesses, and succession profile categories. From this, each team member is identified for career advancement candidacy, which ensures that qualified "internal" candidates are not overlooked for key organizational positions.

Technical expertise relates to specific experience (e.g., sensing circuit design and applications). For team members lacking any significant industry or business experience (e.g., recent hires directly from a higher education institution), expertise is likely based on educational background.

Innovation traits relate to the demonstration of five key attributes:

1. Go-Getter (i.e., independent/energetic spirits)
2. Aggressive (i.e., determined pursuit of goals)

3. Ambitious (i.e., defines visionary goals)
4. Accomplisher (i.e., achieve/attains desired goals)
5. Risk-Taker (i.e., willing to stick one's neck out)

Leadership strengths and weaknesses relate to what the team member excels at and where they may need development. For example, a strength may be "strong at project management" and a weakness may be "a poor listener." This area fundamentally requires doing a traditional SWOT analysis (strengths, weaknesses, opportunities, and threats) for each team member.

Succession profile codes [30] are used to create a picture each team member relative to their current career level or position. Four profiling categories are typically used and each consists of several stages:

1. **Learning Agility**: The ability and approach to learning with subsequent practical application of the knowledge.
   a. **Active**: Creates on own; has sense of urgency for the right issues; seizes opportunities; original; entertains ideas; projects how ideas would play out; practical use of learning experiences including positive changes in behavior and attitude
   b. **Passive**: If given, works at an issue; a positive force but not the first choice; inconclusive as to whether understands results of learning; may practically apply learnings including possible positive changes in behavior and attitude
   c. **Reactive**: Reacts to issues if forced by dictum; maintains theme "not invented here:" can learn but requires investment and presently drains energy to get this accomplished; a tough sell/has trouble extracting meaning from experience; slow at picking up need to change behavior and attitudes
   d. **Blocked**: Fights need and approach to issues; habit bound; does not like to go against the grain; tendency towards perfectionism; over controlling; makes quick decision or studies problems to death; seen as an impediment to change; thinks they are leading change but are not; closed to feedback; defensive; not good at evaluating experiences to extract anything affecting change in behavior and attitude
2. **Potential**: The career level that an individual is currently judged capability of attaining (and have a high probability of success) relative to their present level within the next 5 years.
   a. **Up Two Career Levels**: recognized broadly as a real talent and leader
   b. **Up One Career Level**: needs rounding out to know the business
   c. **Correct Career Level and Job**: technically highly-valued resource; a significant loss if person leaves; handles current assignments competently; not currently managerial moveable at this time
   d. **Wrong Career Level/Wrong Job**: limited ability for expanded contributions and being a change leader; poor match of skill set and requirements
   e. **Too Early to Tell**: less than 3–6 months in position

3. **Readiness**: The time frame within the necessary competencies will be developed to ensure success for next logical organizational position
   a. **Immediate**: Able to move into position with minimal added development; a good portion of the job has been done before...solid results expected; no business risk now
   b. **One to two Years**: Needs some development as position provides a personal development challenge/stretch; coming up learning curve with a strong development plan...strong results expected; longer learning curve; little business risk now
   c. **Three to five Years**: At maximum tolerable development risk point; requires strong managerial coaching/monitoring and a strong development plan...good results expected; acceptable business risk now from a results perspective
4. **Position Life Cycle Stage**: The performance mode within the currently position and career level
   a. **Transition**: Becoming bored with primary role; no challenge; spending significant time on assignments outside currently role; much interaction with higher peer group
   b. **Expansion**: Starting to expand one's horizons; looking for opportunities to do a bigger role, a broader role, a more complex role, or a role with more variety
   c. **Mature**: Current job at this level is firmly in hand; all major responsibilities regularly handled; widely viewed by others to be a solid contributor in current role
   d. **Focus**: Has the job pretty much in hand; lack of experience in the role at this level still showing around the edge; things fall into a crack from time to time.
   e. **Orientation**: Getting a handle on the role and the resources; still in the honeymoon period; still feeling out the boundaries of role relationships
   f. **Introduction**: Brand new in position, unsure of ground rules; many basic work elements to acquire quickly

Together with the performance appraisal, it becomes possible to develop a well-rounded organizational succession plan for each team member. This includes defining individual team member development needs and goals, as well as their succession candidacy and whom their successor(s) may be.

TMA Case Study: Application of the Reaction To Group Situations Test

OVERVIEW

The executive leadership team (ELT) of the organization decided to evaluate their individual emotionality-work profile. Each team member completed the RGST in a group setting with a limited time to ensure quick subconscious answers to the statements. The group then in the same setting each scored their RGST and evaluated the results.

ISSUE

Historically, teams established with executive leaders have not always been able to effectively make decisions. It is evident that depending on the composition, leadership teams may or may not be able to make critical, time-sensitive decisions.

STRATEGIC OBJECTIVE

The strategic objective is to establish the basis for composing specific leadership teams with persons that are best able to provide desired results in an effective and efficient manner.

CASE BACKGROUND

From historical research, findings suggest that work and pairing are associated positively; that work is in greatest opposition with flight and fight; that the interpersonal closeness implied by pairing, and to a lesser extent dependency, is in opposition to fight and flight; and that flight and fight are associated positively.

Several ideas may be applicable to leadership team composition to maximize team effectiveness. If it is desired to build a strong leadership team, the leadership team should be composed of individuals scoring high on work and pairing. If it is desired to avoid forming a leadership team which would adversely affect effectiveness in a simple task situations, team composition should not combine individuals with dominant predispositions for fight, flight, or dependency.

For more complex tasks, it may be desirable to have leadership teams composed of executives with a more aggressive group relevant trait (i.e., fight). Note, that fight was not completely rejected as a preferred response for any of the stimulus situations. For simple tasks, fight could be disruptive, but in a complex situation, it could provide the aggressiveness needed to cut through the frustrations involved in completing more difficult tasks.

RESULT/CONCLUSION

For the leadership team-as-a-whole, work was the dominant emotionality-work predisposition, which may imply that the team members are oriented to completing specific tasks. Fight was the weakest emotionality-work predisposition, which may imply a low leadership team ability to deal with highly stressful situations and retain high team effectiveness. Pairing and dependency were similar in predisposition strength, which may imply that pairing would be a routine part of leadership teams, but that the leadership teams remain dependent on the direction of a leader. That is, although leadership team members may be inherently motivated to pair and work together, the leadership team may not have, or seek, a proactive decision-making capability or empowerment to engage in independent decision-making.

Together, pairing and dependency appeared to be opposites of fight and flight, which again may imply a leadership team pairing orientation but dependent on the leader or someone to take charge. Flight was reflected by a low predisposition strength, which may imply a more serious orientation, and engagement, to leadership team work and activity completion. Likewise, the low fight scoring may imply that there would be little resistance to the team and/or leader, and that there would be a team behavioral preference to wait for someone to provide explicit guidance or direction, or, simply, to go along with the group. This again may have some basis in the dependency variable scoring similar to that of the pairing variable. Finally, work opposed flight and fight, which may imply that flight and fight would not facilitate leadership team effectiveness.

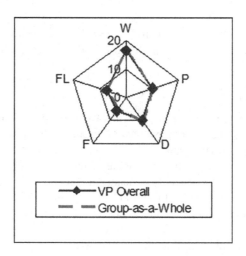

EW Predisposition: Work is dominant and in opposition to Fight and Flight. Pairing is slightly preferred over Dependency.

Individual EW Emphasis: Deliberative, rational, orientation to completing specific tasks. Look to pairing to develop intimate relationships in conjunction with a dependency on external authority.

Team Preferred EW Operating Mode: Focus on the consciously determined, deliberative, reality-bound, goal-seeking aspects of the ELT's activities.

Emphasize providing support for other team members through expressions of warmth, friendliness, or partiality for other team members or their ideas.

Need to find support or direction from a leader or some source external to itself, which may appear as an expression of weakness or inadequacy.

EW Associations: (positive, +; negative, −; null, 0)

	W	P	D	F
P		+		
D		+	−	
F		−	−	−
FL		−	−	−

				+

In addition, to have highly effective leadership teams, it would be desirable to have team members scoring high in work. Leadership teams with members scoring high on pairing would probably be effective, although their main emphasis would be on team building. Generally speaking, it may be wise to avoid composing a leadership team with members scoring high in fight, flight, and/or dependency, particularly in simple task situations, since such teams would probably steer away from being effective. The RGST resultant star diagram indicates that specific team composition is possible based on the diversity of RGST emotionality-work scoring. However, the low overall low fight score may prove to prolong decision-making longer than needed for critical, time-sensitive organizational decisions.

Case Study Question: *What are the business situations that may be difficult with executive leaders having EW predispositions for teaming favoring pairing and dependency?*

7.4 SUMMARY

This chapter addressed the organizational leadership required as part of TMA. Leading the organization is a fundamental, and significant, role of management. If this is effectively accomplished, the consequence is motivated and empowered people which provide innovation and excitement in support of the overall organization.

All departments must realize that together they serve a common purpose and goal. That is, corporate profitability and survival. It is imperative that all walls between departments be eliminated and a broad focus on the manufacturing activity be established by each. A free flow of information must exist throughout the organization. Obviously, the organizational structure and culture must facilitate this to occur.

Organizational structure is important, but the proper motivational environment (i.e., culture) must also exist. Depending on the situation, the appropriate leadership style must be used. However, leadership style flexibility must always be governed by the overall organizational motivational strategy developed and implemented.

To ensure that the organization remains robust in its ability to attract and retain the best talent available, a rigorous approach to performance appraisal, professional development, and succession planning must be in place. Additionally, the ability of the organization to advance forward is dependent on the use of effective teams. The RGST serves as an insightful tool to aid in team composition to accomplish decision-making in the appropriate time frame to support the organization.

QUESTIONS

1. Does TMA require an increase or decrease in management size? In management responsibility? In management integration? Why?
2. Identify several "walls" that exist in your organization. How might they be eliminated? Discuss several possible methods for each situation you can identify.

3. Chart out the information flow existing for an environment that you are exposed to (at work, at school).
4. What are the major factors involved in leadership?
5. Discuss the appropriateness of leadership flexibility.
6. What are the four major organizational behavior models? Discuss the characteristics of each.
7. For technical employee development, what are examples of training that would be beneficial that was not part of any educational curriculum?
8. For personnel with a performance appraisal falling below the 25% quartile, what actions should be taken?
9. Take the RGST to determine your emotionality-work profile. Do the results make sense?…why or why not.
10. How do Vroom's expectancy and valence theory support leadership effectiveness?

REFERENCES

1. Bowers, D. & Seashore, S. (1966). "Predicting Organizational Effectiveness with Four-Factor Theory of Leadership." *Administrative Science Quarterly*, Vol. 11, pp. 238–263.
2. Tannebaum, R. & Schmidt, W. (1958). "How to Choose a Leadership Pattern." *Harvard Business Review*, Vol. 36, pp. 95–101.
3. Vroom, V. (1974). "Decision Making and the Leadership Process." *Contemporary Business*, Vol. 3, pp. 47–64.
4. Davis, K. (1985). "Evolving Models of Organizational Behavior." In *Organizational Behavior*. New York: McGraw-Hill.
5. Vroom, V. (1964). *Work and Motivation*. New York: Wiley & Sons.
6. Deci, E. & Gilmer, B. (1977). *Industrial and Organizational Psychology*. New York: McGraw-Hill.
7. Lawler, E. & Porter, L. (1967). *Antecedents of Effective Managerial Performance*. (Vol. 2, pp. 122–142). ScienceDirect, Elevier B.V.
8. McClelland, D. (1961). *The Achieving Society*. Princeton, NJ: Van Nostrand.
9. Maslow, A. (1970). *Motivation and Personality*. New York: Harper & Row.
10. Vroom, V. & Deci, E. (1970). *Management and Motivation*. Baltimore: Penguin.
11. Davis, K. (1985). *Organizational Behavior*. New York: McGraw-Hill.
12. Cohen, S. G. & Bailey, D. (1997). "What Makes Teams Work: Group Effectiveness Research from the Shop Floor to the Executive Suite." *Journal of Management*, Vol. 23, pp. 239–290.
13. Jackson, S. E., Brett, J. F., Sessa, V. I., Cooper, D. M., Julin, J. A. & Peyronnin, K. (1991). "Some Differences Make a Difference: Individual Dissimilarity and Group Heterogeneity as Correlates of Recruitment, Promotions, and Turnover." *Journal of Applied Psychology*, Vol. 76, pp. 675–689.
14. Barry, B. & Stewart, G. (1997). "Composition, Process, and Performance in Self-Managed Groups: The Role of Personality." *Journal of Applied Psychology*, Vol. 82, pp. 62–78.
15. Banet, A. G. & Hayden, C. (1996). "A Tavistock Primer." In J. E. Jones & W. Pfeiffer (Eds.), *The 1996 Annual Handbook for Group Facilitators* (pp. 155–167). La Jolla, CA: University Associates, Inc.
16. Bion, W. R. (1961). *Experiences in Groups and Other Papers*. New York: Basic Books.

17. Stock, D. & Thelen, H. (1958). *Emotional Dynamics and Group Culture*. New York: New York University Press.
18. Thelen, H. (1954). *Methods for Studying Work and Emotionality in Group Operation*. Chicago: Human Dynamics Laboratory, University of Chicago. The Office of Naval Research (contract: NR 170-176).
19. Karterud, S. W. (1989). "A Study of Bion's Basic Assumption Groups." *Human Relations*, Vol. 42, No. 4, pp. 315–335.
20. Thelen, H. A., Hawkes, T. H. & Stratner, N. S. (1969). *Role Perception and Task Performance of Experimentally Composed Small Groups*. Chicago, IL: University of Chicago Press.
21. Bandura, A. (1986). *Social Foundations of Thought and Action*. Englewood Cliffs, NJ: Prentice Hall.
22. Hackman, J. R. (1987). "The Design of Work Teams." In J. W. Lorsch (Ed.), *Handbook of Organizational Behavior* (pp. 315–342). Englewood Cliffs, NJ: Prentice Hall.
23. McCrae, R. R. & Costa, P. (1987). "Validation of the Five–Factor Model of Personality Across Instruments and Observers." *Journal of Personality and Social Psychology*, Vol. 52, pp. 81–90.
24. Guzzo, R. A. & Dickson, M.W. (1996). "Teams in Organizations: Recent Research on Performance and Effectiveness." *Annual Review of Psychology*, Vol. 47, pp. 307–338.
25. McCrae, R. R. (1989). "Why I Advocate the Five-Factor Model: Joint Factor Analyses of the NEO-PI and Other Instruments." In D. M. Buss & N. Cantor (Eds.), *Personality Psychology: Recent Trends and Emerging Directions* (pp. 237–245). New York: Springer-Verlag.
26. Goldberg, L. R. (1993). "The Structure of Phenotypic Personality Traits." *American Psychologist*, Vol. 48, pp. 26–34.
27. Barrick, M. R. & Mount, M. K. (1991). "The Big Five Personality Dimensions and Job Performance: A Meta–Analysis." *Personnel Psychology*, Vol. 44, pp. 1–26.
28. Thelen, H. A. (2000). "Research with Bion's Concepts." In M. Pines (Ed.), *International Library of Group Analysis. Bion and Group Psychotherapy* (Vol. 15, pp. 114–138). London: Jessica Kingsley.
29. Brauer, D. (1997). "Developing the Working Engineer." *ASEE Conference Proceedings*, Indianapolis, IN.
30. Brauer, D. (1997). *Engineering Career Development: Global Succession Planning*. Design Assurance Sciences.

Section III

Manufacturing System Control

MEN ARE BORN WITH TWO EYES, BUT WITH ONE TONGUE, IN ORDER
THAT THEY SHOULD SEE TWICE AS MUCH AS THEY SAY.

—COLTON

DOI: 10.1201/9781003208051-10

8 System Definition

As in any major undertaking, no great structure is built without first laying a firm foundation. For our task of attaining Total Manufacturing Assurance (TMA), this translates to the solid base of a well-defined operating environment. Men, machines, and materials, as well as information and instructions, must be organized in an efficient manner, according to the best standard industrial engineering techniques. Without a sound basic structure, with all parts smoothly interacting with each other, it may be a moot point in striving towards TMA.

This chapter examines the basics of manufacturing system definition. This includes looking at basic management and organizational functions typical of a modern industrial engineering curriculum, plus new innovations that industry is currently adopting. While exploring this chapter, keep in mind the hierarchy involved in the manufacturing-based company: manufacturing system, subsystem, equipment, module, and part.

8.1 SYSTEM DEFINITION ACTIVITIES

The activities necessary to define a manufacturing system generally include asking a series of open-ended questions and carefully considering the answers. These answers will guide all subsequent design decisions. These questions may include:

- What is the level of product *volume* (how much is planned to make) and what is the level of product *variety* (how different are products from each other)?
- What manufacturing processes will be utilized?
- What are the process inputs and outputs?
- What information is needed to control the processes?
- What scarce resources are necessary to enable the processes? And how many of them will be available?
- What sort of flow will be utilized? (Single part, batch, mixed model?)
- Will fixed routing (all parts follow the same path) or variable routing (different parts follow different paths though the factory) be used?
- Will the system be moving work in process to the various stations, or will moving tooling and labor to the parts? And what will our transportation methods be?

The answers to these and other questions will help to define the type of system that will be created and managed.

DOI: 10.1201/9781003208051-11

8.2   PROCESS PLANNING

Process Planning is the task of deciding exactly how a designed product will be manufactured. It is a creative activity, and as such, different process planners may come up with very different process plans for the same manufactured part. This can be both good and bad, depending on what is most important. It can be good, because different process planners may come up with a variety of ideas, which can then be competed against each other. However, it can be bad if very similar parts end up using very different processes, because excessive and redundant equipment, tooling, and fixturing may end up being acquired. In the end, management must decide how much consistency is necessary in process planning. The use of Automated Process Planning systems can help to enforce consistency of methods.

At its core, Process Planning begins by looking at the features and tolerances of a designed product and deciding which sorts of manufacturing processes will be used to create it. But these decisions are not made in a vacuum. They must be made while remaining cognizant of several important issues:

- What production processes are already performed? (There is no need to purchase new equipment when current, possibly underutilized equipment, can do the job.)
- What core competencies are already possessed? (There is no need to learn an entirely new set of skills when current skills can suffice.)
- What are the organizational strategic goals and objectives in terms of processes? (It might be worth investing in new equipment and skills if it takes us in a direction with long-term benefits.)
- How is the factory currently laid out? (e.g., if currently a cellular layout is used, which manufacturing cells can best handle this new part? If a functional layout is used, which machines will provide the best process flow? Is it worth reconfiguring the manufacturing layout for this new part?)
- Which parts of the process will be done in house versus subcontracting or purchasing off the shelf?

Once these basic questions have been considered, a Process Planner will create the "recipe" for production of the part. This is frequently documented in a "routing sheet" or "shop routing" which lists the steps, the work stations where they are to be performed, and possibly the standard times for both setup and operations of each step. Some routing sheets also list material handling moves between stations and how the parts will be transported.

These routings can be created from scratch by an experienced Process Planner, but often it is expedient to use some sort of automated Process Planning System. These come in two varieties: Generative Automated Process Planning Systems and Variant Automated Process Planning Systems.

A *Generative Automated Process Planning System* uses an Expert System to generate the Process Plan. The Expert System is a computer program that asks various questions about the new product to be planned, questions such as its overall dimensions, the types of features it will possess, any symmetry, and the tolerances needed.

It will then run these *input facts* against a series of canned rules in its *knowledge base* to enable it to output the steps needed to produce the part. A human expert will verify the output and make sure that it makes sense and meets company standards.

A *Variant Automated Process Planning System* uses a database of current parts in the company's catalog and searches for the nearest match to the new part. The process plan for that existing part is then called up and edited as necessary for the new part. This not only saves considerable time on the part of the process planning personnel but also helps to enforce consistency in process plans.

In any case, the result is a complete *technical package* for the new part, including part drawings, specifications for raw materials, dimensions, tolerances, sequence of fabrication and assembly steps, inspections required along the way, and time standards. At this point, the part is ready to be released into production.

8.2.1 PROCESS PLANNING FOR SPECIFIC INDUSTRIES

Naturally, process plans will look very different for different industries. The following paragraphs briefly consider some process planning issues for various industrial domains.

Metal Working: In traditional subtractive metalworking industries, process plans will largely consist of tasks at various machine tools. They often start with a sawing station, where bar stock is cut into blanks. Alternately, they could start with a forging, made either in-house or by a contractor. Forgings tend to have superior mechanical properties over stock metal and can save considerably on metal removal times if they are *near net shape* when they arrive.

Although there is generally an ideal machine tool to use for any processing step that needs to be performed, sometimes it is more efficient overall to use a less-than-ideal machine. For example, a cutting operation can usually be most efficiently performed on a saw. However, if the part is already mounted on a lathe when the cutting step is needed, it is often more efficient to use a cut-off tool on the lathe rather than dismount the part, send it to the saw department, and then return it to the lathe for further turning. This is the sort of trade-off that the Process Planner needs to be cognizant of while developing the process plan.

Molding: In molding industries, the Process Planner must first decide what type of molding operation is most appropriate in terms of material properties and production volume. It must also decided how many parts will be molded in one setup, as it is often more efficient to create many identical units at the same time in a multi-cavity mold. But, the production volume must support such economies for them to make sense, and the molding machine must be capable of holding the mold with sufficient pressure to fill all cavities in the mold.

Micromanufacturing: Both microelectronic systems (MES) and microelectromechanical systems (MEMS) devices are produced by very small-scale operations which are largely photographic and chemical in nature, as well as highly automated. These processes are governed by a large body of technologies foreign to most engineers from other industries and are beyond the scope of this text. The interested reader is referred to other texts on the topic [1].

Assembly: Assembly steps, as distinct from fabrication steps which physically transform materials, are sometimes integrated into the same process flow with fabrication. But, they are more often considered a separate process and organized as such. Assembly can be highly automated, highly manual, or a blend of both. In any case, assembly process planning differs from fabrication process planning in the nature of the process flow: whereas fabrication flow tends to be one-in, one-out, assembly by its very nature is many-in, one out. So, the process flow will consist of several (possibly many) converging flows of material coming together to create one product at the end of the "pipeline."

8.3 PLANT LAYOUT AND FLOW

A manufacturing system is fundamentally composed of a variety of different equipment. This may include equipment for machining, assembly and material handling, as well as storage, loading docks, repair facilities, computers, and personnel offices. Where are these to be located? Plant layout, or in broader terms *facilities design*, enables answering this question via analytical and design techniques aimed at maximizing efficient organization and interaction between various assets.

It is important to establish something more than a mere floor plan. There must be adequate power supply, lines of communication for both voice and data, and, most importantly, room to grow and adapt to the inevitable changes of the future.

Many of the techniques used in facility planning involve minimizing necessary evils. For example, minimizing the amount of equipment needed to facilitate economical production, or decreasing distances to be traveled by material in process to save time and energy, as well as lowering the inventory cost of having material tied up in production. Obviously, a layout which allows fewer workers to attend a greater number of tasks results in manpower savings. A sample list of items to minimize includes:

1. Investment in equipment
2. Production time
3. Material handling
4. Work in process
5. Floor space
6. Variation in equipment
7. Inflexibility
8. Manpower
9. Employee inconvenience, discomfort, and danger

8.3.1 PLANT LAYOUT ANALYSIS

Defining and knowing the major objectives of plant layout design, the fundamental question arises: how to arrive at the layout which achieves those strategic objectives? Many techniques exist, both new and old, which all follow one of two basic schemes:

From \ To	Receiving	Storage	Lathe 1	Lathe 2	Vertical Mill	Horizontal Mill	Packaging	Inspection	Shipping
Receiving		40		2		5			
Storage			8	3	10	12	5		10
Lathe 1				9	22	13	6	7	
Lathe 2			7		3	11	15	9	
Vertical Mill	5	11	6			23	6	1	
Horizontal Mill			5	4	17		11	21	
Packaging		8						10	30
Inspection		6					11		27
Shipping									

FIGURE 8.1 From-To chart for part flow between departments.

1. Select a number of candidate layouts. *Analyze* each layout according to some grading method. Select the layout with the highest score.
2. Define a list of goals, constraints, and requirements for the target layout. *Synthesize* the best possible layout using some optimization technique.

These two basic schemes are referred to as the *analytic* and *synthetic* methods of plant layout.

8.3.1.1 The From-To Chart

The From-To chart, also called a travel chart, is a matrix which tracks flow between the various locations in a plant [2]. Each activity location (machine, storage center, entry or exit point, etc.) is listed across the top row of the chart, as well as down the first column of the chart. Each element in the interior of the chart lists some fact concerning the relationship between the corresponding row and column locations (see Figure 8.1).

One use of a From-To chart is to list distances. An element in row 4 and column 2 would show the distance *from* location 4 *to* location 2. Naturally, the diagonal would have all zeros, and the chart would be symmetrical about this diagonal.

Another From-To chart could be used to tabulate traffic volume. The numbers in the interior of the chart would indicate the flow of parts from each location to each other location. In this application, of course, there would not necessarily be diagonal symmetry.

Combining these two charts, the distance From-To and the traffic From-To, gives the total material handling burden of the proposed layout. Slight changes in the

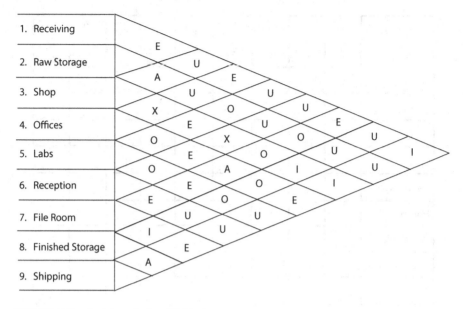

FIGURE 8.2   Activity relation (REL) chart.

layout could be easily documented by altering one row and column. In this way, many candidate layouts can be compared.

8.3.1.2   The REL Chart

Another analytic tool is the REL chart, or *activity relation* chart [3]. This is a more subjective technique than the From-To chart and is useful for considering the more qualitative influences, as well as distance and traffic volume. The REL chart uses a *closeness rating* instead of a numerical score for each location pair (see Figure 8.2).

Unlike the From-To chart, which listed facts about a specific layout, the REL chart is a method for expressing desired distance relationships. The closeness rating for each pair of locations is filled in based upon a qualitative judgment of how important it is that the two locations be situated near, or even adjacent to, each other. The standard ratings are:

A = Absolutely necessary
E = Especially important
I = Important
O = Ordinary closeness OK
U = Unimportant
X = Undesirable

These ratings are selected intuitively, based on the subjective judgment of the plant layout engineer. Often, a reason code is listed as well, indicating the rationale for the closeness rating. Typical reason codes might be sequence of work flow, sharing

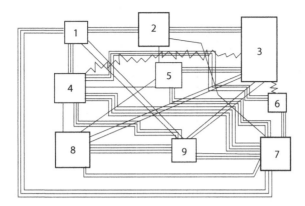

FIGURE 8.3   Space-relationship diagram.

of records, personnel, or equipment, similarity of function, or ease of supervision. Undesirable ratings may arise between noisy operations and those that need quiet surroundings, or particularly dirty or sooty operations and those that require a clean environment.

8.3.1.3   Space-Relationship Diagram and Block Plan

It is now time to start constructing candidate layouts and evaluating them. The REL chart is the starting point for generating the candidates and the From-To chart will enable determining which is best for maximizing manufacturing system efficiency and effectiveness.

The space-relationship diagram is the first step. This combines two sources of information: our REL, or activity relation chart, and our space requirements. For each activity listed in the REL chart, one must calculate the necessary amount of floor space. This can be derived from anticipated production volume, time standards, and machine capacities.

Next, a square is drawn on a piece of paper for each activity. The size of each square is proportional to the area needed for that activity, and this area is listed in the square, along with the activity number (see Figure 8.3). Each square is connected to each of the other squares with a line.

The type of line used is based on the closeness rating assigned in the REL chart: An A pair is connected by a quadruple line, an E pair by a triple line, an I pair by a double line, an O pair by a single line, and a U pair by no line at all. Activity pairs with an X rating are connected by a zig-zag-type line.

This space-relationship diagram is the tool for generating the candidate layouts. Think of the lines connecting each activity as attractive forces. The more lines between two blocks, the stronger the attraction, like so many tightly stretched rubber bands. The zig-zag lines, of course, must be thought of as repulsive forces.

The next step is to try to assemble the activities into a block plan. Using your imagination, let the attractive forces bring the activities as close together as the relationship lines indicate. Think of the process as piecing together a jigsaw puzzle with rubber bands connecting the pieces. The final shape should approximate the shape of

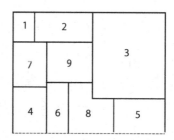

FIGURE 8.4   Example block plan.

the plant or building that will house the activities. An example block plan is shown in Figure 8.4.

Naturally, many block plans are possible from any one given space-relationship diagram. Try to generate as many reasonable plans as possible. The more candidates that are generated, the greater the probability that the best possible design is among them.

When a sufficient number of candidate plans are generated, the From-To Chart can be used to evaluate them. The traffic-volume from-to chart will be the same for each candidate, but the distance chart will not. Measure the distances for each pair of activities in each candidate, and construct the appropriate distance chart. Combining this with the (constant) traffic volume chart, each candidate can be given a score for material handling burden.

A simple final step would be to merely select the candidate with the lowest material handling score, but this can sometimes lead to undesirable results. It is better to select the best three or four candidates and scrutinize each closely. Are the closeness relationships implemented as desired? Are the A pairs and X pairs situated as desired? It is possible, especially in complex systems, for one or two important relationships to be sacrificed in favor of many small improvements in the less important relationships. This results in a better numerical score, but it introduces an "Achilles's heel" slip into the design. Use sound engineering judgment to select the final winner from the top three or four candidates, based on overall adherence to closeness ratings, and any other subjective considerations not programmed into the ratings scheme.

8.3.2   PLANT LAYOUT SYNTHESIS

The preceding subsection focused on *analysis* of candidate layouts, with only minimal effort directed at the generation of those candidates. This is a tried-and-true method of plant layout, and works well for small plants with few departments. Where large, complex facilities are concerned, however, the method of creating these candidates becomes more important.

Synthesis of candidate layouts requires many complex decisions to be made as each function or activity is placed into a layout. Consequently, computer methods are often used. Many programs for plant layout exist, three of the most popular being CORELAP [4], ALDEP [5], and CRAFT [6]. Some programs employ a *generative* approach, building up each candidate layout "from scratch," while others use a *derivative* approach, which continually modifies an existing layout, making incremental improvements at each step. Some programs are capable of using either approach or a combination of the two.

While different programs use different methods of generating layouts, they all have certain functions in common. Most of them require as input the REL chart, as described in the preceding section. They also require a *grading scale*, which translates the codes in the REL chart to a score for each candidate layout. Some programs assume a constant grading scale, others allow the user to supply one.

The programs must also be told how much area is required for each department, and what the size and shape is of the available facility. Some programs can handle multiple floors at one time, and some allow the user to fix certain activities or departments in specific locations before synthesis begins.

Most generative programs begin by selecting one activity from the REL chart. The method of selection of the first activity varies from one program to another but is usually based on some function of the REL chart which allows greatest flexibility in subsequent selections. This activity is placed in the layout, and the REL chart is searched again for an appropriate second activity. Again, each program uses its own criteria for selection of the order of placement based on some function of the REL chart and the grading scale. Each successive activity is placed in the layout at a location which will minimize or maximize some program-specific function. The program will generate a large number of possible layouts, giving each a score based on the grading scale.

The best of these "first cut" layouts will be stored as a benchmark, and the program will then attempt to do better. Some programs will start over with a different first activity. Others will go into a derivative mode, and attempt to make improvements on the benchmark, using various internal functions.

Finally, after several iterations, the program will report its latest and greatest layout. It must then be decided if this layout meets all the non-numeric requirements that were not programmed into the analysis. There is also generally the option of supplying a different grading scale, or a modified REL chart, and seeing if a preferable layout results. As always, one should take into account past layout experience and engineering intuition in making the final decision and not take any computer program's recommendation as gospel.

8.4   MATERIAL HANDLING

The previous section provided some techniques for laying out the facilities in a manufacturing plant. These techniques were based on several principles, one of the most important being the minimization of material handling burden.

Is material handling cost worth worrying about? Absolutely, especially in a TMA environment. For one thing, material handling cost can be up to and beyond one-half of total manufacturing cost. Furthermore, in many applications, material handling is inextricably linked with other manufacturing elements. Conveyors and transfer mechanisms in assembly lines, part feeders and orienters, and robotic manipulation equipment are examples of devices that handle raw material and work in process at the same time that they are being converted to finished products [7].

What materials are handled in a manufacturing system? Basically, everything from raw materials to work in process to finished products. Also, tools, replacement parts, energy, information, and sometimes even personnel are moved by the material handling systems.

What are the goals in the handling of these materials? First of all, materials need to reach the designated destination, and at the proper time, or earlier. They must arrive safe and undamaged and in an economical manner. Further, it is desirable, using readily available technology, to always be aware of each item's location while it is in transit, in case a change in plans makes it necessary to reroute.

Material handling equipment covers a broad spectrum of sophistication, from trivial to highly automated and intelligent. In the most basic sense, manpower can be considered a material handling technique: workers can trot over to a storage area and retrieve the materials they need. This material handling technique has the lowest investment cost, but one of the higher unit costs. As such, it is only advisable for very low production rates.

Moving up the evolutionary scale are human-powered equipment such as dollies and hand carts. Next come human-controlled, but auto-powered, devices such as forklifts and motorized carts. Hoists, gantries, and conveyors require human control only for starting and stopping at desired times. Finally, at the top of the spectrum, are intelligent and automated devices such as automated guided vehicles (AGV) and conveyors with feedback control.

The choice of the best material handling method for each application depends on a combination of factors. Quantity of material to be handled is obviously an important consideration, both from a standpoint of how many units per day, and from the view of how many days in the production run. High material flow rates require large capacity systems, and longer production runs will justify a larger investment in faster and more sophisticated equipment.

Scalability is also worth considering. Is the work volume expected to change significantly in the foreseeable future? If so, by how much? It is easy enough to add an additional forklift or two, or an additional AGV to the fleet. But, increasing a conveyor system's capacity may require an entire redesign and reinstallation.

Another critical factor to consider is flexibility. Will the handling system remain constant for a long time? Will rerouting become necessary often? Are changes in flow rate expected? Naturally, greater flexibility comes at a greater cost and must be justified by both anticipated changes in material handling requirements and enhancement of profitability resulting from the flexibility.

Certain guidelines should be followed when selecting and designing the handling system. These concepts are basically common sense but are still worth discussing briefly. The first guideline is the "bee-line" principle: whenever possible, use straight lines to minimize distance traveled. Of course, practical considerations may overrule this concept in many situations.

Next is the "pallet" principle: unless dealing with extremely large items (e,g, locomotives), it is a waste to drop one unit onto a conveyer. This wastes conveyor capacity, as well as loading and unloading motions. It is better to palletize parts onto a single holder, or pallet, and manipulate them as a single package. The pallets should be as large as possible and should be of consistent size, arrangement, and orientation.

Along with the pallet principle comes the "buffer" principle: build buffers into the production line so that full pallet loads can accumulate before transportation becomes necessary. This not only makes palletizing more efficient, but allows different workstations to work at slightly different rates without blocking the flow lines.

Another guideline is "minimum dead-time" principle, which means to keep the material moving, not sitting around, so, the efforts are to speed up the loading and unloading of parts onto and off the handling equipment. This means not only well-designed pallets and fixtures but also intelligent scheduling of loading and unloading operations during our overall process.

The "two-way" principle is easily overlooked but can double handling efficiency: never send anything back empty if it can be avoided. Use the same pallet that arrived at station n to take station n's output to station $n + 1$. Don't send it back to station $n - 1$ for another load! Arrange flow so that minimal amounts of motion are wasted. Here is where uniform pallet design pays off.

Finally, the "information" principle: whenever possible, integrate the information about the material into the flow with the material. Rather than saving a list of part numbers and their associated pallet number in a computer file for later retrieval by a downstream station, print the part numbers onto the parts or pallet itself. Bar code printers and readers, or even less sophisticated methods for low volume, make this possible.

8.4.1   MATERIAL HANDLING EQUATIONS

The relationships describing various quantities in material handling systems are fairly simple. One basic concept is the rate of flow, R_f, usually expressed in parts/minute. For a one-way conveyor or other handling system, it is given by:

$$R_f = (n_p)(V_c / S_c) \leq n_p / TL$$

where
  n_p is the number of parts per pallet
  V_c is the flow velocity (feet/minute)
  S_c is the spacing between pallets (feet)
  TL is the loading time at each end (minutes)

A material handling system also has a capacity or number of parts that can be accumulated in it. This can be thought of as a temporary storage of work in process. Letting n_c be the number of carriers (pallets) in the system, then:

$$n_c = L_f + L_r / S_c$$

where
  L_f is the length of forward (full) part of route
  L_r is the length of return (empty) part of route

Including a return portion (L_r) makes this equation applicable to recirculating types of systems, such as belt conveyors. Total number of parts in the system, N, is given by:

$$N = (n_p)(L_f / S_c) = (n_p)(n_c)(L_f / (L_f + L_r))$$

One other relationship of importance should be noted here. That is the relationship between travel time and loading/unloading time. If TL is time to load a pallet, and TU is time to unload a pallet, then:

$$V_c / S_c \text{ must be } \geq 1 / TL \text{ and } TL \text{ must be } \geq TU$$

to avoid backups in the system.

Note that other material handling relations, both specific to individual systems, and applicable in general to all types, are found in the literature [8].

8.4.2 AUTOMATED GUIDED VEHICLES

The highest end of the material handling spectrum is occupied by AGVs. These devices are by far the most expensive and therefore not applicable to all situations. Their vast capabilities and inherent flexibility can lead to large payoffs for installations that can justify their cost.

AGVs are basically independent vehicles, rolling along from station to station under their own power and control. Some are guided by wires or painted lines on the floor, others by dead reckoning with periodic alignment checks at known landmarks. They are generally powered by on-board batteries, lasting at least an 8-hour shift before needing recharging. Often, they will have a "safety bumper," a large loop of some flexible material on the front surface which will detect collisions, and stop the vehicle, before the main mass of the AGV does any damage to itself or other objects in its way. Some AGVs use optical or sonar devices (e.g., LIDAR) for collision detection.

Different control schemes can be used with AGVs. In some systems, the AGV is controlled by typed instructions entered at an on-board control panel. This is a simple and straightforward method but requires large amounts of human intervention. In other systems, a decentralized control scheme is used: a station in need of an AGV will issue a general "service call" via radio frequency. Any AGV unused load capacity can detect and answer the call and service the station. This is a useful scheme in large factories with "islands of automation," but no centralized control of the entire floor. If a central computer is used to direct flow in the entire factory, it will generally make the decisions as to which AGV should service which stations and when. This method offers greatest overall efficiency but requires extreme levels of planning and programming.

Equations to plan an AGV system are similar to those given above for any material handling system, with the addition of a "traffic factor" to account for time wasted in waiting for obstacles to pass, avoiding collisions, adjusting a path, etc. This factor is usually somewhere between 0.8 and 1.0 and is used to reduce the number of parts per minute that could be handled by an AGV with no need to worry about traffic congestion. Naturally, the more AGV and other traffic in the system, the lower the traffic factor will become.

8.5 AUTOMATION

Automation, from the Greek words for self and moving, is one of the most common buzz words in manufacturing today. This section looks at some of the basic concepts of automation and examines some of the strengths, limitations, and trade-offs involved.

The word automation can be applied to all parts of the manufacturing enterprise where direct human control has been minimized or even eliminated. This includes material handling, machining, assembly, inspection, storage and retrieval, and any other tasks that are carried out in the process of converting raw materials to a useful and salable product.

Automation is almost universally accepted as a good thing. The reasons for this acceptance are compelling. Productivity increases are an obvious advantage of automation, both in terms of production per hour and production per worker. Other advantages of the automation trend are decreased unit cost, increased safety, increased product quality, reduction of work in process, and greater predictability of the manufacturing function.

Basically, automation can be broken down into two broad classifications: hard and soft [9]. Hard automation is the older of the types, and is based on machines which do only one thing, but do it extremely well. The auto industry was dominated by hard automation a few decades ago and continues to use the approach extensively. Examples of hard automation include automatic presses and die-casting machines, automated transfer lines, and any other heavy machinery that performs the same task over and over. It is characterized by a lack of flexibility, long production runs, and high production rates.

Soft automation, on the other hand, is characterized by machines which can perform a variety of tasks. Soft automation is also called flexible or programmable automation and is dominated by robots and other general purpose, programmable devices. The idea behind flexible automation is that a single investment in equipment will pay dividends in the production of many different products and processes. One of the costs of this flexibility is that the tasks are sometimes performed at a lower production rate than if dedicated, single-purpose machines were used.

Which type of automation is best for a given application? The answer to this question depends on many factors, including the type of industry, the anticipated production rate and run length, frequency of product changes, and future business plan. For any one given manufacturing task, it is much cheaper to acquire hard automation equipment that can perform the task than soft automation equipment.

However, if the task becomes unnecessary, the hard automation equipment will require extensive modification, if not outright scrapping. The flexible equipment can be easily modified for a new task via software alone, with no further investment in hardware. Furthermore, the flexible equipment can be instructed to perform the new task along with the old task, as each one is needed, if a variable product mix is desired.

As a final note on this topic, it should be mentioned that the distinction between hard and soft automation is not as clear cut as it is often made out to be. In reality, a sort of continuum exists, where any level of flexibility that is desired can be achieved. There are individual devices that exhibit certain degrees of flexibility, and there are entire factories that contain elements of the programmable along with the inflexible. The amount of flexibility that is selected for any specific installation should be based on the amount that is needed and anticipated. To pay for excess flexibility that will never be utilized would be an unwise waste of funds.

8.5.1 AUTOMATION ELEMENTS

All automation is based on certain types of equipment, capable of performing standard automation tasks such as material handling, assembly, inspection and so on. Each of these, with the exception of robotic devices, is available in varying degrees of programmability and as such should be considered key elements of both hard and soft automation systems. Robots and similar devices, of course, are inherently flexible and should be considered elements of soft automation only.

Transfer mechanisms move a workpiece from one station to the next in an automated fashion. Several configurations exist. An in-line transfer mechanism consists of a conveyor or overhead-chain system to move the workpiece along in a more or less straight line, while a rotary mechanism is more like a turning table, around which the necessary processing stations are arranged like guests at a dinner table. The rotary configuration is useful only for situations with a small number of workstations, whereas an in-line system can accommodate as many stations as are desired by merely increasing its length. Both types can move in either a continuous or indexed fashion, depending on the types of operations to be performed at each station.

Automated machining devices perform the same functions as traditional machine tools but are directed automatically with some sort of computer control, generally Computer Numerical Control (CNC). These machine tools can consist of traditional tools such as drill presses, vertical or horizontal milling machines, lathes, presses, and the like, or more modern devices, which combine several of these functions. The machines may operate upon a stationary workpiece positioned by the transfer mechanism or may have a built-in multiaxis positioning table to hold the work.

With regard to the newer multipurpose machine tools, many types and configurations are available. These are often called machining centers, and may have several spindles driven by a common motor, and capable of coming to play on the workpiece in whatever order is required by the process plan. A typical machining center may combine a drill press, vertical and horizontal mill, and tool holder.

PLC and CNC refer to methods of utilizing computers to control automated machine tools. However, it should be kept in mind that new hardware and techniques are constantly being developed, with the result that clearly drawn distinctions between these types of control are becoming more difficult to make.

PLC stands for programmable logic controller, the original device used to control automated machine tools. A PLC is a dedicated device attached to one tool and often is supplied by the machine tool manufacturer as a part of the tool system. It is basically a digital computer which can store a sequence of operations for the tool and send the appropriate commands to the tool at the appropriate times. The input console of the PLC often consists of a collection of buttons corresponding to the standard commands for the tool. A small screen is often included so that the program can be reviewed and edited.

The PLC bears little resemblance to a general purpose computer but is more of an intelligent command console with a memory. Some are equipped with disk drives so that programs may be saved, although more recent trends utilize network connectivity so that programs can be downloaded from a central computer. PLCs have been popular since the late 1960s and continue be important industrial components today.

CNC was a natural outgrowth of an older technology (i.e., DNC, or Distributed Numerical Control) where a centralized computer controlled a number of machine tools across the factory. A CNC computer only controls one machine, or possibly a small number of machines working together. The advantages of CNC over DNC arise from the localization of control. An engineer on the factory floor has direct access to the computer at the site of the tool. Also, since CNC computers control a single machine, there is no need for time sharing or other resource allocations, and the complete power of the processor can be utilized in complex control algorithms.

There is no danger of a single computer failure crippling the entire factory, as in the DNC implementation. CNC computers are usually networked to a central computer, so that there is no loss of the DNC ability to monitor all processes and keep all elements of the factory in communication with each other element.

Robots, or more precisely, programmable manipulators, are the final element of manufacturing which help to pull everything together. Robots have many applications in an automated factory, including loading and unloading pallets, applying sealants and adhesives, cutting, inspection, and assembly. Popular uses of robots currently are for assembly, stacking, inspection, painting, and welding. Any task which involves the manipulation of some material or device through a precisely defined trajectory can be accomplished with some type of robotic device.

As with any programmable automation element, a robot must be able to justify its high price tag through savings in manufacturing cost. Tasks which are performed many times per hour, perhaps for two or three shifts per day, are especially deserving of robotic attention. Also, tasks performed under hazardous conditions, requiring many expensive protections and precautions for human operation, may be much cheaper and safer for a robot.

Several issues must be kept in mind when considering a robot for a task, based on the robot's capabilities and the task requirements. Chief among these is accuracy: can the robot position and orient the object to be manipulated with the required accuracy? Related to accuracy is repeatability. Repeatability is often defined as a robot's ability to reach the same location several times in a row, without drifting as time goes by. Another issue is resolution, the smallest increment of motion possible for the robot. Is it small enough to perform the most delicate of the required motions?

Other, more practical considerations include such things as load carrying capacity, reach of the arm, and type of grippers (also called end effectors) available for the robot. Another is the capability of the robots controller, how programmable it is, and what kind of sensors it can be connected to. These capabilities vary widely from one type of robot to another, and the options available on the market today are quite vast.

8.6   FLEXIBLE MANUFACTURING SYSTEMS

Flexible Manufacturing System (FMS) is an industry standard term for a highly customized, highly automated machining, material handling, and possibly inspection device [9]. It can replace a large quantity of individual machine tools and eliminate the need for mounting and dismounting parts on the various separate machines. It is also capable of running unattended for extended periods of time, often limited only by its capacity to store raw material and finished parts.

The FMS is generally custom designed for each individual installation, being assembled of standard components such as milling, drilling, and turning workstations, inspection stations, and material manipulating devices. They require a large investment to bring online and are justified only under specific situations. Specifically, they make most sense when:

- Parts are complex and require many operations.
- Product mix has a medium to high level of variety, requiring ability to use different CNC programs for each part.
- Production volume is sufficiently high to justify long-term, multi-shift production.

8.7   SIMULATION

Assume that there has been developed a tentative layout of our new manufacturing system, proposed changes to our old system, or perhaps just a new control scheme for a physically unchanged system. Before implementing our new ideas, which could cost millions of dollars, it would be nice to have some evidence that the design ideas are sound. Ideally, it would be desirable to conduct a scale-model experiment, a dry run of the new plan, before committing to a specific design.

That is exactly what the concept of simulation allows. General purpose computer-based simulation is a programming technique, which allows system analysis with random variables in a manner, which gives fairly good estimates of the candidate system's performance.

The concept of random variables is a crucial one. In most complex systems, we do not know exactly what the future holds. Sales volumes, consumer demand; breakdown times, frequency, and severity; randomly varying production times; and many other system variables are not precisely known beforehand. If they were known, we could develop mathematical relationships between these inputs and our system outputs, although these relationships may be extremely complex.

However, with simulation techniques we do not need the exact inputs. All that is needed is a mathematical expression for the distribution of the inputs. Typical distributions are:

1. **Normal**: with known mean and standard deviation
2. **Exponential**: with known mean
3. **Poisson**: with known mean
4. **Triangular**: with known minimum, maximum, and mode
5. **Uniform**: with known minimum and maximum

These distributions are described by fairly simple mathematical expressions and can be derived from historical data of the process in question. As long as no anticipated change is expected in the overall distribution of a random variable, the distribution developed from the historical data can be considered an excellent prediction, over time, of the future events. These predictions become the inputs to our simulation.

How is the simulation developed? First and foremost, the manufacturing system must be precisely defined. That is, it must be decided exactly what parts of our factory are to be modeled. Essentially, this requires drawing a "dotted line" around that part of the factory which will be included in the model. Scrutinize this dotted line and determine all inputs and outputs to the system. These include everything that crosses the dotted line, either in or out. Identify all material, personnel, products, energy, and information interfacing with the design. Anything which happens completely inside the dotted line will become part of the internal workings of the model; those things which cross the lines are modeled as inputs and outputs.

What exactly is a model? In a general sense, a model is an abstract description of some real system. It is an analogy, whose behavior can be used to infer the behavior of the system which it models. Many types of models exist in the world, some of which are used in computer simulation. Scale models, such as miniature trains, cars, or army men, are useful to architects and generals. Graphical models are useful in other situations; a classic example of a graphical model is a road map.

Additionally, math models and logical models are common. An example of a math model is an equation, such as $F = ma$. This model describes the behavior of the quantities of force, mass, and acceleration, and can be used to predict future events, if the inputs are correctly supplied. Another example of a math model is a look-up table: given row and column values as inputs, the output can be looked up in the interior of a table of values. This type of model is most applicable to functions which cannot be described by a simple equation.

A logical model is a description of inputs and outputs in terms of logical statements. For example, a simple IF-THEN rule could constitute a logical model. A truth table would also fit the definition of a logical model. A complex computer program, with many branching IF statements, is a large logical model of some decision process. An expert system (see Chapter 6) is another example.

The system model will include both math and logical models in its attempt to mimic system behavior. It will include those portions of the system which are considered to be important and will describe them to a level of detail, or resolution, that considered necessary. It will expect to be supplied with those quantities identified as system inputs and will return the system outputs.

When the model has been completed, it will enable do four things that could not have been accomplished as easily, if at all, before:

1. If the model can be made to perform exactly like the current system, it can be used to EXPLAIN the real system. The math and logical models we selected must be actually functioning in the real system.

2. The model can PREDICT how the system would react to a new set of inputs or to a change in the system itself.
3. This allows one to ANALYZE the system and its performance.
4. Iterative analysis of progressive changes to the system model allow one to DESIGN a new system, or changes to the old system, in accordance with new goals.

8.7.1 SIMULATION EXAMPLE

What does a typical simulation look like? Consider the example of a simple system with random variables can be demonstrated with by the operation of a fast-food restaurant. Assume the restaurant has one server, and customers wait in one line to be served. This line is usually called a queue.

First, define the state of the system. The state is a set of variables which completely, yet without redundancy, describes the condition of the system. Any variable not included in the state description can be derived from those which are included. Many possible state descriptions exist for any given system. For the fast-food system, the state can be described by two variables: the server status (either busy or idle) and the number of customers waiting in the queue, not counting the customer, if any, being served by the server. For convenience, it is assumed that the server works diligently at serving any waiting customers and will only become idle when the queue is empty.

Next, identify the ways in which the state can change. In this example, there are two ways: a new customer can arrive, or the server can finish serving a customer, who then leaves. Identifying these state-changing activities is something of an art and requires a bit of practice.

Next, examine how these state-changing activities change the state. Generally, the manner in which the state changes is dependent not only on the activity that occurs but also on the state of the system just before it occurs. The first state-changing activity was that of a new customer arriving. If the teller is idle, the new customer gets immediate service. This means the new customer does not get in the queue, so the queue length stays at zero. The server status, however, changes to busy. If the server had been busy when the new customer arrived, the customer would have to wait, increasing the queue length, but not effecting the server status.

The other activity was the server completing service of a customer. If there are no customers in the queue, server status changes to idle. If there are customers in the queue, server status remains at busy, and queue length decreases by one.

Now, having completely defined the system and identified how it will function, to set it in motion, the simulation must be supplied with some inputs. The inputs will be arrival times of each customer and time required for service by each customer. Each of these values will be supplied from some random distribution, supposedly derived from a study of past data concerning customer interarrival time and service time.

Output statistics are completely up to the simulation user. That is, the simulation can be programmed to collect whatever statistics we are interested in, as long as they are based on variables in the model. For this example, one might be interested in things like average time spent in the queue by the customers, average queue length, and percentage of the time that the server is idle.

Now the simulation is ready to begin. It is given an initial state (usually server=idle and queue=empty), and let the program run. The program selects a customer arrival time from the appropriate random distribution, as well as a service time. It selects the next arrival time, and if this is sooner that the first service is completed, it increases the queue length by one. As each customer arrives and leaves, the state is noted, and statistics are recorded. After sufficient "simulation time" has passed, there will be a full set of output statistics on the values that requested. These outputs, of course, will be statistical in nature, mostly averages and standard deviations. But there is no way one could easily have calculated them from average input statistics; the relationships, in general, will have been far too complex.

What about the actual coding of this model into a computer program? Even the relatively simple example of the single-server/single-queue system could result in a large programming task. Fortunately, commercial packages are available which make this task fairly simple. Most systems are composed of basic building blocks, which tend to look the same, from the viewpoint of a model, regardless of the physical system they are based on. Simulation packages supply ready-made program chunks for each of these building blocks. The user needs only to define the blocks they need, define the interconnections between them, and supply actual values for the variables.

Examples of these building blocks, also known as "network elements," are:

- **Arrival Elements**: these model customers, orders, workers, etc., as they enter the system. Must be supplied with interarrival times for the arrivals.
- **Queue Elements**: these model a queue in the system. Must be supplied with an initial queue length, and possibly with a maximum allowable queue length.
- **Activity Elements**: these model an activity which takes a certain amount of time. Must be supplied with a duration, and how many customers can be served at the same time.
- **Conditional Elements**: allow decisions to be made in the system, based on system state.
- **Statistical Elements**: allow the program to collect statistics based on current model state.
- **Departure Elements**: these model the departure of customers, orders, etc., from the system.

Other, more complex elements also exist, but these tend to vary from one simulation package to another. Many exist, and each has its own strengths and weaknesses. Some are more geared towards specific situations or industries than others. Some have built-in animation and graphics packages, which can display a schematic diagram of the system model on a computer screen as it operates, allowing the user to watch customers arrive, get serviced, wait in queues, etc. Some popular simulation packages currently in use are Arena, FlexSim, Simul8, GPSS, and WITNESS [10–15].

TMA Case Study: Molding Manufacturing System

OVERVIEW

The factory layout provides an efficient use of space and access to material supply, additional manufacturing system integration, and ergonomic productivity gains. There are two basic layout patterns: process layout and product layout. In the former, the basic organization of technology is around processes, and, in the latter, it is the product and its flow that dictates the layout design. Combinations between the two are common.

ISSUE

A new product line is to be launched and a product layout is desirable in providing low production cost per unit. Production control is to be relatively simple, since product designs are stable, and routes through the overall factory manufacturing system are standardized.

STRATEGIC OBJECTIVE

The strategic object is to implement a product layout design that overcomes the high investment in single-purpose equipment by maximizing system reliability to avoid system stoppages due to equipment failure and corrective maintenance.

CASE BACKGROUND

The system for producing the item is evaluated to explore system design and layout alternatives. The engineer assigned the project has modeled the system to predict its reliability. To facilitate the evaluation, it was assumed that the system components are in their useful-life phase and the system is repairable. From this, the reliability was predicted for an 8-hour production shift.

The manufacturing system for this product includes:

A. Molding System
  i.   Mold
  ii. Injection Molding Machine
B. Molded Product Trimmer
C. Product Transfer Loader
D. Product Bagger Equipment
E. Labeling Equipment

A feature of the system functional layout enables one production team member to monitor multiple production cells producing similar products with varying features. Taking advantage of the overall factory product layout design, if the production cell experienced a failure causing catastrophic downtime accumulation, the product manufacture could continue by bringing in applicable system components from other product manufacturing systems (except the injection molding machine). However, that would sacrifice the manufacture of another product.

The engineer developed a system reliability block diagram (RBD) and reliability math model (RMM) reflecting the system functional diagram. The system was designed without any redundancy built in, and system availability, in addition to reliability, was evaluated to assess the impact of corrective maintenance to minimize system downtime.

Data was collected as summarized in the following.

Total Assessment Time: 14.12 hours or 847 minutes
Total Assessment System Cycles: 3,404
Total # Verified Events: 139
Corrective Maintenance (CM) Downtime: 120.4 minutes
Preventative Maintenance (PM) Downtime: 0.2 minutes
Molding System Events: 0
Trimmer Events: 1
Transfer Loader Events: 55
Bagger Equipment Events: 71
Printer System Events: 12

Using the data collected, the RBD and RMM provided the following reliability prediction equation for the 8-hour production shift.

$$R_S = R_1 \times R_2 \times R_3 \times R_4 \times R_5$$

Since all manufacturing system components are serial and the failures are assumed to be random (i.e., reflecting useful life), the failure rates can be added together to ease the calculation. This enables quickly calculating reliability for the 8-hour (or 480 minutes) production shift.

$$R_S = e^{-(1/\mathrm{MTBF})(8)}$$

$$\text{Uptime} = \text{Potential Uptime} - \text{CM Time} - \text{PM Time}$$

$$\text{Uptime} = 847\,\text{minutes} - 120.4\,\text{minutes} - 0.2\,\text{minutes} = 726.4\,\text{minutes}$$

$$\#\,\text{Failure Events} = 139$$

$$\text{MTBF} = (726.4\,\text{minutes}) / 139 = 5.23\,\text{minutes between failure}$$

$$\text{MTBF} = (3,404\,\text{cycles}) / 139 = 24.5\,\text{cycles between failure}$$

$$\text{Mean Time to Repair} = (\text{CM time}) / \#\,\text{events} = (120.4\,\text{minutes}) / 139$$

$$= 0.87\,\text{minutes per event}$$

RESULT/CONCLUSION

The system reliability and system availability are calculated as follows:

$$R_S = e^{-(1/5.23)(480)} = 0.0$$

$$A_{inherent} = MTBF/(MTBF + MTTR) = (5.23 \text{ minutes})/(5.23 \text{ minutes} + 0.87 \text{ minutes})$$

$$= 0.86$$

$$A_{achieved} = Uptime/(Uptime + Downtime) = 726.4/(726.4 + 120.6) = 0.857$$

The reliability of the system is zero over the 8-hour production shift. That is, the system will fail causing downtime. Making additional reliability calculations considering run periods shorter than the 8-hour shift time reveals that reliability equals 0.12 for running 11 minutes without failure, and the reliability value lowers to 0.0008 making it 39 minutes without failure.

Due to the system being repairable, the availability of the system is 0.86. That is, the probability that the system will be running at any given time is 0.86, or it will be running 86% of the time during the shift period.

Case Study Question: *How does system functional layout relate to system reliability block diagramming and reliability math modeling? What is revealing about maintainable systems?*

8.8   SUMMARY

This chapter addressed the need to define the manufacturing system with TMA in mind. It is the first chapter in the second major element of TMA, namely, Manufacturing System Control. The system needs to be defined in a manner that facilitates controllability.

The manufacturing area is laid out to enable maximum efficiency. It is important that manufacturing subsystems be arranged together in a design that reduces distances, inconvenience, and inflexibility. Methods available to achieve this include From-To charting, REL charting, and Space-Relationship Diagramming. Computer programs are available that expedient such analyses and allow for comparative assessment of several candidate layout designs.

Material handling is an always-present issue. Obviously, the desire is to move materials and work in process through the facility in a cost-effective manner. The best material handling method is a function of how much and how fast material needs to move to an intended destination. How these issues are answered guides the handling investment decision and the sophistication of the handling equipment.

System efficiency is also enhanced via the use of automation. This technology enables repetitive tasks to be performed faster and more consistently. Automation is generally classified as either hard or soft. The latter distinguishes itself by being flexible in its manufacturing intent. Many factors (e.g., programmability) are involved in selecting the most cost-effective automation approach, if any, for a manufacturing system.

In this day and age of computer technology, a major system definition advantage exists, that is, simulation. Before we invest lots of money in a manufacturing system design that may or may not be what is desired, the candidate system design(s) can be evaluated via uniquely developed or commercially available computer simulation programs. This enables the optimal system design to be selected for implementation.

QUESTIONS

1. Think of several pairs of departments in your organization that deserve an "A" closeness rating. Are they located as close together as they should be? Also, consider pairs of departments that deserve E, I, O, U, and X ratings.
2. Select a list of 5–10 departments in your organization that work together frequently. Develop a From-To chart to describe the material handling burden between them.
3. List several examples of material handling equipment used in your organization. Rank them in order of initial investment, from most expensive to least expensive. Now rank them according to unit operating cost, from maximum to minimum. What does this tell you?
4. Reconsider the material handling "principles" discussed in this chapter. Think of an example of how each is utilized in your organization. Think of an example of how each is violated.
5. List several examples each of hard and soft automation in your factory. If there are none, what tasks can you identify that would be suitable for each type of automation?
6. Discuss the differences between DNC and CNC. What are their relative advantages and disadvantages? Which would be most appropriate for each task in your organization?
7. Identify several tasks in your plant which could be handled by robots. What characteristics of these tasks make them especially suited for robotics?

REFERENCES

1. Kalpakjian, S. & Schmid, S. (2020). *Manufacturing Engineering and Technology*. New York: Pearson.
2. Francis, R. L. & White, J. (1974). *Facility Layout and Location; An Analytical Approach*. Englewood Cliffs, NJ: Prentice-Hall.
3. Muther, R. (1955). *Practical Plant Layout*. New York: McGraw-Hill.
4. Muther, R. (1961). *Systematic Layout Planning*. Boston: Industrial Education Institute.
5. Sepponen, R. (1969). "CORELAP 8 User's Manual." In *Department of Industrial Engineering*. Boston: Northeastern University.
6. Seehof, J. M. & Evans, W. (1967). "Automated Layout Design Program." *The Journal of Industrial Engineering*, Vol. 18, No. 12, pp 136–158.
7. Buffa, E., Armour, C. & Vollman, T. (1964). "Allocating Facilities with CRAFT." *Harvard Business Review*, Vol. 42, No. 2, pp 690–695.
8. Apple, J. (1977). *Plant Layout and Material Handling*. New York: John Wiley and Sons.

9. Muther, R. & Haganas, K. (1969). *Systematic Handling Analysis*. Kansas City, MO: Management and Industrial Research Publications.
10. Azadivar, F. (1984). *Design and Engineering of Production Systems*. San Jose, CA: Engineering Press, Inc.
11. Kelton, W., Sadowski, R. & Zupick, N. (2015). *Simulation with Arena* (6th ed.). New York: McGraw-Hill.
12. Beaverstock, M. (2018). *Applied Simulations: Modeling and Analysis Using FlexSim*. Orem, Utah: FlexSim Software Publishing.
13. Concannon, A., (2003). *Simulation Modeling with SIMUL8*. Tucson, AZ: Visual Thinking International.
14. Schriber, T. (1991). *Simulation Using GPSS*. New York: John Wiley & Sons.
15. Al-Aomar, R. (2015). *Process Simulation Using WITNESS*. New York: John Wiley & Sons.

9 Product Manufacturing Degradation Control

This chapter addresses controlling and assessing product performance degradation during manufacture. Specifically, the focus is on controlling reliability, safety, and quality degradation. These three performance parameters are critical to a product's strategic success in the market place, as discussed in Chapter 3. Admittedly, investment cost often strongly influences the user's expectations regarding how long a product or system will last without failure. However, the bottom line is that a product or system must be able to stand the test of time, and high reliability, safety, and quality are the key ingredients to make this happen.

To control the degradation of these performance attributes, it is necessary to understand and know the design-in inherent level of reliability, safety, and quality. It is important to note that all too often, product reliability, safety, and quality are secondary issues in designing and developing new products for manufacture.

It is not uncommon that a basic product design itself is fundamentally flawed. Or, the product may be too difficult, or impossible, to manufacture because of its complexity or from assembly requirements surpassing the state-of-the-art manufacturing technology. It would be hard to imagine a product from either one of these scenarios coming out of a manufacturing system without some type of problem or defect.

When talking about "control," two key thoughts must be kept in mind. First, product degradation during manufacture occurs as a result of defects induced during the process. Second, no matter how hard one tries to remove and eliminate defects, a probability always exists for products with latent and/or patent defects being shipped to customers.

Therefore, in order to eliminate, or minimize, degradation, it is necessary to ensure, first, that products are designed with optimal performance safety margins and, second, that materials are specified with proper characteristic tolerances. And, finally, that manufacturing systems (or subsystems or equipment) are designed to be reliable and maintainable, capable of outputting consistent products, and integrated with special inspections, screens, and tests, as appropriate, to detect and remove defective products and components. It is the manufacturing system equipment(s) that enable a process to be performed.

It is easy to accept the fact that a product must be designed to inherently exhibit the market place-required performance levels of reliability, safety, and quality. In the same light, it must be understood that the manufacturing equipment will not necessarily maintain the required desired parameter levels. The playing field doesn't remain the same. That is, the product raw materials change, the operational environment changes, and the equipment parameters change.

This chapter consists of three main sections addressing reliability engineering and control, safety engineering and control, and quality engineering and control. It is

DOI: 10.1201/9781003208051-12

emphasized again that in order to control these product performance parameters, they must be present in the design prior to initiating manufacture. The attainment of TMA ensures that they remain present in the product design upon shipment to the customer.

9.1   RELIABILITY ENGINEERING AND CONTROL

Reliability Engineering is the act of putting together the right combination of design features and support approach to achieve a specified level of successful operation that minimizes design life-cycle costs. To do this requires understanding fundamental reliability engineering concepts, knowing what reliability things to do, and applying reliability engineering techniques.

This classical definition of reliability, "the probability of performing a required function under stated conditions for a define period of time," stresses four elements:

- **Probability**: A quantitative term expressing the likelihood of an event's occurrence.
- **Performance Requirements**: Criteria clearly describing or defining what is considered to be satisfactory operation.
- **Time**: The measure of the period for satisfactory performance.
- **Use Conditions**: The environmental conditions in which an item functions.

There are a couple key ideas from this definition worth remembering. First, reliability is a probability taking into consideration a lot of variables, and, second, reliability changes over time.

In essence, reliability is a moving target. So, no matter what value of reliability is calculated or demonstrated, its accuracy is never an absolute. It is critical to always to view reliability values as figures of merit and not absolutes.

Figure 9.1 illustrates how this reliability figure of merit changes over time. Starting at an inherently designed-in level, reliability decreases upon transition to the production environment, grows with the production learning curve, and then decreases again once in use. The various life-cycle components are:

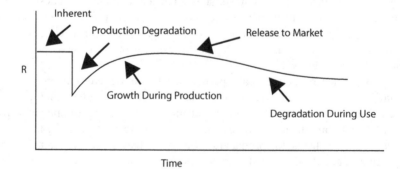

FIGURE 9.1   Reliability degradation over time.

1. **Inherent Reliability**: An upper limit of reliability established by the product design configuration, the component technologies, the component quality levels, and the component derating.
2. **Product Degradation**: The reliability resulting from initial or early manufacturing processes before manufacturing learning, inspection efficiency, and test procedures are optimized.
3. **Reliability Growth during Production**: Improvements in the manufacturing process resulting from trained personnel and inspection and test designed to force out latent defects.
4. **Reliability as Released for Actual Field Operation**: Represents the level of reliability as the product leaves the factory prior to deployment or purchase.
5. **Reliability Degradation during Use**: The degradation resulting from operational and maintenance deterioration factors. The extent of degradation is dependent on the design's hardware and software components selected, as well as the effectiveness of servicing, the maintenance program, inspection and control tasks, and the familiarity and experience of maintenance personnel.

The value of Reliability Engineering comes from its ability to provide practical insight into the potpourri of issues addressed by design, modeling and prediction, and testing and assessment. These include personnel safety, operational success, personal comfort, minimized downtime, minimized maintenance costs, improved customer confidence, and business sustenance. Since there is such breadth to reliability engineering application, reliability engineering management is key to getting the greatest return on investment [1].

There are three major areas of reliability engineering requiring "management" attention. These are design, modeling/prediction, and testing/assessment. Accepting that there is great benefit to be derived from the application of Reliability Engineering, it necessary to put to bed the myth that the subject is an ambiguous pool of mathematical theory subject area that drowns people in mathematical theory. Quite frankly, drowning in mathematics is a real possibility…in any engineering, financial, or other subject. But, the need for detailed mathematical calculations for the majority of cases is avoidable simply by following one golden rule…don't imply more accuracy and precision than is possible or needed.

Reliability in everyday life is described efficiently and effectively by the reliability bathtub curve. Depicted in Figure 9.2, this bathtub curve is also commonly referred to as the life-characteristic curve. The importance of this reliability bathtub curve is that it describes the life of everything in this world, including products, manufacturing systems, and people.

A product's inherent, designed-in level of reliability can easily be degraded during manufacture. However, reliability actually grows (see Figure 9.1) through experience and data collection as problems are discovered and corrected; both design and process. High inherent reliability does not negate the effects of the typically dynamic nature of the manufacturing environment. Both incoming materials and manufacturing system/process parameters change. Such change may be unnoticeable

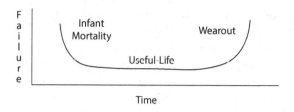

FIGURE 9.2 Reliability bathtub curve.

or catastrophically apparent. The thing to remember is that all changes impact product performance in some manner, whether good, bad, or indifferent.

A common method of detecting product changes is via tests or inspections. However, a single product passing an inspection or test does not imply that all items in the production lot have the same reliability. Inspections and tests are primarily effective in removing patent defects. The sources which degrade product reliability are latent defects.

Examples of patent and latent defects are:

Patent Defects
a. Broken or damaged in handling
b. Wrong or no part installed
c. Faulty part installation
d. Electrical overstress or electrostatic discharge damage (ESD)
e. Missing parts

Latent Defects
a. Partial damage through electrical overstress or ESD
b. Partial physical damage during handling
c. Material or process-induced hidden flaws
d. Damage from soldering operations (excessive heat)

Latent defects are identified through the application of various operational stresses. They are not readily obvious and they may not exist in all products. This means that they likely will move into the market place undetected and result in early product failure upon the experience of an uncharacteristically low overstress situation. The trick is to transform latent defects into patent defects via some type of environmental stress screen. That is, force the defects to be readily identified and removed by special inspection and/or test.

This highlights a significant difference in the philosophical objective of inspections and tests and stress screens. Inspections and tests are intended to pass products. Whereas, stress screens are intended to fail products.

In dealing with latent defects, it is necessary to implement a manufacturing product reliability control program which complements the design/development reliability program. From such a properly planned degradation control effort, it is possible to maintain product reliability during manufacture.

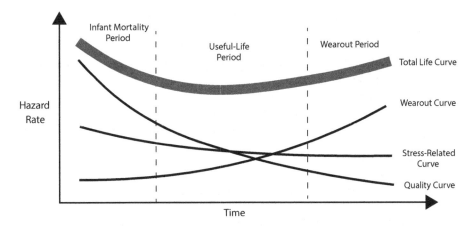

FIGURE 9.3 Product reliability life periods.

9.1.1 RELIABILITY FUNDAMENTALS

As stated previously, reliability is most easily explained via the famous "reliability bathtub," or life characteristic curve. Figure 9.3 illustrates this curve with more detail [2]. The figure illustrates the three failure components that comprise the overall product life characteristic curve. These are (1) quality failures, (2) overstress failures, and (3) wear-out failures. The distributions of these failure components come together to form the infant mortality, useful-life, and wear-out life periods.

The bathtub shape is formed by the decreasing, leveling off, and then increasing hazard rate, or instantaneous failure rate, associated with a product over its lifetime. The high hazard rate during infant mortality is due to latent and patent defects induced during manufacture. The useful-life period reflects a constant hazard rate, or failure rate, which is due to random overstress occurrences, faulty maintenance practices, and/or remaining latent defects. In the wear-out period, an increasing hazard rate is evident, which results from the gradual physical or chemical change of an item over time causing a decrease in strength. Keep in mind that the "bathtub" curve applies to all items (e.g., an airplane, manufacturing system, and a shoe string).

Determining reliability involves the concept of *failure rate* varying as a function of time. A failure rate is a measurement of the number of malfunctions occurring per unit of time and varies over three life periods which comprise the reliability bathtub curve.

1. **Early Failure**: due to design and quality-related manufacturing flaws (a decreasing failure rate).
2. **Stress-Related Failure**: due to random application stresses (a constant failure rate).
3. **Wear-out Failure**: due to aging and/or deterioration (an increasing failure rate).

During each period of the life-cycle, a specific failure component predominates which influences the shape of the "bathtub" curve.

- **Infant Mortality**: This period is characterized by a high but rapidly decreasing failure rate that is comprised of:
  - A high-quality failure component
  - A constant stress-related failure component
  - A low wear-out failure component
- **Useful Life**: This period is characterized by a constant failure rate that is comprised of:
  - A low (and decreasing) quality failure component
  - A constant stress-related failure component
  - A low (but increasing) wear-out failure component
- **Wear Out**: This period is characterized by an increasing failure rate that is comprised of:
  - A negligible quality failure component
  - A constant stress-related failure component
  - An initially low but rapidly increasing wear-out failure component.

"Failure" is the result of an over-stress situation. If the load is greater than what the design is capable of dealing with then failure occurs. Failure can either be due to a random overstress or an application overstress. The quality and wear-out failures result from an application overstress caused by a performance capability less than that expected either due to manufacturing induced problems or aging in the use environment. Whereas, random failures result from a probabilistic stress that exceeds the expected and exhibited performance capability. The reliability objective is to avoid quality and wear-out failures and only experience random, stress-related failures... that is the useful-life part of the reliability bathtub curve.

Note in the figure that product reliability is inversely related to the hazard rate. This means that as the hazard rate decreases, reliability increases. Although reliability itself is a probability (of success), it is often expressed in terms of a mean-time-to-failure (MTTF) or more commonly a mean-time-between-failure (MTBF) (if the product is repairable). The MTBF is derived from the following:

$$MTBF = \int R(t)dt$$

where: R(t) is the reliability function

During the useful-life period, the relatively constant hazard rate enables reliability to be described by the exponential distribution. From this, the inverse relationship between the constant hazard rate, or failure rate, and MTBF is derived.

$$MTBF = \int R(t)dt = \int exp(-\lambda t)dt = 1 / \lambda$$

Note that this relationship only applies during the useful-life period experiencing random failure.

Obviously, it is desirable to be in the useful-life period. This is achieved by growing reliability during product development and implementing effective inspections, tests, and screens (ITS) during manufacturing to minimize the infant mortality period. In addition, wear-out effects are minimized by a comprehensive maintenance program, which addresses short life parts and components. Proper preventive maintenance actions pull the product back into the useful-life period and delay wear out (discussed in Chapter 10).

Accordingly, the focus is on making the useful-life period as long as possible. From a design perspective, this involves increasing the separation between the probability distributions of stress and strength by derating items (i.e., the intentional reduction of stress to strength). From a manufacturing perspective, this involves reducing the scatter of product strength through tighter processes and assurance controls. Finally, from the product user's perspective, this involves controlling application stress variations.

It is apparent that product reliability is affected by (1) applied load variability, (2) environmental conditions, and (3) method of manufacture. Also, material suppliers or maintenance procedures adopted can make a significant difference in the reliability experienced. Again, failure results when the load that the component or assembly is being asked to carry, whether it be dynamic or static, exceeds design capability.

In general, there is little to be gained from reliability activities if the material properties are known precisely but the loads applied are known only approximately. Keep in mind that it is often difficult to establish the distribution of applied loads accurately. It is important to understand how material properties apply to the operating environmental conditions.

If the design specification does not clearly define the limits of the loads to be applied, then all the reliability theory in the world is not going to be of much help. In such a case, either some large safety factor must be applied or else it is a question of build one and see what happens. In either case, the item is likely to be heavier or bigger than is really necessary.

This is when "deep" mathematics is applied in order to gain an understanding of the probabilistic distribution of material strengths with the caveat being that it is critical to ensure that the desired level of reliability is achieved or achievable. However, the key to any application of mathematical theory still relies on the fundamental collection of failure data that ultimately reflects the applicable mathematical situation. It takes real data to confirm the correctness of a theoretically defined situation.

9.1.1.1 General Approach to Reliability Improvement

There is a general approach to reliability engineering. First, minimize initial failures by doing screens and inspections at the part and assembly levels. Second, minimize wear-out and aging effects through inspection and replacement of short-life components. And third, concentrate on eliminating stress-related failures via a focus on design and reliability engineering action.

Linking this back to the reliability bathtub curve, the objective is to get out of the infant mortality period (quality failures), avoid going into the wear-out period (aging failures), and stay in the useful-life period (random, stress-related failures).

The language of associated with reliability evaluation and ongoing improvement is pretty straightforward.

As discussed earlier, the most common terms used are "failure rate" and "MTBF." Failure rate is the number of random failures per unit interval (e.g., time or cycles) for a specific item. MTBF is the unit interval per the number of random failures. The reciprocal of the failure rate, λ, during the useful life period is defined as the MTBF. The MTBF is primarily a figure of merit by which one hardware item can be compared to another.

Expressed mathematically, $MTBF = 1/\lambda$. It is the MTBF, where all failures that result in an unscheduled maintenance action are included. MTBF, as a reliability parameter, serves as a popular index of failures in the useful-life period. Though it does not indicate what will fail or when it will fail, it can indicate the expected number of unscheduled maintenance actions during a given time period, which facilitates preventive maintenance planning.

It is important to make a distinction between "life" and MTBF. They are not the same thing, but are often confused as such. Life, or end of life, is the point in the overall life cycle when it is no longer cost-effective to restore an item to a useful state. Whereas, the MTBF defines an interval between failure. The MTBF by definition implies a reparable item which may or may not be the case. But again, note that the two numbers are not the same. The MTBF may be smaller or larger than the life simply as a result of the design and its intended use.

9.1.2   Design for Reliability

As a design moves off the drawing board and through manufacturing learning, field service, and life, the reliability experienced degrades. One way to off-set this degradation is to obviously create a design that can withstand the stresses it will face. To do this, a clear focus on designing-in and improving reliability is necessary.

There are a host of reliability design analysis and improvement techniques. The short list includes part selection and control; part derating; design simplification; and redundancy.

9.1.2.1   Part Selection and Control

Reliability ultimately is a function of the design's components. To get the highest reliability, use standard, high-quality parts with proven reliability and long-life characteristics wherever possible. Many standard parts (particularly, electronic) are available to serve as the building blocks of a design and, as such, greatly impact the actual reliability of the hardware. The importance of selecting and applying high-quality, proven parts cannot be overemphasized.

If a standard part is not available, special attention should be given to select the best non-standard part. This involves evaluation of the candidate part, its reliability history, construction, and its potential failure modes. Don't forget to keep in mind the life cycle cost curve discussed in Section 9.1, which begs the need to pay attention to individual part cost and its future availability. Avoid the use of expensive, hard-to-support parts.

9.1.2.2   Part Derating

Part derating is the intentional reduction of stress to strength ratio in the item's application.

$$\text{Derating} = \frac{\text{Maximum Stress}}{\text{Minimum Strength}}$$

The basic of the concept of reliability is that a given item has certain stress-resisting capacity, and if the stress induced by the use conditions exceeds this capacity, failure results. This concept expresses item reliability as a function of applicable stress, and strength is generally assumed to be represented by a Normal Frequency Distribution for the scatter of the stresses applied to the item. The overlapping area between the two distributions indicates the extent to which a simultaneous occurrence of high stress and low strength can occur and result in failure or in other words the unreliability of the item.

For various components, failure rate versus stress data may be obtainable from the manufacturer or users, but time rate data may not be available. In using a manufacturer's rating and single design stress values, keep in mind that they are really distributions, not single values. Either the worst-case "tolerances" for both stress and strength or a plot of the distributions must be utilized. When there is time dependency for the distributions (e.g., degradation, wear out), the stress and strength distributions must be related to the cyclic or time operation in the intended environment.

Engineers usually think in terms of safety factor and margin of safety.

$$\text{Safety Factor} = \frac{\text{Minimum Strength}}{\text{Maximum Stress}}$$

$$\text{Margin of Safety} = \frac{\text{Minimum Strength} - \text{Maximum Stress}}{\text{Minimum Strength}}$$

Since failure is not always related to time, the designer needs techniques for comparing stress versus strength and determining the quantitative reliability measure of the design. The traditional use of safety factors and safety margins is inadequate for providing a reliability measure of design integrity.

The concept of stress/strength in design recognizes that loads or stresses and strengths of particular items subjected to these stresses cannot be identified as a specific value but have ranges of values with a probability of occurrence associated with each value in the range. The range of values (variables) may be described with appropriate statistical distributions for the item. Stress/strength design requires knowledge of these distributions. After the strength and stress distributions are determined, a probabilistic approach can be used to calculate the quantitative reliability measure of the design.

The stresses acting on components can generally be grouped into two categories:

1. Environmental stresses include humidity, atmospheric pressure, radiation, chemical content, impurities of the atmosphere, microbes, and environmental temperature. These stresses normally do not cause chance failures

of inactive components, but they may cause a deterioration of component strength under prolonged impact. Items may also be exposed to mechanical stresses such as vibration, shock, and acceleration. These stresses are still environmental stresses which act on the component when it is in operation or not, but they differ from the other stresses in that they can cause a chance failure of the component, also causing gradual fatigue and thus, deterioration of component mechanical strength.

2. Operating stresses (such as voltage, frequency, current, and self-generated heat) appear when the component is in active operation and if the item performs mechanical motions. As in relays, actuators, or rotating devices, self-generated mechanical stresses such as friction and vibration are also added. These stresses normally lead to chance failures, but at the same time, they also contribute significantly to component strength deterioration which shows up in wear out.

Eliminating a stress impacting a design, then the potential failure(s) caused by that stress are eliminated. Thus, the key is to eliminate or minimize stress. There are four basic ways to deal with stress:

1. Increase the strength
2. Decrease average stress
3. Decrease stress variation
4. Decrease strength variation

The failure rate model of most parts is stress and temperature dependent, and a good method for achieving high part reliabilities is to reduce temperature and use parts of higher ratings, as depicted in Figure 9.4. The Arrhenius Law rule-of-thumb is that component failure rates double with each 10°C temp rise. However, as a general rule, derating should not be conservative to the point where costs rise excessively. Neither should the derating criteria be so loose as to render reliable part application ineffective. Optimum derating occurs at or below the point on the stress–temperature curve

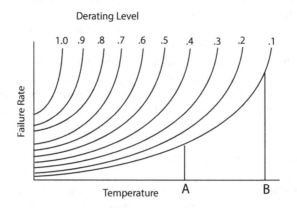

FIGURE 9.4 Arrhenius law.

where a rapid increase in failure rate is noted for a small increase in temperature or stress.

There are two steps for determining the required environmental strength.

Step 1: Identify and describe the application environments
Step 2: Determine the performance of parts and materials, comprising the design, when exposed to the stresses imposed by the application environments

In order to fully realize the benefits of a reliability-oriented design, consideration must be given early in the design process to the required environmental strength of the equipment being designed. The environmental strength, both intrinsic and that provided by specifically direction design features, will singularly determine the ability of the equipment to withstand the harmful stresses imposed by the environment in which the equipment will be operated.

9.1.2.3 Design Simplification

Since reliability is a function of complexity, anything that can be done to reduce complexity will, as a rule, increase reliability. If a component can be removed from a design, the effects of its failure have been eliminated. During design reviews, attention should be directed towards a determination that all items are required to perform the intended function(s), that is, design simplicity. Design simplicity contributes to optimal reliability by making system success dependent on fewer components and the resultant decrease in potential failure modes.

Figure 9.5 illustrates how the number of components affects reliability of the "whole."

9.1.2.4 Redundancy

Redundancy is the existence of more than one means for accomplishing a given task, where all means must fail before there is an overall system failure. Depending on the specific application, a number of approaches are available to improve reliability through redundant design. These approaches are classified on the basis of how the redundant elements are introduced into the system to provide a parallel signal path.

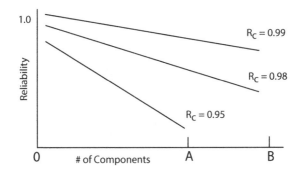

FIGURE 9.5 Reliability versus # of components.

In general, there are two major classes of redundancy:

- **Active Redundancy**: External components are not required to perform the function of detection, decision, and switching when an element or path in the structure fails.
- **Standby Redundancy**: External elements are required to detect, make a decision, and switch to another element or path as a replacement for a failed element or path.

Redundancy Advantages
1. Improves reliability
2. Enables repair without system downtime
3. Prolongs operating time

Redundancy Disadvantages
1. Increases item weight, complexity, cost, etc.
2. Decreases unscheduled maintenance reliability
3. Increases maintenance costs

There are several common redundancy techniques that can be integrated into a basic product or system series design to enhance reliability and safety. Examples include (1) parallel, (2) bimodal parallel-series, and (3) bimodal series-parallel, which will consist of active or standby redundant components. These arrangements are depicted in Figure 9.6.

The decision to use redundant design techniques must be based on a careful analysis of the trade-offs involved. Redundancy may prove the only available method when other techniques of improving reliability have been exhausted or when methods of part improvement are shown to be more costly.

In general, as shown in Figure 9.7, the reliability gain for additional redundant elements decreases rapidly for additions beyond a few parallel elements. There is a diminishing gain in reliability and MTBF as the number of redundant elements is increased. For the simple parallel case, the greatest gain achieved through addition of the first redundant element is equivalent to a 50% increase in the system MTBF.

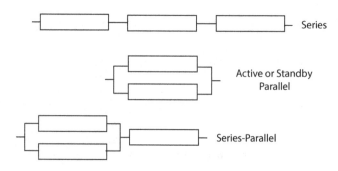

FIGURE 9.6   Basic reliability configurations.

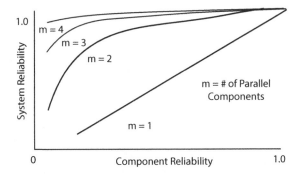

FIGURE 9.7   Reliability versus redundant components.

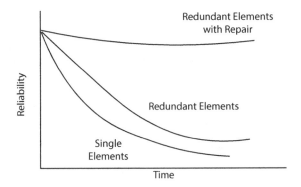

FIGURE 9.8   Design redundancy and repair policy.

Note that in addition to maintenance cost increases due to repair of the additional elements, reliability of certain redundant configurations may actually be less. This is due to the serial reliability of switching or other peripheral devices needed to implement the particular redundancy configuration. If at all possible, it is generally more advantageous to use higher reliability components.

Figure 9.8 illustrates that the effectiveness of certain redundancy techniques (especially standby) can be enhanced by repair. Standby redundancy allows repair of the failed unit (while operation of the good unit continues uninterrupted) by virtue of the switching function built into the standby redundant configuration. The switchover function can readily provide an indication that failure has occurred and operation is continuing on the alternate channel. With a positive failure indication, delays in the repair are minimized. A further advantage of switching is related to built-in-test (BIT) objectives. BIT can be readily incorporated into a sensing and switchover network for ease of maintenance purposes.

Design Analysis Example

From Figure 9.9, a study of the reliability curves shows that of the configurations, only (e) meets the reliability specification of a minimum of 0.995 for 10 hours. In practice, however, configuration (e) might not be used because of

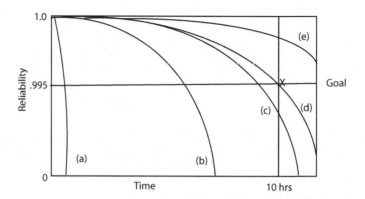

FIGURE 9.9   Design analysis example.

weight and size considerations with the three-unit system. Therefore, efforts would be made to increase the reliability of configurations (c) or (d) until the system reliability goal is met. It may be that from further analysis, the design reliability curves of these two design configurations indicate that only moderate efforts will be needed to upgrade reliability to the required level.

9.1.3   RELIABILITY PREDICTION

A key activity during product during design and development is to predict the product's inherent reliability. This typically comes after conceptualizing the designed in reliability (e.g., reliability block diagrams). It is from this conceptualization that probabilistic reliability mathematical modeling evolves [3]. This provides a means for assessing any discrepancy between the specified reliability requirement (and lower allocations) and the designed-in level. This activity also provides a means by which design improvements and alternatives are judged in regard to reliability. In addition, reliability prediction data are useful input for other activities such as reliability growth and qualification tests, design review, logistics and support cost estimates, and general product improvement programs.

Reliability predictions are made at various times during development. A prediction's accuracy depends on the detail and quality of the design information available. As a design progresses from early to detailed stages, more rigorous prediction methods and models are used to reflect the greater level of definition.

Remember that reliability failures are stress-related failures. These occur during a products useful-life period and are based on three key assumptions: (1) they occur randomly, (2) they occur independent of each other, and (3) they occur at a constant mean rate.

Also, keep in mind that all failures are mechanical in nature. This applies to both "mechanical" parts (such as gears) and electronic parts. The latter part type is more commonly associated with reliability methods since a standardization of part types exists. This enables large data collection efforts subsequently leading to the development of generic reliability prediction algorithms.

TABLE 9.1

Hierarchy of Prediction Techniques

Similar Equipment: Broad comparison of similar items of known reliability.

Similar Complexity: Comparison similar type items based on design complexity.

Similar Function: Evaluate based on previously demonstrated correlations between operational function and reliability.

Part Count: Evaluate based on the number of parts, in each of several part classes, included in the item.

Stress/Strength: Evaluate based on the individual part failure rates and considering part type, operational stress level, and derating characteristics of each part.

Accepting the fact that all failures are actually mechanical in nature (since electrons do not fail), it is a good idea to some understanding of the strength characteristics of an item or material. In general, there are two general theories for viewing the strength probability distribution for materials [4].

The first theory states that the strength of a material is determined by its weakest point. The distribution of strength is then determined by the lowest value of samples of points in the material. This is represented by the Extreme Value Distribution. The second theory states that the weaker points in a material receive support from surrounding stronger points (an averaging effect occurs). The distribution of strength is the mean value of all points. This is represented by the Normal Distribution.

Both theories state that materials (can and often due) exhibit strengths well below their theoretical capacity. This occurs due to the existence of imperfections, or defects, induced by manufacturing. The probable strength is related to the effect of the imperfection creating the greatest reduction in strength. A reliability failure results when a random overstress occurs exceeding the probable strength. In other words, when stress exceeds strength, failure occurs. Via reliability prediction, one can obtain an estimate of how often (on the average) an overstress failure may occur (i.e., MTBF).

Table 9.1 identifies the hierarchy of reliability prediction techniques used to assess reliability based on the strength–stress interactions. The techniques accommodate different analysis requirements and the availability of detailed data as the product design progresses. It is noteworthy that reliability predictions are best used for comparison purposes. Use and look at reliability prediction data for what it really is, a figure of merit of performance. However, the more accurately a prediction model presents the actual operational application, the more credible is the figure of merit.

The basic reliability prediction methodology consists of five unique steps:

Step 1: Compile System Data

This involves aggregating the data and information needed to structure a Reliability Block Diagram (RBD).

Step 2: Develop the RBD and Reliability Math Model (RMM)

The RBD reflects the series/parallel connectivity of the systems functional groups and the components/assemblies. The RMM is a mathematical representation of the RBD.

Step 3: Compile/Compute Part Failure Rates
    The failure rates inputted to the RMM.
Step 4: Compute Reliability
    The failure rate data is inputted into the RMM and a baseline system reliability estimate is computed.
Step 5: Perform Sensitivity Analysis

The sensitivity analysis optimizes the reliability relative to identify significant terms of design, test, application, and environment.

Of great benefit is the development of the RBD to illustrate the reliability connective of design components and to calculate reliability. The structure of a block diagram depends on the definition of success. RBDs are iterated as design is iterated and better defined.

The reliability modeling formulae given below do not account for repair.

$$\text{Series Configuration: } R_s = R_1 \times R_2 \times \ldots\ldots R_n$$

$$\text{Parallel Configuration: } R_s = 1 - (1 - R_1) \times (1 - R_2) \times \ldots\ldots (1 - R_n)$$

They account only for the single event failure.

Series
- Solved using Product Rule
- R_s decreases as the number of series blocks increases
- $R_s \leq \text{minimum } \{R_i\}$
- Reliability is improved by:
    - Decreasing the use of series components
    - Increasing component reliability

Parallel
- **Active**: All systems are activated when system is activated; failures do not influence the reliability of surviving subsystems.
- **Standby**: Most commonly used – Standby component is not activated unless inline component fails
    - Activated by an automated sensing device or by operator (consider $R_{sw} = 1.0$) [e.g., spare time].
    - **Shared Parallel**: As a redundant element fails, the λ of the other elements increases (e.g., lug nuts on wheel).

When modeling product or system design that is repairable, then "availability" is the value to be calculated.

$$\text{Series Configuration: } A_s = A_1 \times A_2 \times \ldots\ldots A_n$$

$$\text{Parallel Configuration: } A_s = 1 - (n!/(k-1)!) \times Y_n - k + 1$$

where
$$Y = MTTR/MTBF$$
n = # of redundant channels
k = # of channels that must be operable

Chapter 10 provides a detailed discussion of the relationship between reliability, maintainability, and availability as it pertains to maintenance strategy to maximize design "uptime" and minimize "downtime." The formulas following are for operational availability (A_o) and inherent availability (A_i).

$$A_o = (\text{uptime})/(\text{uptime} + \text{downtime})$$

$$A_i = (MTBF)/(MTBF + MTTR)$$

It is seen from the equations that system availability increases as the uptime or the mean time between failure increases and the downtime or mean time to repair decreases. So, designing in high reliability or making the item very easy/fast to repair maximizes availability.

In using the availability formula to account for repair, the following assumptions and limitations must be kept in mind.

1. The redundant channels are identical where units are initially operating, one online – the other offline; each unit fails at the rate, λ, and are repaired at the rate, u.
2. Repair is done on only one element at a time, and that any element under repair would remain so until the complete repair has been accomplished.
3. Each element is completely independent. This means that redundant channels are designed such that they are completely isolated from each other, that failures do not propagate from one channel to the other and common failure modes do not exist.
4. Perfect switchover exists, that is, the reliability of any components used to sense failure of the online channel and to switch to the offline channel is assumed to be 1.0.
5. Both logistic time and repair time are considered random. This means that the repair is completed at a random time after it is started and also that the repair is initiated at a random time after the failure occurs. It should be noted that although the assumptions of random failure and repair (constant λ and μ) are not always correct, both assumptions are necessary to avoid extensive mathematical complications. Furthermore, assuming constant repair rates does not usually introduce serious limitations when availability calculations are performed.
6. The MTTR of each element (serial and redundant channels) is identical.
7. The failure rate of a redundant channel is less than or equal to ten times its repair rate.

There are many standard statistical distributions which may be used to model the various reliability parameters. For continuous distributions, the Normal, Lognormal, Exponential, Gamma, and Weibull Distributions are used. For discrete distributions, the Binomial and Poisson Distributions are used. It has been found that a relatively small number of statistical distributions satisfies most needs in reliability work. The particular distribution used depends upon the nature of the data in each case.

In practice, one is usually forced to select a distribution model without having enough data to actually verify its appropriateness. Selection is based on either previous experience or knowledge of the particular physical situation which caused the failures.

Accepting the fact that all failures are actually mechanical (i.e., electrons do not fail), it is a good idea to have some understanding of the strength characteristics of an item or material. There are two general theories for viewing the strength probability distribution for materials. The first states that the strength of a material is determined by its weakest point. The distribution of strength is then determined by the lowest value of samples of points in the material. This is represented by the Extreme Value Distribution.

The second theory states that the weaker points in a material receive support from surrounding stronger points (an averaging effect occurs). The distribution of strength is the mean value of all points. This is represented by the Normal Distribution.

Both theories state that materials can and often do exhibit strengths well below their theoretical capacity. This happens because of the existence of imperfections, or defects, induced by manufacturing. The probable strength is related to the effect of the imperfection creating the greatest reduction in strength. A reliability failure results when a random overstress occurs that exceeds the probable strength. In other words, when stress exceeds strength, failure occurs. Using reliability prediction, one can obtain an estimate of how often (on the average) an overstress failure many occur (i.e., MTBF).

During an item's useful-life period its failure rate is considered to be constant as described by the exponential failure distribution. This is based on (1) the stress/strength relationship and varying environmental conditions result in effectively random failures, and (2) over a period, the system failure rate oscillates, but this cyclic movement diminishes in time and approaches a stable state with a constant failure rate.

The exponential distribution can be used as an approximation of some other function over a particular interval of time for which the true failure rate is essentially constant. This distribution is concerned with the useful-life period and a constant failure rate due to stress-related failures.

As discussed earlier, there is a hierarchy of reliability prediction techniques to accommodate different reliability study and analysis requirements and the availability of detailed data as the system design progresses. Figure 9.10 reiterates the prediction approaches relative to design level (i.e., system, subsystem, assembly, and part).

1. **Similar Equipment**: The design under construction is compared with similar items of known reliability in estimating the probable level of achievable reliability.

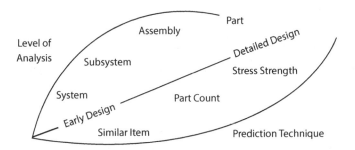

FIGURE 9.10   Reliability prediction throughout design.

2. **Similar Complexity**: The reliability of a new design is estimated as a function of the relative complexity of the subject item with respect to a "typical" item of similar type.
3. **Similar Function**: Previously demonstrated correlation between operational function and reliability is considered in obtaining reliability predictions for a new design.
4. **Part Count**: Design reliability is estimated as a function of the number of parts, in each of several part classes, to be included in the equipment. The advantage of this method is that it allows rapid estimation of reliability in order to quickly determine the feasibility (from the reliability standpoint) of a given design approach.
5. **Stress/Strength**: The item failure rate is determined as an additional function of all individual part failure rates, and considering part type, operational stress level, and derating characteristics of each part. The stress/ strength reliability prediction technique is applicable during the design phase of hardware development.

Of course, in order to predict reliability, maintainability, or availability, there needs data to be available. Several data sources are delineated in Table 9.2.

9.1.4   GROWING PRODUCT RELIABILITY PRIOR TO MANUFACTURE

An important consideration during product development is reliability growth. As stated earlier, one of the first tasks performed during product development is reliability prediction. However, the actual product reliability level will be lower than that predicted. Thus, the need to perform adequate testing and assessment prior to reaching manufacture stages.

Figure 9.11 provides an overview of the placement of reliability development testing to grow reliability through the development and manufacturing process. As illustrated in the figure, there are five key points in the transition from design through manufacture that are of reliability interest.

1. Represents the reliability of the design as estimated by the reliability prediction task. This value is the upper limit of reliability established by the design-in reliability and maintainability attributes.

TABLE 9.2

Data Sources for Reliability Prediction

Reference	Available From	Contents
IEEE Std 500-	Institute of Electrical and Electronic Engineers New York, NY	Electrical, electronic, and sensing component reliability data for nuclear power generating stations
MIL-HDBK-217	Naval Publications and Forms Center 5801 Tabor Avenue Philadelphia, PA	Hazard rate data on military electronic equipment
NPRD (Nonelectronic Parts Reliability Databook)	Reliability Analysis Center Rome Air Development Center Griffiss Air Force Base, NY	Failure rate data on devices intrinsic to computer peripherals; other data not available in MIL-HDBK-217
GIDEP (Government-Industry Data Exchange Program)	Department of the Navy GIDEP Operations Center Corona, CA	Reliability-maintainability data interchange; failure experience data interchange
NPRDS (Nuclear Power Plant Reliability Data System)	American National Standards Institute New York, NY	Failure-reporting data bank for systems and components
FEED (File of Evaluated and Event Data)	Electric Power Research Institute Palo Alto, CA	Personnel, component, and system failure rates
System Reliability Service	United Kingdom Atomic Energy Authority, Wigshaw Lane, Culcheth, Warrington, Lancashire, England	Failure rate assessments derived from UK and other available European sources
FARADA (Converged Failure Rate Data Handbooks)	Fleet Missile Systems Analysis and Evaluation Group Annex, NWS, Sea Beach, Corona, CA	Failure rate assessments derived from Army, Navy, Air Force, and NASA sources

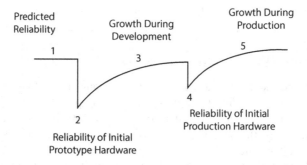

FIGURE 9.11   Reliability from design to manufacture.

2. Represents the value of reliability for the initial assembled engineering test hardware. This value usually falls within the range of 10%–30% of the predicted reliability.
3. Represents the reliability growth of the test hardware as a result of reliability development and growth testing (RDGT). The rate of growth, depicted

by the slope of the curve, is governed by the amount of control, rigor, and efficiency by which the failure recording, analysis, and corrective action (FRACA) program is conducted.

4. Represents the reliability degradation incurred when the final qualified design configuration is transitioned to production.
5. The subsequent growth that can be expected due to manufacturing learning and the application of Environmental Stress Screens (ESS).

Reliability development testing includes planning and implementing design verification tests, design approval tests, reliability growth tests, screening and burn-in tests, qualification tests, and acceptance tests, as well as other special safety tests which are conducted during product development. It includes preparing a detailed and fully coordinated test plan tailored to project needs and applicable engineering test item configuration. Reliability testing includes the following:

- *Design verification tests* consist of limited environmental tests performed on early engineering test units (ETU) to identify design deficiencies. These tests include mechanical stresses, climatic stresses, and safety stresses.
- *Design approval tests* consist of a comprehensive sequence of environmental tests performed on advanced ETUs.
- Reliability *growth* tests consist of a planned *test-analyze-and-fix* process in which environmentally qualified ETUs are tested under repeated actual, simulated, or accelerated environmental test cycles to disclose design for process deficiencies and defects.
- *Screening and Burn-in* consists of a series of stress tests where selected equipment or components are subjected to environmental and operating stresses for the purpose of eliminating early life defects on pilot or production units prior to delivery for field test and/or commercial use.
- Qualification and acceptance tests are conducted on pre-production ETUs representative of the final design configuration to determine compliance with specified requirements and acceptability for field testing.

Table 9.3 provides recommended test requirements and parameters for the development test types.

As reflected in Figure 9.11, a RDGT program enables a design's reliability to grow to the level required. Growth occurs during the test which consists of operationally stressing the product to force out problems. This results from analyzing failures to determine root cause, developing and implementing sound corrective action, and verifying that the failure has been effectively eliminated.

Formal RDGT is performed for a full-scale preproduction article that reflects the final product configuration. It is also manufactured (as close as possible) via the process(s) associated with the planned manufacturing system. The intent is to get the product's reliability level at or near that predicted. This occurs as result of eliminating or minimizing failure sources via a rigorous test program. Supporting this RDGT effort must be a comprehensive FRACA system (discussed in Chapter 12).

TABLE 9.3

Test Requirements and Parameters

Test Type	Test Unit Configuration	Length	# Units	End Criteria	Acceptance Criteria
DVT	Model	1–100 hours[1]	1–2	Failure	Design Spec.
DAT	Prototype	1–1,000 hours[1]	1–2	Failure	Design Spec.
DQT	Production	T_L[2]	1–5	Time	Se ≥ 0.9[b]

[1] or cycle equivalent

[2] $T_L = (MTBF_s) (M)$[a]

Confidence

Level % = 99 97.5 95 90[a] 85 80 75 70 60 50 40 30 20 10

      M = 4.1 3.4 2.8 2.3 1.9 1.7 1.5 1.3 1.0 0.75 0.5 0.4 0.2 0.1

[a] Recommended minimum CL %.

[b] System Effectiveness per H1-1562.

Knowing that reliability *will* degrade during manufacturing, it makes sense to apply RDGT to ensure that the product design itself meets reliability requirements. Attention can then be focused completely on controlling reliability degradation occurring as a result of the manufacturing system and raw material variation.

The RDGT planning and implementation activity is driven by the objective of meeting a specified level of product reliability at the minimum cost. This requires that the RDGT program reflect a proper balance between desired hardware reliability, test cost, and test time [5].

As with all projects, a detailed management plan is essential. The RDGT plan defines the cumulative test time required to grow to the specified or targeted MTBF, the number of test units, and the anticipated test time per unit. The plan also defines details regarding the growth model, reliability starting point, growth rate, and test schedule.

A popular growth model for planning the reliability growth process is the Duane Model [6]. The Duane Model assumes that if a uniform RDGT program is implemented, then reliability growth maintains a linear relationship with time. Also, the rate of the growth is dependent on the rigor of the test, analyze, and fix (TAAF) program implemented. Figure 9.12 illustrates this model.

Two key variables depicted in the figure are the reliability starting point and the growth rate. The reliability starting point, $MTBF_0$, is typically between 10% and 20% of the predicted MTBF. A conservative value should be used to assure adequate funding and time to complete the RDGT.

The value of growth rate typically varies between 0.1 and 0.6. A growth rate of 0.1 is expected in those programs where no specific consideration is given to reliability. A conservative growth rate of 0.3–0.4 is typically planned in all cases. Although rates of 0.5 and 0.6 are theoretically possible, they are seldom achieved.

Note that the calendar time required for a RDGT can be prohibitive (a year or more). This does not mean avoid RDGT but rather explore the many RDGT program trade-offs possible to reduce test time while not reducing test integrity. Common trade-offs to reduce RDGT schedules involve increasing FRACA program rigor and the number of dedicated test units.

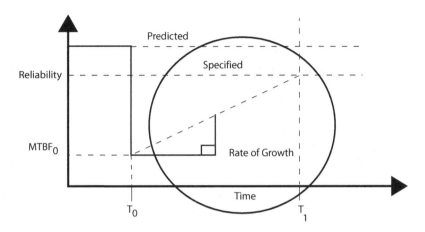

FIGURE 9.12 Duane model illustration.

9.1.4.1 Test Data Collection and Analysis

As with all design and manufacturing activities, it is critical to analyze and correct failures through a formal "closed-loop" recurrence control process. Data collection is facilitated via a FRACA activity. FRACA facilitates isolating and correcting the failures and verifying corrective action effectiveness. Use the actual or simulated end-use environment to enable experiencing all performance-impacting conditions that apply such as mechanical changeovers, operator changes, and maintenance (preventive and corrective).

Data and information include (as applicable):

1. Time to failure/stoppage event.
2. Reason for each failure/stoppage event (e.g., preventive maintenance).
3. Total successful operation time (or uptime).
4. Total unsuccessful operation time (or downtime).
5. Corrective action taken for each failure/stoppage event.
6. Identification of and the time used for various activities contributing to the downtime for any one downing incident (i.e., failure or other) (e.g., 1 hour in tool and die area or 1 hour for lunch break).
7. Classification of each failure/stoppage event as relevant or nonrelevant.

Relevant events include:

a. Design defects.
b. Manufacturing defects.
c. Functional degradation below design specifications.
d. Intermittent failures.
e. Failures of known limited life parts prior to specified life.
f. Failures not attributable to a specific cause.
g. Failure of built-in test features.

Nonrelevant events include:

a. Failures resulting from improper handling or installation.
b. Failures of instrumentation or monitoring devices external to the qualification subject.
c. Failures due to stress beyond design specifications.
d. Failures resulting from procedural/human error.
e. Failures induced by maintenance actions.
f. A secondary failure which is the direct result of a failure of another part within the equipment.

Note the following for availability/reliability assessment conducted in accordance with the sample test parameters and criteria presented in Table 9.3.

1. Intermittent failures are relevant. If several such failures are due to a specific defect and the defect is eliminated during the test, then only one failure is counted.
2. Failures occurring during test setup, troubleshooting, or maintenance are not considered as reliability test failures.
3. Only operational cycles occurring during normal "power on" portion of the test are considered as test time. Operating time accumulated as a result of "checkout" or "verification" activities are not considered test time.
4. Test time restarts from time zero after implementation of a *major* design change.
5. Test time may be interrupted to make a minor design change. Test time resumes from the point of stoppage and the design change has no effect on the classification of previous failures.

In evaluating data, Weibull analysis provides a simple graphical tool. Figure 9.13 illustrates the process which consists of plotting a curve and analyzing it. The horizontal scale is some measure of life, perhaps start/stop cycles of operating time. The vertical scale is the probability of the occurrence of the event. The slope of the line

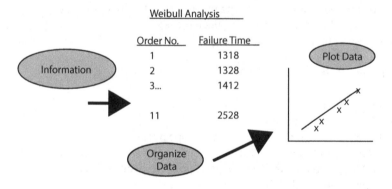

FIGURE 9.13   Weibull analysis.

(β) is particularly significant and may provide a clue to the physics of the failure in question. This type of analysis relating the slope to possible failure modes can be expanded by inspecting libraries of past Weibull curves.

Most applications of Weibull analysis are based on a single failure class or mode from a single part or component. An ideal application would consist of a sample of 20–30 failures. Except for material characterization laboratory tests, ideal data are rare; usually the analysis is started with a few failures embedded in a large number of successful or censored units.

The age of each part is required. The units of age depend on the part usage and the failure mode. For example, if low and high cycle fatigue produce cracks leading to rupture, which indicates that the age units are fatigue cycles. Or, for an aero-jet starter, the age unit may be the number of engine starts. Another example might be that burner and turbine parts fail as a function of time at high temperature or as the number of excursions from cold to hot and return. In most cases, knowledge of the physics-of-failure will provide the age scale. When the units of age are unknown, several age scales must be tried to determine the best fit.

The first use of the Weibull plot is to determine the parameter β, which is known as the slope, or shape parameter. Beta determines which member of the family of Weibull failure distributions best fits or describes the data. The failure mode may be any one of the types represented by the familiar reliability bathtub curve, infant mortality with slopes less than one, random with slopes of one, and wear out with slopes greater than one [7]. The Weibull plot is also inspected to determine the onset of the failure.

9.1.5 Manufacturing System Modeling

As an integral part of designing and evaluating a manufacturing system, it is desirable to get an estimate of the product reliability degradation that is going to occur. This is done by performing product reliability manufacturing degradation analysis (PRMDA) (see Figure 9.14).

The analysis is composed of two fundamental activities. The first activity consists of modeling the manufacturing system via a detailed flow diagram depicting each step of the overall manufacturing process which is performed the system. This includes the various ITS integrated into the overall process, as appropriate.

The second activity involves exercising the model to estimate the number of defects introduced into the overall process, removed from the process, and outgoing from the process. These estimates are then used to derive a degradation adjustment factor which is multiplied with the predicted product MTBF to provide an estimate of system outgoing product MTBF.

The analysis enables knowledgeable trade-offs to be made between the type and quantity of ITS to be integrated into the manufacturing system and process. This makes it possible to optimize the overall effectiveness of the manufacturing system design and minimize the product defectivity rate transferred to the consumer markets.

At this point, it is prudent to make a distinction between what we are doing here and what is done as part of quality control (discussed later in this chapter). With

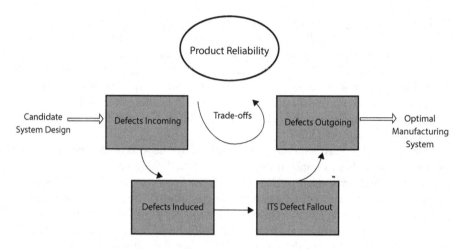

FIGURE 9.14 PRMDA approach.

reliability control, the objective is to minimize the number of failures causing (particularly, near term) product defects that make it to the market place. For the most part, these defects are not readily observable (i.e., latent defects) and require special ITS to be discovered and removed.

With quality control, the objective is to minimize the number of readily observable product defects that make it to the market place. The key word here is "minimize." Statistical process control, for example, is designed to minimize the number of customer observable defects. Certainly, zero defects is the goal. By addressing reliability degradation control, that goal is made more practical.

Integrating ITS into the manufacturing system requires knowledge of the quantity and type of defects expected to reside in the product (at various levels). It is the reliability degrading defects (i.e., latent defects) that are of importance to discover and remove. The quantity and type of latent defects which are introduced into a product are dependent upon several factors [8]. These factors include:

- **Design Complexity**: The quantity and type of parts and interconnections used in the product. Increased complexity creates more opportunities for defects.
- **Part Quality Level/Grade**: The quality levels of parts or materials.
- **Operational Environment**: The stress conditions to which the product will be exposed in the actual use environment.
- **Manufacturing System/Process Maturity**: New systems and processes require learning curve time to identify and correct planning and process problems, train personnel, and to establish vendor and process controls.
- **Packaging Density**: Product assemblies with high part and interconnect density are more susceptible to system/process, workmanship, and environmental-induced defects.
- **Manufacturing System/Process Control**: Good control and monitoring reduce the number of defects which are introduced into the product.

- **Workmanship Standards**: Stringent and enforced workmanship quality standards minimize workmanship defects induced into the product.

Even though manufacturing systems are typically, optimistically, viewed as a product reliability-conscious designs, it is imperative to remember that ITSs are not perfect. At each step of manufacture, where ITSs are applied, defects will escape to the next process step and new opportunities for introducing defects are created.

9.1.5.1 Defect Removal Planning

As with other engineering activities, defect removal planning for the manufacturing system must begin early in the design phase. Successful defect removal is strongly dependent on knowledge of the product and the manufacturing system and overall process. The following should be kept in mind:

- The type of potential defects in the product.
- Experience data available for similar products (e.g., in composition, construction, and maturity).
- Information sources including:
  1. Item historical performance.
  2. Supplier/vendor performance and certification.
  3. Qualification test data.
  4. Supplier provided test data.
  5. Incoming inspection data.
  6. ITS records for previous manufacturing programs.
  7. Reliability growth test results.
- The need for cost-effective integration of ITS into the manufacturing process.
- Project management activities:
  1. Establishing objectives/goals
  2. Preparing a written plan
  3. Obtaining planning estimates of defect density
  4. Selecting and placing of ITS

9.1.5.2 PRMDA Technique

The PRMDA technique enables various manufacturing system configuration alternatives to be evaluated. This means that the system configuration providing the best relative benefit-cost is identified prior to any significant resource investment. The following provides a formula to determine this benefit-cost figure of merit.

$$RBC = DAF \times (ECS - EIC)$$

where RBC is the relative benefit-cost
DAF is the reliability degradation adjustment factor
ECS is the expected cost savings (rework or repair costs x #defects removed)
EIC is the expected implementation cost (hardware costs + inspection or test costs)

The DAF is derived (later in this section) based on the removal of expected latent defects. Several factors which are key in assessing the benefit-cost include:

- Outgoing defectivity rate
- Incoming defectivity rate
- Types of ITSs relative to product characteristics
- Screen effectiveness
- Test effectiveness
- Inspection effectiveness
- Product environmental design limits
- ITS facility costs
- ITS monitoring costs
- Material scrap quantities
- Product rework costs
- Product field repair costs
- Failure analysis costs
- Product volume

Many of the above are fixed for a given corporation, particularly those associated with basic costs. The PRMDA adjusts the multipliers for these "fixed" costs via exercising the model in light of various trade-off scenarios (i.e., manufacturing system ITS configurations).

Earlier in this chapter, PRMDA was defined as consisting of two fundamental activities. The first activity focuses on modeling the manufacturing system and the process(es) supported thereby. The second activity focuses on exercising the model to estimate product reliability degradation. These activities are addressed in the following paragraphs.

PRMDA Activity One

In developing a PMRDA model, a system engineering approach is used. The manufacturing system is viewed as essentially a "black box." There are defects incoming to the box, induced in the box, falling out of the box, and outgoing from the box to the customer. Figure 9.15 illustrates the highest level of this modeling approach.

The level at which a system is modeled is a judgment call. However, it must be to a level which depicts individual steps of the overall manufacturing process(es) and ITS locations integrated therein. Figure 9.16 depicts some lower level of manufacturing system modeling. Note the identification of the PRMDA parameters. These are the number of latent defects in materials entering the system (D_i), the individual ITSs' effectiveness $(E_1, E_2...E_n)$, the number of latent defects in materials entering each process (or step) (D_{in}), the outgoing number of latent defects for each process (D_o), the latent defect fallout from each ITS (D_f), and the system outgoing number of latent defects (D_r).

PRMDA Activity Two

In order to exercise the model, it is necessary to estimate the values of the various PRMDA parameters. The first parameter to estimate is D_i. This is done via manipulation of the Chance Defective Exponential Formulation [9]:

$$\lambda_{s(t)} = \lambda_o + (D_i/N)(\lambda_{ld})(\exp(-t\lambda_{ld}))$$

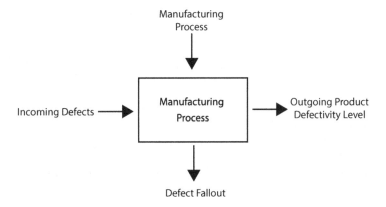

FIGURE 9.15 PRMDA systems modeling approach.

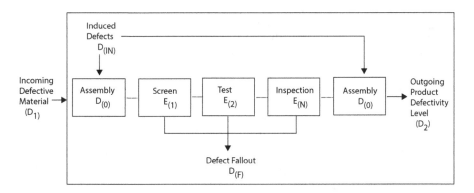

FIGURE 9.16 Detailed PRMDA systems model.

where

$\lambda_{s(t)}$ is the actual item failure rate
λ_o is the predicted item failure rate
D_i is the number of incoming latent defects
N is the quantity of material or number of parts
λ_{ld} is the assumed latent defect failure rate
t is the amount of exposure (e.g., time, length, area, and volume)

The parameter D_i is derived for this formulation via the application of several key (and conservative) assumptions. First, the average failure rate of a defect in the marketplace is greater than one failure per thousand t in order to be considered an early failure, or latent defect. From this, λ_{ld} is estimated as 0.001 defects per t. Second, for products undergoing RDGT, the actual product failure rate is at least 50% of that predicted. For products not undergoing RDGT, the product failure rate is at most 30% of that predicted.

Applying these assumptions allows D_i to be derived as:

$$D_i = (1,000)(\lambda_o)(N)(\text{with RDGT})$$

$$D_i = (2,300)(\lambda_o)(N)(\text{without RDGT})$$

The second parameter to estimate is E for each the inspection or test in place to detect the latent turned patent defect. Inspection effectiveness is a function of two possible implementation scenarios: manual and automated. Obviously, automated inspection methods are typically the most cost-effective and values for E of 0.95 and higher are appropriate. For manual methods (i.e., performed by a human), values for E of 0.7–0.9 are generally appropriate.

A more exact formula for determining E value is derived from the following [10]:

$$E = (E_b)(E_t)(E_c)(E_e)$$

where
 E is the inspection effectiveness level
 E_b is the baseline error probability under ideal conditions
 E_t is a factor to account for level of training or experience
 E_c is a factor to account for inspection complexity
 E_e is a factor to account for the inspection environment

Test effectiveness is defined as the fraction of patent defects detectable by a defined test procedure to the total possible number of patent defects present. While stress screens are effective in transforming a latent defect into a detectable failure, removal of the failed condition is dependent on the capability of the test procedures used to detect and localize the failure.

It is important to ensure that tests have detection efficiencies as high as is technically and economically feasible. The difficulty in accurately simulating functional interfaces or the inability to establish meaningful acceptance criteria reduces test effectiveness. To simplify determining the E value, Table 9.4 provides example values of test effectiveness [9].

Screen effectiveness is defined by the ability of screen to precipitate a latent defect to a detectable state given that a defect susceptible to the screen stress is present. A basic premise of stress screening is that with specific stresses applied over time, latent

---

TABLE 9.4

Test Effectiveness, E or DE

Test Type	Detection Effectiveness, DE
Manual Go/No Go	0.85
Operational	0.90
Automated functional	0.95

defect failure rates of material are accelerated over that which would occur under normal field operating stress conditions. For example, by subjecting an electronic product assembly level to accelerated stresses (such as rapid temperature cycling and random vibration), latent defects are precipitated to early failure.

More severe stresses accelerate failure mechanisms and the rate of defect failure. However, care must be taken to not implement destructive screens but to only stress the subject item within its design limits.

The most popular screens are vibration and temperature cycling. In general, vibration screens are more effective for precipitating workmanship-induced defects, and thermal screens are considered more effective for part defects. There are also classes of defects responsive to both vibration and thermal screens.

It is advantageous to operationally exercise the product during the screen. Also, performing the screen at the lowest possible product level (i.e., part, intermediate, or product) reduces the cost of rework or repair. However, finding problems at higher levels of assembly generally costs less.

An important issue to remember is that the applicable product level must be tested prior to entering a screen. If this does not happen, it cannot be determined whether the defects were precipitated by the screen or present in the product level (as patent defects) before the screen.

The screen effectiveness value is derived from the following:

$$SE = (SS)(DE)$$

where
SE is screen effectiveness
SS is the screening strength (see Table 9.5)
DE is the detection effectiveness (this value is derived from either detailed inspection criteria or Table 9.4)

Once the above PRMDA parameter values are defined, it is possible to estimate the number of defects outgoing to the market place. This involves summing the D_i values moving into each process and the outgoing D_o value outgoing to each defined ITSs, as applicable. The D_f is determined by multiplying ITS E value to derive the outgoing D_i into the next process, as applicable. These calculations continue through the model until the D_r value is estimated.

It is now possible to estimate the product reliability degradation occurring during manufacture [1]. The following provides this estimation.

$$MTBF_a = (MTBF_i)(DAF)$$

where
$MTBF_a$ is the achieved reliability (resulting from the degradation forces)
$MTBF_i$ is the predicted reliability
DAF is the degradation adjustment factor

The degradation adjustment factor is derived from:

TABLE 9.5

SS Values (Adapted from DOD-HDBK-344)

Temperature Stress Screening

# Cycles	Temp Rate of Change, °C/min.	80	100	120	140	160	180
6							
	5	0.7	0.75	0.79	0.82	0.84	0.86
	10	0.9	0.93	0.95	0.96	0.97	0.98
	15	0.97	0.98	0.99	0.99	0.99	0.99
	>20	0.99	0.99	0.99	0.99	0.99	0.99
8							
	5	0.8	0.84	0.87	0.9	0.91	0.93
	10	0.96	0.97	0.98	0.99	0.99	0.99
	>15	0.99	0.99	0.99	0.99	0.99	0.99
10							
	5	0.87	0.9	0.92	0.94	0.95	0.96
	10	0.98	0.99	0.99	0.99	0.99	0.99
	>15	0.99	0.99	0.99	0.99	0.99	0.99

Temperature Range (°C)

$$DAF = D_i / (D_i + D_r)$$

This value can be used in the relative benefit-cost equation defined previously.

PRMDA Example

Consider the manufacture of a handheld electronic product that costs $1,000 and 10,000 units are planned to be manufactured. Each unit consists of some several integrated circuits, semiconductors, and resistors. The predicted MTBF for the item is 26,315 hours. An automated functional temperature cycling screen having six cycles with a range of 80°C and a 10°C per minute rate of change and power cycling is to be performed at the board level (SS = 0.9). From Table 9.4, test effectiveness is 0.95. The cost of a repair is $200 (considers inspection time, repair time, materials, field service costs, etc.) and the cost of implementing the screen into the manufacturing system and process is $10,500 (considers thermal cycling chamber, test equipment, accessories, inspection time, handling time, etc.). Figure 9.17 illustrates the model.

The following are derived:

$DAF = 825/(825 + 120) = 0.87$

$MTBF_a = (26,315\, \text{hours})\, (0.87) = 22,894\, \text{hours}$

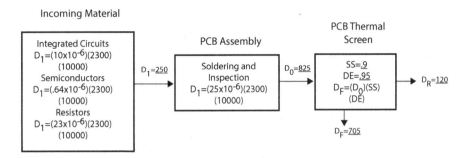

FIGURE 9.17 Example system model.

ECS = ($200) (705) = $141,000
EIC = $10,500
RBC = (0.87) ($141,000 − $10,500) = $113,535

So, the relative benefit-cost, RBC, is greater than $100,000.

9.2 SAFETY ENGINEERING AND CONTROL

Controlling product safety during manufacturing is a critical issue. Obviously, product designs should never reach the manufacturing floor if they exhibit a reasonable risk to the customer. The legal liabilities can prove very costly; both in dollars and reputation. This section also addresses the manufacturing system as a product since it too should be designed to minimize product safety degradation and present no impending danger to its operators.

As part of TMA, safety is an important function. Safety engineering is a broad discipline; however, it focuses primarily on three fundamental areas: (1) products, (2) systems, and (3) occupational environments. Within these areas, extended focus is on the customer and on the personnel working in the manufacturing environment.

Product safety is important to the customer as is system and occupational safety to the workers in the manufacturing facility. All three are a function of design and must receive priority attention from management. The rationale is simply the fact that company liability exists for each of these areas. Furthermore, there is absolutely no justification for knowingly endangering persons using your product, or making your product. The key is to know who you are designing for and then use this information to develop a design which can be used safely and effectively.

To ensure that safety is adequately controlled, it is imperative that engineering design evaluation techniques addressing both product and manufacturing system safety be applied. In the same breadth, it is important that a company safety program plan be developed to improve worker morale and productivity relative to their perspective of manufacturing system and occupational safety. This links into worker motivation discussed in Chapter 7.

9.2.1 SAFETY ASSESSMENT

One can generally expect a large number of safety issues/concerns to be present during the development of new products or manufacturing processes and systems. Discovering and addressing these issues requires performing a formal risk assessment.

A risk assessment, or hazard analysis, provides insight to the risks, or potential hazards, involved with the subject item. From this, responsible management decisions are made concerning the perceived risk(s). For example, knowledgeable decisions to answer questions such as; is the perceived risk(s) reasonable/acceptable? Is it reducible to acceptability via the use of protective measures? Or, is it completely unacceptable dictating redesign or discontinuing the development effort?

Obtaining the information required to answer these questions involves several activities. First, it must be determined whether or not probable risks are acceptable. Second, it must be determined what is the justifiable level of funding for accident prevention measures. Third, comparisons must be made between risks and hazard rates for similar existing products or processes.

Many techniques exist to identify safety hazards and rank them in terms of their criticality, fault tree analysis (FTA) and failure mode, effects, and criticality analysis (FMECA) are two popular and powerful techniques. These analyses provide rationale for addressing those hazards which when eliminated, or reduced, provide the greatest impact in reducing risk. In addition, these analysis techniques facilitate:

- Selection of optimal designs
- Assessment and documentation of failure modes and their effects
- Early visibility of hardware and human–machine interface issues
- Ranking of potential failures relative to severity of effects and probability of occurrence
- Identification of single failure points critical to operation or performance
- Performance criteria for test planning
- Quantitative and qualitative data and information for use in reliability, maintainability, and logistic analyses
- Design and location of performance monitoring and built-in test capabilities
- Evaluation of design, operational, or procedural changes

From an overall cost and time project life-cycle perspective, it makes sense that design hazards should be eliminated, or reduced, as early as possible during the design/development cycle. This requires initiating the risk assessment during the design concept phase and then updating the analysis with current information as the development effort progresses.

By definition, a risk assessment/hazard analysis identifies the existence of potential hazards and their probability of occurrence. Also, the adequacy of proposed hazard controls is determined. To actually determine whether or not hazard safeguards provided are adequate, a verification effort is required. The verification effort provides assurance that:

1. The subject item and corresponding operating procedures mitigate the potential for a safety incident,

2. The subject item exhibits no hazardous characteristics not foreseen by analysis,
3. All potentially hazardous characteristics are controlled, and
4. The requirements of industry codes and standards are satisfied.

Verification may be performed during design/development or on a random basis during production or operation.

Typically, design safety verification is achieved via one or more of the following: technical evaluation, inspection, demonstration, or test. Technical evaluation involves taking an in-depth look at the theoretical basis for a design, as well as the reference documentation (e.g., codes(s) or standard(s)) which served to guide the basic design formulation. Technical evaluation is often substituted for other means of verification which may be too difficult, time-consuming, or costly.

Inspection serves to provide verification through visual, or other senses (possibly enhanced), detection of workmanship of material characteristics which effect the existence of a hazardous condition. Hazards may also be detected through some form of measurement or simple manipulation. Demonstration is a trial conducted to show that a specific operation can be accomplished safely, that a product operates safely, or that a material contains or lacks a certain property. Tests are more detailed than demonstrations in that specific measurable parameters must be met. A test is often used to verify that values for a certain operational parameter fall (or do not fall) within specified limits, and that various operational parameters do not cause a failure, damage, or hazardous condition.

To be completely safe, a candidate product or manufacturing system or process must have no potential for causing injury or damage under any circumstances. Unfortunately, no such product or process exists. Therefore, some risk must always be expected and accepted during the normal use of any design. How much risk is acceptable depends on the benefits derived from usage. Therefore, when risk is greater than that deemed acceptable, the subject design must be considered unsafe. Subsequently, it is necessary that the design be made safe through some compensative/corrective action before being introduced for use.

9.2.1.1   Fault Tree Analysis

FTA is a deductive graphical technique applied to products and systems for modeling the various faults that lead to the occurrence of a defined, undesired hazardous event. The analysis allows a product or system to be analyzed in the context of its environment and operation to find all possible ways in which the undesired event can occur.

The fault tree itself provides a visual representation (see Figure 9.18) of the logical interrelationships between a specific failure event, or basic fault, occurring and the ultimate effect it has upon the subject item. A basic fault reflects a specific part or component hardware failure(s), human error(s), or any other pertinent event(s) which can lead to a higher undesired event. Note that this analysis technique distinguishes itself from FMECA (to be discussed later in this section) in that it addresses human error. FMECA is limited to evaluating hardware and functional failures.

A key part of the analysis is the definition of the top undesired event to focus on. The top event generally reflects a complete or catastrophic failure of the item under consideration. Careful choice of the top event is important to the success of the

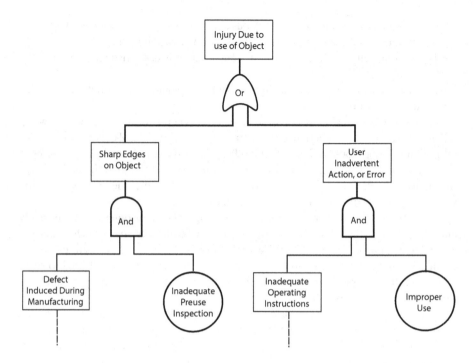

FIGURE 9.18 Example fault tree diagram.

analysis. If it is too broad, the analysis loses its ability to adequately address a specific problem. If it is too specific, the analysis loses its ability to adequately address
all significant issues contributing to a specific problem.

The FTA process is an interactive approach in identifying basic faults and establishing their criticality. The power of the fault tree comes from its qualitative and
quantitative ability to assess criticality. A fault tree structure reflects a series of logic
gates which serve to permit or inhibit the passage of fault logic up the tree to the top
event. Figure 9.19 presents the variety of logic symbols used in a fault tree diagram.

Thus, through qualitative and/or quantitative evaluation, the fault tree provides an
assessment of the probability of reaching the top undesired event. For example, visually examining the fault tree diagram for the types of logic gates predominating gives
us an indication of the relative ease with which the top event can occur.

This analysis power enables corrective action recommendations (both design
and operational) to be readily formulated and prioritized to cost-effectively enhance
product or system safety (and/or safety). The analysis technique is most beneficial
when implemented early during design conceptualization and updated throughout
design/development phases in light of design changes.

FTA involve four major steps [11]:

Step 1: *Construct the Fault Tree Diagram*: This involves initially defining
the top event of the tree using terminology that will encompass the lesser
events, individually or collectively. Next, beginning with the top event,

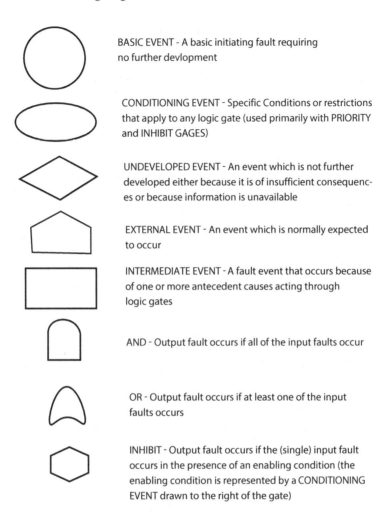

FIGURE 9.19   Fault tree logic symbols.

paths leading to each succeeding lower level of detail are developed. This requires having a thorough enough understanding of the product or system to visualize all the events that could conceivably take place as a result of malfunctions, failure, or human error. The diagram expands as a progression of events through the logic gates.

Step 2: *Collect Basic Fault Data*: This involves compiling statistical data based on experience (either specific or industry-wide generic) to develop a failure probability for each basic fault.

Step 3: *Evaluate Diagram*: This involves determining which combinations of failures are most likely to cause the top event of the tree (i.e., which failure paths are dominant). In doing this, the criticality of each failure is determined by calculating its failure impact on the top event. A secondary evaluation goal is to determine the probability of the top event occurring;

however, the use of this value should be limited to serving as a benchmark to assess design and operational improvements.

Step 4: *Formulate Corrective Action Recommendations*: This involves prioritizing the basic faults relative to their criticality and defining corrective action to reduce the most critical faults' impact on the occurrence of the top event of the fault tree diagram.

9.2.1.2   Failure Mode, Effects, and Criticality Analysis

FMECA is an inductive technique for systematically evaluating and documenting the potential impact of each functional or hardware failure on operational success, personnel and system safety, system performance, maintainability, and maintenance requirements. Note that human factors are not considered in this analysis technique.

As with FTA, the result is a ranking of each potential failure mode relative to the severity of it effects. This subsequently enables appropriate corrective actions to be taken to eliminate or control high risk items.

The analysis is inductive in nature. This means that the analyst induces failure to each item at the lowest applicable level of product, system, or process hierarchy. It actually serves as a paper test prior to expensive hardware assembly and configuration in a test fixture. From a knowledge of the predominant failure modes of each element, the analyst traces up through the various levels of the subject item to determine the effects a failure will have on the subsequent indenture levels and ultimately output performance.

This differs from FTA, which is a deductive technique. As stated previously, FTA uses a top-down approach, which assumes that the top hazardous event has occurred. From this, all the failure modes and basic faults contributing the occurrence of the top event are defined.

The FMECA is initiated early in the conceptual phase to support the evaluation of candidate designs and to provide a basis for establishing corrective action priorities. The analysis plays a key role in design reviews from concept through final hardware development and, ultimately, technology transfer. In addition to the obvious benefits the FMECA provides to actual hardware design, the FMECA aids in defining special test considerations, quality inspection points, preventive maintenance actions, operational constraints, useful-life factors, and other pertinent information and activities necessary to minimize failure risk. Note that all recommended actions resulting from the FMECA should be evaluated and formally dispositioned by appropriate implementation *or* documented rationale for no action.

The analysis consists of two fundamental parts. First, there is the identifying of the failure modes and effects of each indenture level component and, secondly, there is the calculating of the criticality of each failure mode.

The FMECA documents all probable failures in a product, system, or process within specified ground rules. These include the analysis approach, the lowest level of indenture to be analyzed, sources of data and information describing the subject item and failure of components within, and include general statements of what constitutes a failure. Every effort should be made to identify and record all ground rules and assumptions prior to initiating the analysis, and expanded thereafter as necessary.

The following steps are performed during the analysis [12].

Step 1: *Define the item to be analyzed*: This involves gathering comprehensive information including identification of internal and external interface functions, expected performance at various indenture levels, item restraints, and failure definitions.

Step 2: *Construct Functional and Reliability Block Diagrams*: This involves developing diagrams which illustrate the operation, interrelationships, and interdependencies of functional components of the design.

Step 3: *Identify and Evaluate Failure Modes*: This involves assessing all potentially significant failure modes of each component at the lowest appropriate indenture level. Next, the effects on subsequent higher indenture levels are determined. Finally, the worst-case end effect is defined relative to the following severity classifications:

- **Category 1**: Catastrophic, a failure which may cause loss of life.
- **Category 2**: Critical, a failure which may cause severe injury, property damage, or loss of function.
- **Category 3**: Marginal, a failure which may cause minor injury, property damage, or degraded function performance.
- **Category 4**: Minor, a failure of low severity resulting in unscheduled maintenance or repair.

Step 4: *Identify Failure Mode Detection Methods and Compensation Provisions*: This involves establishing the means for isolating failures through the use of inspection criteria or design features such as built-in test equipment (BITE).

Step 5: *Determine Design Corrective Action Priorities*: This involves calculating the criticality of each failure mode and ranking them.

Step 6: *Formulate Specific Correction Actions*: This involves identifying specific corrective actions (either design or procedural) for those failure modes most critical.

The FMECA is documented on a worksheet such as that presented in Figure 9.20. The worksheet is straightforward and facilitates conduct of the analysis. The top blocks on the worksheet provide background information. The remainder of the columns are delineated below.

Reference Number – The unique part identification number.
Identification – The part name.
Failure Mode – The part potential failure mode under consideration.
Failure Effects – The failure effects of the failure mode at subsequent higher indenture levels.
Compensation Provisions – The design features or procedural routines in place to offset the effects of the failure mode.
Severity Class – The worst-case failure mode classification
λ_p – The part failure rate.

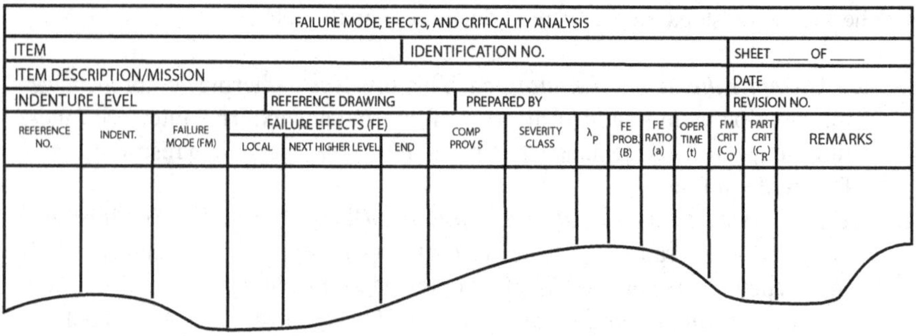

FAILURE MODE, EFECTS, AND CRITICALITY ANALYSIS														
ITEM				IDENTIFICATION NO.								SHEET ____ OF ____		
ITEM DESCRIPTION/MISSION												DATE		
INDENTURE LEVEL			REFERENCE DRAWING		PREPARED BY							REVISION NO.		
REFERENCE NO.	INDENT.	FAILURE MODE (FM)	FAILURE EFFECTS (FE)			COMP PROV S	SEVERITY CLASS	λ_p	FE PROB. (B)	FE RATIO (a)	OPER TIME (t)	FM CRIT (C_Q)	PART CRIT (C_R)	REMARKS
			LOCAL	NEXT HIGHER LEVEL	END									

FIGURE 9.20   FMECA worksheet.

Failure Effect Probability, β (beta) – The conditional probability that the failure effect results in the identified criticality classification. If unknown, use 1.0.

Failure Mode Ratio, α (alpha) – The fraction of the component's total failure rate due to the failure mode under consideration.

Exposure, t – The exposure of the subject item (e.g., hours)

Failure Mode Criticality, C_m – The calculated criticality of the failure mode under consideration,

$$C_m = (\beta)(\alpha)(\lambda_p)(t)$$

Part Criticality, C_p – The cumulative sum of the failure mode criticalities calculated for the part. (Note that there is typically more than one failure mode defined for a specific item, or part.)

9.2.2 SAFETY AND THE WORKER

Aside from product and manufacturing system safety, occupational safety plays a key role in the company. Particularly in regard to worker morale, it is desirable to have a safe and healthy work environment. Establishing a safe work environment enables the production floor personnel to concentrate on their work productivity and quality. This in turn leads to more efficient and effective production and subsequently quality, cost-competitive products.

Obviously, the manufacturing environment does not have to be absolutely safe for products to roll out the door. However, an unsafe work environment may directly affect reduced product throughput and lower quality. In addition, safety may impact product cost to the consumer due to direct and indirect accident costs.

A worse scenario centers around a worker perception that management is not concerned about worker safety. The result is that the worker may not care if the product goes out the door "right." Worker pre-occupation with avoiding injury is removed with implementation of a well-defined occupational safety program. The plan for such a program identifies operational critical safety issues, serves to establish good worker morale, and ultimately, establishes a quality product base and a healthy management operation.

9.2.2.1 Human Factors

The subject of human factors addresses providing work environments which foster effective procedures, work patterns, and personnel safety and health, and minimizing factors which degrade human performance, or human reliability. Thus, the intent is to design manufacturing systems and equipment such that operator workload, accuracy, time constraints, mental processing, and communication requirements do not exceed operator capabilities. In addition, designs should minimize personnel and training requirements within the limits of time, cost, and performance trade-offs.

Both human and machine elements can fail, and just as equipment failures vary in their effect on a system, human errors can also have varying effects on a system. In some cases, human errors result from an individual's action, while others are a consequence of system design or manner of use. Some human errors cause total system failure or increase the risk of such failure, while others merely create delays in reaching system objectives. As the man–machine interface (MMI) becomes increasingly complex, the possibility of human error increases with an accompanying increase in the probability of system failure to due to human error.

One obvious approach to eliminating failures due to human errors is to replace the human by a machine. This approach, however, must consider the complexity, reliability, interactions with other equipment, cost, weight, size, adaptability, maintainability, safety, and many more characteristics of a machine replacement for the human.

An interesting facet of the human factors/reliability relationship (and which also concerns the maintainability engineer) is that the continuation of the system designed-in reliability depends upon the detection and correction of malfunctions. This task usually is assigned to humans. Thus, system performance can be enhanced or degraded, depending upon whether or not the malfunction information is presented so that it is understood readily.

Human behavior is a function of three parameters:

1. **Stimulus-Input (S)**: Any stimuli, such as audio or visual signals, failure indications, or out-of-sequence functions, which act as sensory inputs.
2. **Internal Reaction (O)**: The act of perceiving and interpreting the stimulus and reasoning a decision based upon these inputs.
3. **Output-Response (R)**: The response to stimulus based upon internal reaction. Talking, writing, positioning a switch, or other responses are examples of output responses.

All behavior is a combination of these three parameters, with complex behavior consisting of many S-O-R chains in series, parallel, or interwoven and proceeding concurrently. Each element in the S-O-R chain depends upon successfully completing the preceding element. Human errors occur when the chain is broken.

The intent is to design manufacturing systems and equipment in such a way that operator workload, accuracy, time constraints, mental processing, and communication requirements do not exceed operation capabilities. Designs should minimize personnel and training requirements within the limits of time, cost, and performance trade-offs.

Items degrading human reliability include:

- Visual
  - Obstructed Instruments
  - Poor Illumination
  - Limited Visibility
- Accessibility
  - Obstructed Movement
- Nonstandard Controls
  - Interpretation Error
  - Identification/Location
- Overwhelming Instructions
  - Lack of Clarity
  - Lack of Equipment
- Discomfort
  - Space
  - Temperature/Environment
  - Vibration
  - Sound

In short, it is necessary to optimize the MMI. Numerous examples exist where the safety or operation effectiveness of complex systems have been compromised by defects in the design of the MMI and which have rendered the systems difficult or impossible to operate.

This catapults us into the realm of ergonomics. Simply defined, ergonomics is the science of designing based on the characteristics of the intended users of the design in its final form and in its intended environment. This means applying the "principle of user-centered design." [13].

User-centered design promotes four fundamentals:

1. Direct observations of human beings and their behavior, supported by systematic investigations of human experience
2. Apply both empirical analysis and evaluation
3. Modify the product to fit the user
4. Achieve the best possible match for the greatest number of people so far as is reasonably practical within constraints of cost and time.

The end result is a design that reflects an allocation of functions to personnel, equipment, and MMI to achieve the required sensitivity, precision, time, and safety; required reliability of system performance; the minimum number and level of skills of personnel required to operate and maintain the system; and the required performance in a cost-effective manner.

The following is a generic list of MMI issues to consider, as applicable, in developing a user-oriented design [14].

- Satisfactory atmospheric conditions including composition, pressure, temperature, and humidity.
- Range of acoustic noise, vibration, acceleration, shock, blast, and impact forces.
- Protection from thermal, toxicological, radiological, mechanical, electrical, electromagnetic, pyrotechnic, visual, and other hazards.
- Adequate space for personnel, their equipment, and activities as they are required to perform under both normal and emergency conditions.
- Adequate physical, visual, auditory, and other communication links between personnel, and their equipment, under both normal and emergency conditions.
- Efficient arrangement of operation and maintenance areas, equipment, controls, and displays.
- Provisions for insuring safe, efficient task performance under reduced and elevated gravitational forces with safeguards against injury, equipment damage, and disorientation.
- Adequate natural or artificial illumination for the performance of operation, control, training, and maintenance.
- Safe and adequate passageways, hatches, ladders, stairways, platforms, inclines, and other provisions for ingress, egress, and passage under normal, adverse, and emergency conditions.
- Provision of acceptable personnel accommodations including body support and restraint, seating, rest, etc.
- Provision of nonrestrictive personal life support and protective equipment.
- Provision for minimizing psychophysiological stress effects of fatigue.
- Design features to assure rapidity, safety, ease, and economy of operation and maintenance in normal, adverse, and emergency environments.
- Satisfactory remote-handling provisions and tools.
- Adequate emergency systems for contingency management, escape, survival, and rescue.
- Compatibility of the design, location and layout of controls, displays, work areas, maintenance accesses, storage provisions with the clothing and personal equipment to be worn by personnel operating or maintaining systems or equipment.

The most effective way to avoid MMI problems is to utilize a set MMI standardized design criteria. Such standardization should encompass controls, displays, marking, coding, labeling, and arrangement schemes. Remember to base internal design standards on a recognized national or international source(s).

9.2.2.2 Safety Program Planning

Utilizing knowledge gained from safety management theory and practice, it is possible to define the basic elements of a successful safety program. This includes addressing the key elemental motivators having the strongest impact on employee morale and attitude [15].

Keep in mind that in defining a tailored program plan, it is always easier to remove task elements than it is to add elements at a later date. Therefore, include items in the initial plan even if you strongly suspect their removal from the final program plan. Basic elements of a safety program include the following components.

Management Commitment: Program commitment is displayed in two ways. First a written management statement of safety policy is displayed to all corporate employees and prepared and signed by top management (i.e., the president or chief executive officer). Second, the commitment is backed by actions. This requires that top management takes part in safety-related activities to display interest and assure that the policy is adhered to.

Assignment of Responsibility: A single person is assigned responsibility for overall management and direction of the program. This includes having the authority to make decisions and direct actions to positive effect. Typically, this is a safety director in a large organization; however, it can be any effective manager having adequate training.

Supervisor Responsibility: Supervisors are held responsible and accountable for the safety activities and accident experience within their respective departments and they are given the authority to meet this responsibility. As necessary, the supervisor must seek assistance to solve safety problems from either in-house or out-of-house sources.

Supervisor Training: Supervisors are educated to create a safe environment. They learn how to deal with people, recognize hazards and safety concerns, perform accident investigations and identify corrective actions, train employees, and know their importance to the overall success of the safety effort.

Hazard Identification: Implement a project for hazard identification. There are several techniques to accomplish this, and the simplest is probably periodic inspection of a manufacturing operation in accordance with a customized checklist. When a hazard is identified, a corrective action is formulated and followed up on. Employee input is another good way to communicate hazards. Also, state-of-the-art technology involving video recorders provides another means to observe hazardous operations.

Safety Committees: A committee comprised of management and labor representation is developed and given specific responsibilities and objectives to achieve. Top management involvement enhances the effectiveness of such a group significantly. The objective of the committee can be general in scope, such as ongoing review of loss history, hazardous input, and hazard control input. It can also be strictly on a per project basis, such as to research and develop new procedures, operational controls, etc. The committee focuses on product issues, liability, or even vehicular functions of the operation.

Employee Training: Safety training is provided to employees and periodically reinforced. Initial new employee orientations include at least the basic safety concerns of not just one job but of the whole corporation. Specific safety training as to individual jobs is presented by a competent trainer. Ongoing training of employee groups is needed to reinforce specific concerns and may be related by loss history or degrees of operational exposures. A brief safety talk given once per week is a viable alternative. This is quick, effective, and takes a minimal amount of time. Its effect on reinforcing management concern is consistently effective.

Accident Investigation: Establishing a specific procedure to investigate and document each accident and incident (close-calls with no loss) is critical. Corrective action is then formulated to prevent accidents from reoccurring.

Accident Analysis: Data analysis involves studying past loss history to identify trends of accidents by body part, cause, equipment, location, department time of day, shift, etc. This aids in focusing on those areas which are most significantly impacted by corrective action.

First Aid/Medical Facilities: Having the ability and means to take care of injured persons is basic to any safety program. At the minimum, a first aid kit is required along with persons available who are trained in first aid. Specific plans for addressing a major medical emergency are necessary.

9.3   QUALITY ENGINEERING AND CONTROL

The control of product quality naturally encompasses many important concepts. First of all, the product must be designed and engineered for a desired quality objective. Next, the process by which it is manufactured must be designed and controlled for maximum product quality. Finally, the entire process must be managed for adherence to the quality objectives defined for the product.

When one thinks of quality, the traditional silo focus of quality control comes to the forefront. But the concept of designing a product to strategically meet quality standards that make it viable from consumer and business perspectives is often not given the attention required. That is, manufacturing a successful product goes beyond the statistical monitoring and control of process quality and adherence to standards.

However, it is critical to ensure prior to getting to manufacture that the product and manufacturing processes and system have been adequately designed for the purpose to be served. This means for the product and process that design tolerances have been adequately specified, and a manufacturing process has been selected which can produce the item correctly and economically. Only then does the task become to make sure that the manufacturing process(es) and system continue to function properly to produce product as designed and if it deviates, to get an understanding as to the nature of the problem and corrective action.

9.3.1   DESIGN OF EXPERIMENT

It is important to define and monitor the system indenture level at which product manufacture actually occurs in order to minimize degradation that *will* occur. However, once the overall manufacturing process(es) which the system (and its equipment) will perform is defined and, subsequently, developed, it is necessary to determine the optimal system operational conditions. How fast (i.e., throughput) should the machine run? What temperature should be used for the hot bath? How often should maintenance be done? Where should the crimp be on the product? How high should that thing be? And how low should that other thing be in the system equipment or in the product itself? These are all questions that must be answered, and all can be answered via formal design of experiment (DOE).

By developing and implementing the optimal product and manufacturing system and process design, numerous TMA benefits are derived. Leading the list is increased long-term profitability. Others include:

- Design
  - Improve manufacturability
  - Reduce sensitivity to variability
  - Implement value analysis/engineering
  - Maximize functional performance
- Testing
  - Factor significance (critical vs unimportant)
  - Engineering priorities
- Capability
  - Optimal procedures
  - Meeting production goals
  - Optimal operational parameters

Figure 9.21 provides an overview of the DOE process. Depicted are three major steps: (1) planning the experiment, (2) implementing and collecting the data, and (3) performing analysis and interpreting the experimental results. Each step itself has a lot of concerns.

In planning, perhaps the most significant concern is the experimental objective. A major benefit of formal DOE is its introduction of structure and discipline, and its requirement of defining exactly what is the question to be answered. Upon having an objective defined, it is necessary to get the proper people involved who can provide

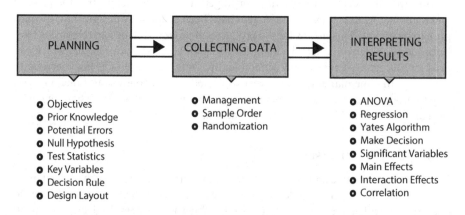

FIGURE 9.21   Design of experiment steps.

valuable input based on their industry experience. It is also important to define all the statistical ground rules to be used in the study, such as the decision rule, test statistic, and significance level.

In collecting data, it is important that good management practice be used in running the experiment. By not running and collecting the data as defined, one is almost certain to bias or invalidate the results. The experimental run should reflect a randomized order to minimize data variability due to introduced biases. *This is not a trivial task!* Frequently, an engineer has specified an experiment to be run a certain way, only to have shop personnel reject the requirements as too inconvenient, resulting in flawed data. Worse yet, the data is flawed but nobody knows it.

In interpreting the results, it is critical to properly identify significant effects impacting the experimental outcome. This includes using proper statistical techniques to determine the statistical significance of results and combining this information with engineering intuition and experience. If practical to the situation, replication of results is desirable.

DOE often provides answers to questions that somebody thought they knew the answers to. In accomplishing this, it characteristically saves valuable time lost to "shooting in the dark."

Obviously, this information only briefly touches on the whole DOE arena. There are an abundance of texts providing excellent insight to DOE [16–18]. The reader is strongly encouraged to explore the literature available and find a text which is easy to understand and at the appropriate level of mathematical detail.

Even better than DOE, for many manufacturing purposes, are the use of Taguchi Methods of quality engineering. These include a subset of DOE theory, wrapped up in sound engineering principles. One of the advantages of Taguchi Methods is its use of deeply fractional factorial experimental designs, allowing investigation of a great many factors, without requiring a huge experimental burden. Another advantage is the inclusion of noise factors, enabling the experiments to include deliberate attempts to increase variability in the outcomes, thereby identifying designs which are "robust" (insensitive) to those noise factors.

9.3.1.1 Variable Identification

Whether using either classical DOE or Taguchi Methods, the first step is always to determine what is to be optimized. Often, this will be something like "maximize process throughput" or "minimize out-of-spec output," or some similar target. Sometimes this optimization determination will be obvious; other times it will not.

With design optimization goals determined, the next action is to identify the design variables of interest. These are also called control factors, as they are factors in the production process that are under our control. Temperature and duration of a heat-treating process, speeds and feeds in a machining process, pressures and speeds in a forging process; these are all examples of control factors. These are best brainstormed in a meeting with process experts, who have a pretty good gut feeling for what is likely to be important in achieving design optimization. Some of these control factors may turn out to be relatively unimportant, but some will be critical. Formal DOE or Taguchi analyses are used to reveal which are factors important [19,20].

Another component of DOE is doing noise factor analysis. These also are commonly identified via brainstorming with manufacturing process experts. These are factors which should be controlled as best as possible to not have inflated impact on experiment results and analysis. A classic example of a noise factor is ambient temperature. The air temperature can be controlled during the experiment, but once the production process goes live, it will not likely be absolutely controllable on a factory floor, which is subject to a wide range of temperature variances during the course of a typical day and/or over the course of a typical year. If the process is optimally designed, noise factors will have minimal impact on its process performance.

9.3.1.2 Analysis of Variance

Analysis of variance (ANOVA) is a mathematical technique for determining which of defined control factors and noise factors have significant effect on the performance attribute of interest [21]. For those manufacturing process attributes, or variables, that do have significant effect, the value of those factors, or variables, is selected which optimize desired performance. For those that have minimal effect, the factor values are selected that are the cheapest and/or easiest to implement. In this way, the optimal performance is obtained with minimal cost and effort.

The ultimate objective of DOE is to establish a model that predicts the optimal performance of a subject item interest based on the operating level of significant variable(s) (e.g., part strength as a function of injection molding screw speed, mold temperature, or holding time). Such a model, or prediction equation, is derived from data collected regarding the performance characteristic of interest and manipulating the variable of interest. A line is fit to the data that best predicts the performance based on variable adjustments. The fitting of this line to the data is the result of ANOVA, which analyzes the variability. This is essentially using the data for regression analysis to develop a performance prediction equation. The less variability in the data, the better the prediction equation.

The variability is analyzed using sum of squares (SS). Fundamentally, the objective is to minimize the distance the data points are above or below the prediction line. The data is squared to make all variability evidenced positive (i.e., no negative numbers). The sum of squares got its name because it is calculated by finding the sum of the squared differences.

In regression analysis, the three main types of sum of squares are the total sum of squares, regression sum of squares, and residual sum of squares.

The total sum of squares reflects the variation of the values of a *dependent variable* from the sample mean of the dependent variable. The total sum of squares (TSS) quantifies the total variation in a *sample*. It can be determined using the following formula:

$$TSS = \sum_{i=1}^{n} (y_i - \bar{y})^2$$

where:

y_i – the value in a sample

$\bar{y}$ – the mean value of a sample

The factor or regression sum of squares (SSR) describes how well a regression model represents the modeled data. A large regression sum of squares indicates that the model does not fit the data well (i.e., there is a lot of variability in the data). This is also known as the explained sum of squares. The formula for calculating the regression sum of squares is:

$$SSR = \sum_{i=1}^{n} (\hat{y}_i - \bar{y})^2$$

where
$\hat{y}_i$ is the value estimated by the regression line
$\bar{y}$ is the mean value of a sample
n is the number of data points

The residual sum of squares essentially measures the variation of modeling errors. In other words, it depicts how the variation in the dependent variable in a regression model cannot be explained by the model. Generally, a lower residual sum of squares indicates that the regression model can better explain the data while a higher residual sum of squares indicates that the model poorly explains the data. This is also known as the sum of squared errors (SSE) of prediction. The residual sum of squares can be found using the formula below:

$$SSE = \sum_{i=1}^{n} (y_i - \hat{y}_i)^2$$

where
y_i – the observed value
$\hat{y}_i$ – the value estimated by the regression line

The relationship between the three types of sum of squares can be summarized by the following equation:

$$TSS = SSR + SSE$$

ANOVA Example

There is an injected molded product that needs to be cured at a high temperature to increase its strength. What is to be determined is the optimal mold holding temperature (i.e., what it should be, or if temperature variability makes a statistically significant difference). Accordingly, an experimental design is developed to evaluate the impact that mold holding temperature has on product performance. First, a sample of parts is manufactured. The experimental design dictates that the process maintains a mold curing temperature at three different temperatures. Upon release from the mold, the cured prototypes are tested for strength. Data collected is as follows:

Curing at 100° F: Strength values of 3, 3, 5, 4
Curing at 150° F: Strength values of 7, 6, 7, 8, 7
Curing at 200° F: Strength values of 10, 12, 10, 9

The control factor, or experimental design variable, curing temperature was held at three defined temperature levels, and at each level there were multiple replications (i.e., samples). Note that the number of replications at each temperature do not need to be the same. From visually looking at the data, it appears that higher temperatures give greater strength. To provide a sound mathematical selection optimal temperature on part strength, an ANOVA is performed. This mathematical analysis starts with a *null hypothesis*: "Strength is the same regardless of curing temperature." This is called a "null" hypothesis because it supposes that there is "no difference" caused by the control factor. The ANOVA will yield statistical mathematical data and indicate if the null hypothesis should be accepted or rejected.

The ANOVA requires determining three SS values: SS due to the control factor, SS due to randomness in the data, and SS of the total. By applying the SS formulae above using all 13 strength measurements, the calculations provide:

$$\text{TSS or SS(total)} = 94$$

$$\text{SSR or SS(factor)} = 84.5$$

$$\text{SSE or SS(error)} = 9.5$$

At this point, the calculations can be verified by that SS(total) equals the sum of SS(factor) and SS(error), which shows that 94 does indeed equal $84.5 + 9.5$. The "degrees of freedom" (df), (i.e., the number of free terms in calculating the mean of the samples) is derived from the sample sizes. The df(total) is the number of measurements minus one. The df(factor) is the number of levels of the control factor minus one.

$$df(\text{factor}) = 3 - 1 = 2$$

$$df(\text{total}) = 13 - 1 = 12$$

$$df(\text{error}) = (4 - 1) + (5 - 1) + (4 - 1) = 10$$

Dividing the SS by the applicable degrees of freedom gives a mean square, or MS, for the error and the factor values.

$$MS(\text{factor}) = SS(\text{factor})/df(\text{factor}) = 84.5/2 = 42.25$$

$$MS(\text{error}) = SS(\text{error})/df(\text{error}) = 9.5/10 = 0.95$$

At this point, two values are compared: MS(factor) and MS(error). Since MS(factor) is much greater than MS(error), this tells us that the temperature difference had a much greater effect on the output of the experiment than randomness did. If the MS(error) had been larger, or if they had been similar in value, it would tell us that the temperature difference was no more important than randomness was in the process.

However, just looking at the two values is not very mathematically rigorous. What is done next is to test the original hypothesis using the F-Test. A F-Statistic for this experiment is calculated by dividing MS(factor) by MS(error), which gives 44.47. This calculated F-value is compared to a critical value of the F-Statistic found in an F-Distribution Table associated with a risk of error, α. The F-Statistic has three input parameters, which are α (i.e., uncertainty, or willingness to be wrong), the df(factor), and the df(error).

Using a 95% confidence level for the hypothesis test, the α is equal to 0.05. From a F-Distribution Table, the Fcrit for $\alpha = 0.05$, df(factor) $= 2$, and df(error) $= 10$, is Fcrit $= 4.10$. Since the calculated F-statistic of 44.47 is much larger than the tabled Fcrit, the test data indicate that the temperature of the curing process has a statistically significant effect on the strength of the part. The null hypothesis is rejected that temperature of curing, via mold holding temperature, makes no difference.

9.3.2 STATISTICAL PROCESS CONTROL

Statistical process control is based on a fundamentally simple concept: (1) make some assumptions about a manufacturing process, (2) collect some data, and (3) calculate some metrics. Then a judgment call is made: is the value of that metric likely to have occurred if our assumptions about the process were true? If yes, it is assumed that the assumptions were reasonable. However, if the value of the metric is very unlikely given the assumptions, it can be concluded that the assumptions may well have been incorrect.

For example, suppose it is assumed that our manufacturing process is providing product that is 95% within specifications. If every hour 20 units are inspected and if zero or one of them is outside of specifications, it makes sense to stay with the assumption that the process is producing product in specification. But, if ten or more of the 20 are out of spec, it would be concluded that the assumption is probably no longer true. Careful use of statistical techniques can take this simple concept and turn it into a powerful decision-making tool.

The statistical analysis of a process is viewed from two different vantage points: online and offline. The online perspective assumes that while manufacturing the product, key process variables are monitored in real-time. One can make conclusions about the process parameters and implement feedback techniques to correct problems as they develop, hopefully before they become significant. Naturally, this online perspective lends itself ideally to the production process. It is usually referred to as process monitoring.

Statistical process control can also be implemented in an offline manner. This involves examining large batches of product, possibly long after they have been produced. The techniques in this case are time-independent, as it is generally not known which items were produced first. Conclusions to be made concern the entire batch and whether or not it conforms to standards. Little can be inferred as to why a particular batch may be inferior. This offline implementation may be used to monitor a manufacturing process if online techniques are not feasible, but it more naturally lends itself to decision-making between independent units in the production process. For example, an inspection department may periodically test a day's worth of product to check up on a production department. More frequently, however, this offline control strategy is employed when acquiring items from a vendor or supplier. Entire shipments may therefore be accepted or rejected (a process called *Lot Sentencing*) on the basis of their overall quality, and the vendor is left to worry about the cause of any degradation. Since these techniques are used to decide whether a batch is to be accepted or not, it is known as acceptance sampling.

The fundamental mathematical science of statistical process control is probability and statistics. In statistical process control, statistical analysis is used to calculate probabilities, and the probabilities enable deriving conclusions as to whether a process is in control (i.e., producing good product).

Probability is a measure of knowledge about some aspect of the world. When knowledge is complete, the probability of any outcome or event is either one or zero, depending on whether the event has occurred or not. When knowledge is incomplete, the probability is somewhere between zero and one.

The classic example is tossing a coin. It will land with either heads or tails up and while it is in the air, spinning in a random-incalculable manner. The perfect coin will have assigned a probability of one-half to each of the possible outcomes (i.e., heads up or tails up), because in a truly random coin flip, there will be no way of knowing which will occur or even if one is more likely to occur than the other.

Of course, with enough good sensors, detailed enough math models, accurate data on air currents in the room, and the precise torques imparted to the coin as it was flipped, it could have been predicted exactly how the coin would land. The probability of the correct outcome would be adjusted to one, the other to zero. Without this knowledge, however, one must be satisfied with a 50% chance of each outcome.

This is how statistics are used in process control. Just like the tumbling coin, a manufacturing process has some exact state, and some precise number of non-conforming units will be produced. Before the fact, however, this number is not known and the probabilities will have to be assigned. Statistics collected on past events are used to predict the probabilities of future events.

If X represents an event which may or may not occur, the designation $Pr(X)$ denotes the probability of X occurring. For an experiment as a process whose outcome is not yet known, the set of all possible outcomes is called the event space of that experiment. Since one of the outcomes definitely must occur, the sum of the probabilities of all events in the event space must be unity.

A probability distribution is a function or rule that assigns probabilities to the events in an event space. Basically, three methods exist for determining the probability distribution of an event space. The first method is based on reasoning. Each possible outcome is analyzed in terms of its causes, and the probability of it occurring is calculated based on the probabilities of its precursors. This approach is generally not feasible in the manufacturing industry, as any process complicated enough to require a statistical treatment will proceed according to largely unknown mechanisms.

A more generally useful second method of determining a probability distribution is to collect data on past experiments. If 1,000 experiments are run, and the number of occurrences of each possible outcome is tabulated, a distribution can be calculated. A simple histogram of results is a form of distribution, although not particularly useful. If an equation can be fit to the data, of the form:

$$Pr(X) = \text{some function } f(X)$$

then a useful tool for further mathematical analysis is at hand. The only problem with this technique is that it requires that the process under consideration does not change with time. That is, the probabilities of each outcome must remain constant into the future, or the function which was fit to the data is no longer valid.

The third method is to force the event space into some pre-selected distribution. By designing an experiment that is guaranteed to have some known distribution, then meaningful analysis can be performed. This approach has intuitive appeal and is a favorite method in process control.

The Central Limit Theorem provides the tool needed to facilitate sound statistical-based decisions. In mathematical terms, this is stated [22]:

For a population having mean, μ, and finite standard deviation σ_p, let X represent the mean of n independent random observations. The sampling distribution of X tends toward a normal distribution with mean, μ, and standard deviation, $\sigma = \sigma_p/\sqrt{n}$.

In plain words, whatever the distribution is of the raw data coming out of our process, the theorem describes the data as a Normal, or Gaussian, Distribution by plotting the averages of the samples of the data.

Suppose the manufacturing process is producing out a stream of data of unknown distribution. The first datum is x_1, the next is x_2, then x_3, x_4, and so on, up to the last piece of data, x_n, where n is some very large number. From a sample, or sub-set, of this data, (e.g., the first n values) the sample mean, or average, can be calculated for the data set, or sample. The mean of the next n values is called x-bar$_1$, and so on. The mean of the means is what adheres to the Central Limit Theorem.

The Central Limit Theorem states that, if n is large enough, the distribution of the sample means will be Normally distributed. The only question left to be answered is, how large is large enough? The precise answer to this question depends on the precise distribution of the original data, and how close we wish to come to a Normal Distribution. The larger the n, the closer we come to true normality. In most engineering applications, four is considered a lower limit on sample size. A sample size of five is common in practice. However, if data are plentiful and measurements are cheap to make, ten is an even better size to use. The sample size in any actual implementation will be a decision based on the cost and availability of the data, and the precision required in the analysis.

9.3.2.1 The Normal Distribution

The famous Normal Distribution, also known as the bell curve or Gaussian distribution, is well known and easy to handle. In probability terms, it is expressed as:

$$Pr(X) = \frac{1}{\sigma\sqrt{2\pi}} \exp\left[-\frac{1}{2}\left(\frac{x-u}{\sigma}\right)^2\right]$$

where
   μ is the process mean
   σ is the standard deviation.

To calculate the probability of any outcome X, only needing to be known is the process mean and standard deviation for insertion in the above formula. Remember, in process control situations, the sample means are the data of interest. Finding μ and σ are a simple matter of collecting data and calculating them, and then assuming that the process remains stable in the future. The formulae for μ and σ are:

$$\mu = \bar{x} = \frac{1}{n}\sum_{i=1}^{n} x_i$$

$$\sigma = \sqrt{\frac{\sum_{i=1}^{n}(x_i - u)^2}{n-1}}$$

where n is the number of samples.

It should be noted that these formulae would only give the exact values of the process mean and standard deviation if infinite data were collected. However, a reasonably large set of data is the best and is generally good enough.

9.3.3 PROCESS CAPABILITY ANALYSIS

Before beginning to monitor a process to see if it *remains* within the required level of quality, capability analysis is used to determine if it *begins* within the required level of quality in the first place.

Any process in the real world is subject to random variabilities. These are minimized to the greatest extent possible, but some will always remain. Raw materials will have some inherent variability, as will the machines that process the material. People operating the machines will perform their duties in slightly varying ways, both from person to person, and from hour to hour for one person. Some amount of variability, of a purely random nature, must be accepted.

If the process was designed correctly, these variations will be kept to a bare minimum, and will follow a random, unpredictable distribution. The cumulative effect of all of the random variations within a process will have an even less predictable effect on the outcome of the process. However, through the use of sample Means, the data will adhere to a Normal Distribution.

According to mathematics of the Normal Distribution, any outcome is possible, which indicates that values infinitely smaller and larger than the Mean are possible. The farther the outcome is from the Mean, however, the less likely it is to occur. Of course, there are practical limits involved. For instance, if tracking the weight of items, a negative value is theoretically possible but will not be practically possible.

TABLE 9.6
Percentages for Sigma Values

Process Limits	Percent of Distribution Included
±1σ	68.26%
±2σ	95.46%
±3σ	99.73%
±4σ	99.9937%
±5σ	99.999943%
±6σ	99.9999998%

But, in general, vast deviations from the nominal value are always possible and will occur, however infrequently.

The task in process capability analysis is to determine just how infrequently these large deviations do occur. Process capability can be defined as that range of variation in a product characteristic that will include "almost all" of the product produced. For instance, if the process operation is to fill a soda can with 12 ounces of soda, what range of values will include the amount of soda in almost all of the cans filled? The smaller this range of variations is, the better.

In practice, it is very common to define "almost all" of a process as anything within plus or minus three standard deviations of the mean. In a normal distribution, the sum of the probabilities of all outcomes between the + and − three sigma limits equals 99.73%. If ±3 sigma limits are chosen and the process is capable, this indicates that 0.27% of product will be outside of those limits. Other definitions are of course possible, and some companies are now using ±6 sigma as their capability limits, which includes 99.9999998% of the product. Table 9.6 lists percentages for other sigma values.

Process capability limits are often confused with tolerances. Although they look alike, both having an upper and lower limit, they are completely independent concepts. Design specifications, or tolerances, are what is needed, while capability limits are what the process delivers. If the capability limits and the tolerances are equal, this is a satisfactory situation: 99.73% of the product will be within tolerance, which is quite good. If the capability limits are farther apart than the design tolerances, that is not good. It means that the process is not precise enough to produce product that is all within the required limits.

The best possible situation is when the capability limits are well inside the tolerances. This means that *more* than our 99.73% of the process output is satisfactory, and even if the process were to degrade, it would still be in good shape. Of course, if the capability range was much smaller than the tolerance range, then quality assurance activities may be reduced, or not warranted. A common rule of thumb is that capability range should be around 70%–80% of the tolerance range.

Sometimes, it will happen that the process cannot be tightened to within the tolerance range. In this case, there are several options, none of them good. One is to use 100% inspection on the output, separating the good from the bad. Another is to redesign the process from scratch, purchasing more precise equipment or hiring workers with greater skill. A last resort is to change the tolerances by redesigning the product. This requires extensive changes in many departments and will not make the process engineer very popular in the overall organization, but sometimes is the only approach.

A commonly used measure of the relationship between process capability and desired tolerances is known as the process capability ratio, or PCR. This is defined as the difference between the upper and lower tolerance specifications, divided by six times the process standard deviation (the range of process values which includes ±3 sigma of the process) which is to say, 99.73% of the process output.

$$PCR = \frac{(UpperSpec) - (LowerSpec)}{6(\text{process } \sigma)}$$

If exactly 99.73% of the process output falls within the tolerances, then the PCR will be exactly 1.0. As discussed above, this represents a good situation, but leaves no margin for error. Similarly, if 99.73% of the process lies within a range smaller than the tolerance range, the PCR will be greater than 1.0. If the PCR is less than 1.0, it means that to include 99.73% of the process output, a range greater than the specified tolerance range is needed. To follow the rule of thumb given above, an ideal PCR would be in the range of 1.20–1.4.

9.3.3.1 Process Monitoring

Process monitoring is the online version of statistical process control, where the collection of data enables making inferences about a process and its quality while the process is running. The goals are to detect process degradation as quickly as possible and correct any problems before an excessive amount of sub-standard product is produced [23–25].

One of the possible degradations to be monitoring for is a change in the process capability, as defined above. That is, attention is given to see if the range of variation which includes some standard percentage of the product increases. An example of this would be if our soda cans, formerly filled to between 11.5 and 12.5 ounces, now contain anywhere between 11.0 and 13.0 ounces of soda. If something of this nature occurs, the process will need to be corrected as soon as possible.

Another of the possible degradations to watch for is a shift in the mean, or average, of the process performance parameter of interest. This can occur without a loss in process capability. That is, the size of the range which includes most of the product is the same, but the location of the range has changed. For example, the soda cans no longer contain between 11.5 and 12.5 ounces, but now contain between 9.5 and 10.5 ounces.

The basic technique of process monitoring is the control chart. This is a graphical tool which tracks product quality over time and gives visual evidence of problems, as well as intuitive clues as to their possible origin. Many forms of the control chart exist, and with the most popular discussed herein.

A control chart is a graphical "hypothesis test." The assertion, known in mathematics as the "null hypothesis," is that the process capability has not changed. Each time a new piece of data is added to the chart, this assertion is tested. The result of

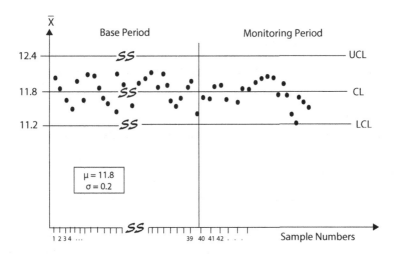

FIGURE 9.22 Typical control chart for a variable.

each test will be either to accept the null hypothesis, or to reject it. Fundamental assumptions for control charts include (1) the Central Limit Theorem is valid and (2) data are collected randomly in an unbiased manner.

In its simplest form, a control chart consists of three horizontal lines and a vertical scale. These lines are called the center line (CL), the upper control limit (UCL), and lower control limit (LCL). The vertical scale reflects possible values of the quality characteristic to be tracked. In the example from the previous section, the consideration was ounces of soda in a soda can. In that case, the scale would go from zero to the volume of the can, so that any possible value of our variable could be plotted (see Figure 9.22).

The horizontal axis of the chart can be thought of as time, or more precisely, measurement number. Measurements will be taken at regular intervals, so that each new entry in the chart will be one unit to the right of the previous entry.

Remember, the measurements that plotted are not individual values, but mean values of samples of some size n, preferably at least four. As long as all points on the chart stay within the UCL and LCL, the process is "in control." That is, nothing is going wrong, the process capability is remaining stable, and the mean is not drifting. However, if points begin wandering outside of the control limits, the process is becoming "out of control," meaning that something has gone wrong.

How are the values of the control limits calculated for the process control chart? Basically, two phases are involved, known as a Base Period and a Monitoring Period. The sequence of steps in control chart development are outlined below:

Base Period:
1. Select a quality characteristic to measure
    Select a "confidence level" (typically, ±3 sigma). Select a sample size, n (at least 4). Select N, the number of samples in the base period (20 is considered a good minimum, 50 is better).

2. Take N*n measurements, and calculate the N sample averages, or means (i.e., the $\bar{x}$)

    Calculate $\bar{\bar{x}}$, the average of the sample means. This is an estimate of the process mean and will be the center line of the control chart. Calculate sigma, the standard deviation of the $\bar{x}$s. Calculate UCL $=\bar{\bar{x}}+3\sigma$. Calculate LCL $=\bar{\bar{x}}-3\sigma$.

3. Draw the CL, UCL, and LCL at the appropriate location on the control chart. Plot the N sample averages, $\bar{x}$ on the chart, in the order in which they were measured.

4. Examine the chart, and decide if the process is in control. In control: no points outside control limits, and no "trends" in the data. If the process is in control, go on to monitoring period.

    Monitoring Period:

5. At regular intervals, take n measurements. Calculate a new $\bar{x}$ from the new measurements. Plot the new $\bar{x}$ on the chart. If it is within the control limit, and no trends are appearing, the process is remaining in control.

6. Repeat.

Notice that there are two different situations that signal an "out of control" process. One is an $\bar{x}$ outside of the control limits, and the other is any sort of "trend" in the data. An $\bar{x}$ value falling outside of the control limits is the easiest red flag to interpret. But what does this actually mean? The control limits are constructed according to our definition of capability. Assuming ± 3 sigma limits, then 99.73% of our $\bar{x}$ should be inside those limits. That implies a finite chance of any one $\bar{x}$ being outside the limits, but that chance is extremely slight. During the base period, with anywhere from 20 to 50 $\bar{x}$, the chance of one falling outside the limits, due merely to chance in the distribution, is negligible. If an $\bar{x}$ is outside the limits, it is far more likely that there is something wrong with the process, that it was out of control from the start, and that there is no point in progressing any further.

During the monitoring period, if the process has been running in control for a long time and suddenly one data point falls outside the limits, what is the probability that nothing is wrong? Since 99.73% of an in-control process should fall inside the limits, only 0.27% can be expected to fall outside. That amounts to about one value in 370. If the experience has been that hundreds of data points fall inside the control limits before one falls outside, it may not warrant stopping the process. But, if two data points in, say, 100 values fall outside the control limits, it will be prudent to immediately investigate for the root cause and take corrective action. If the control chart is based on other than ± 3 sigma limits, of course, these numbers would be different.

What about "trends" in the data? The assumption of a Normal Distribution calls for more than just control limits. It also requires that the data be randomly distributed between those limits. This means approximately equal points above and below the center line, more near the center line that near the control limits, and no discernible pattern to the data. Typical patterns which might appear, signaling an out-of-control process, would include a slow, steady drift away from the center, a cyclic distribution,

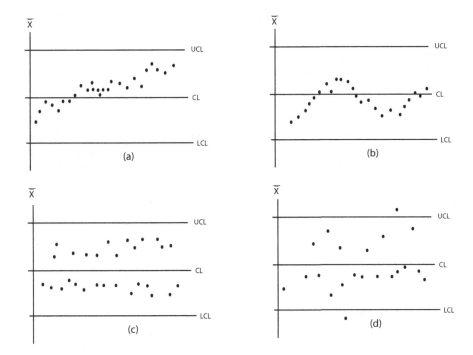

FIGURE 9.23 Control chart out-of-control conditions.

or two distinct populations (see Figure 9.23). Any of these indicate that the process is not as it appears, and should be investigated.

What might cause these trends? The control chart alone cannot pinpoint the problem but can give certain insights. For example, a sudden jump in the distribution of points would indicate a sudden change in the process, perhaps caused by a new supplier of raw materials, a new operator on a machine, or a minor failure of some component in the process. A gradual change in the distribution, on the other hand, would signal possible wear of a tool or component, or some other gradual degradation of a process element.

A cyclic trend in the data would indicate that some sort of cyclic influence is acting on the process that was not anticipated. Environmental effects often fall in this category, such as temperature variations from day to night, or from season to season. If the period of the cycle is weekly or daily, it could be based on personnel factors, such as exhaustion and personal issues. Maintenance schedules may also cause cyclic variations in the process.

Whatever cause is ultimately determined for the trend, it should be eliminated, preferably by removing the source of the variation. If this is not possible, the data should be segregated into groups, each of which will have its own trend-free chart.

The procedure outlined above for developing a control chart is just the bare bones of the process. Many different embellishments exist for specific situations and can be found in any text on quality control. The most important of the specific control charts, and their uses, is discussed here.

9.3.3.2   Control Charts for Variables

A variable can be defined as a quality characteristic with a numeric value, such as weight, length, diameter, resistance, strength, and hardness. Control charts are developed based on variables when the actual value of the quantity is important. If more than one characteristic is to be tracked, separate control charts must be developed for each one.

In tracking variables with a control chart, one is actually trying to understand a random stream of numbers. The mathematics of Stochastic Systems can show that a random number stream has many degrees of freedom and can be viewed in many different ways. In practical terms, this means there is a need to develop more than one control chart for each variable. In normal practice, two charts are used, one to track the central tendency of the variable and one to track the variability of the variable. For the central tendency, the $\bar{x}$ chart is used, and for the variability, either a range chart or a sigma chart is used.

The x-bar chart is the most basic type of control chart and is precisely as described above. The quality characteristic is measured in samples, and each sample's mean, or $\bar{x}$, is plotted against the average and ± 3 sigma values of the $\bar{x}$.

The range chart follows the same basic philosophy, but is calculated quite differently. For each sample, range R is calculated, which is the largest value minus the smallest value. If starting with N samples, there will be N ranges. Next, the average of these ranges, $\bar{R}$, and the standard deviation of the ranges, σ_R, is calculated. A control chart now constructed with $\bar{R}$ as the center line, and (± 3) σ_R as the upper and lower control limits.

Now there are two blank charts, an $\bar{x}$ chart and an R chart. For each sample, both in the base and monitoring periods, plotted are both the $\bar{x}$ and R of the sample. Superficially, the charts look very much alike. In reality, though, they tell us very different things. In control and out of control is defined the same for both of charts: all points should be between the control limits, with no trends. But, an out-of-control indication in the $\bar{x}$ chart tells us that the central tendency of the process is changing, whereas a problem in the R chart tells us that the range is changing. Of course, if the range gets smaller, it could be reflecting an improvement in the process, but this is still something that to be investigated.

The third chart mentioned was the sigma chart. This performs the same function as the R chart: it tracks the variability of the process. The advantage of the sigma chart, though, is that all data are used. A range chart, using only the largest and smallest values in a sample, ignores what the rest of the values are doing. In small samples, say five or less, this is fine. But, in larger sample sizes, valuable information could be ignored. The sigma chart is created in exactly the same way as the R chart, except that it uses the average standard deviation of the samples, rather than the average range, as the center line, and ± 3 sigma of these standard deviations as control limits. In this way, all data is used, and subtler changes can be detected. The only disadvantage of the sigma chart over the R chart is that it requires more calculations.

In actual practice, calculations of $\bar{x}$, R, and sigma charts are performed with tables, and actual standard deviations of range and sigma need not be determined. Tabulated values, based on sample size n, give factors which are used to determine center lines and upper and lower control limits.

9.3.3.3 Control Charts for Attributes

An attribute is defined as a quality characteristic which has one of two values: good or bad. This is a useful way of judging conformance to specifications when the actual value of a variable is of no importance, as long as some maximum and minimum have been satisfied. An example would be the length of a screw that is long enough to mount a nut on but not so long that it causes interference. In other circumstances, the quality characteristic being tracked may have no numerical value at all, such as a component that has either been mounted on a circuit board, or has not.

Attributes are further divided into two classes: defects and defectives. A defect is a negative quality characteristic, such as a scratch in a surface, a tear in fabric, or a missing component. The important point in defect tracking is that any one manufactured item may have many defects. A defective, on the other hand, is a single item which has too many defects to be considered usable. For most attribute situations, either a defect viewpoint or a defective viewpoint may be chosen, depending on the preferences and goals of the quality engineer.

Unlike control charts for variables, the attribute-based control charts do not use the Normal Distribution. Since they are not based on variables with numeric values, the Normal Distribution no longer applies. Control charts for defectives are based on the Binomial Distribution, and charts for defects are based on the Poisson Distribution.

The most common type of control chart for defectives is called a p-chart. In this case, p stands for proportion. Samples of size n are collected, but instead of calculating an average, the calculation is to determine the proportion of defectives in the sample. For example, if a sample of 100 items has 28 defectives, the p value for that sample would be 0.28. The choice of sample size n is based on the proportion of defects expected in the long run. Samples should be large enough so that each sample contains at least one defective; otherwise, the calculations will be too coarse to detect the effects of process variations.

In the base period (initial phase) of a p-chart, N samples of n units are inspected, and a p value for each sample is calculated. The average p for all samples, $\bar{p}$, is the center line of the chart. The control limits, based on the Binomial Distribution, are calculated from:

$$\bar{p} \pm 3\sqrt{\frac{p(1-\bar{p})}{n}}$$

This gives limits that include 99.73% of the population, as before. The p values for each sample are plotted, as in variable control charts, and the p values for each monitoring period sample are plotted similarly. Interpretation is the same as for variables. The only difference is that there is no need for any kind of range or sigma chart, since the focus is on attributes.

Control charts for defects are similar, except that the equations are different. That is because the distribution that governs defects is the Poisson Distribution. As before, samples of n units are inspected. Now, however, rather than counting defective units, the defects per unit are counted. If defects occur frequently, a sample may consist of a single unit, with n equal to 1. However, if defects are rare, it is wise to define a sample as large enough to ensure several defects per sample.

The standard defect chart is called a c-chart, which tracks the count of defects per sample. N samples are inspected, each with some count, c, of defects. The average number of defects in all N samples is $\bar{c}$, which becomes the center line of the chart. Based on the Poisson distribution, 99.73% of the population will be between the limits calculated as:

$$\bar{c} \pm 3\sqrt{\bar{c}}$$

As usual, the N $\bar{c}$ of the base period are plotted on these limits and the process is judged to be in or out of control. As long as it remains in control, the monitoring period continues with a new c for each new sample, which is subsequently plotted. Again, there is no need for any range or sigma chart.

9.3.4 Acceptance Sampling

Acceptance Sampling is the offline version of statistical process control and is used to infer quality information about a large quantity of items which have already been produced. The goal is to infer the overall quality level of the batch without the necessity of examining every item in the batch.

Acceptance sampling is traditionally thought of as an activity performed by a receiving department before accepting a shipment from a supplier. At this point, it is too late to monitor the process which produced the shipment, so that the only question to be answered is: is the quality of the shipment sufficiently high?

This question could be answered by the simple process of examining each and every item in the shipment. This process is called "100% inspection" and in some cases is unavoidable. In general, however, it is too expensive in terms of time, money, and manpower. Furthermore, it is not significantly more accurate than inspecting a small sample of the batch and using statistics to infer the overall quality level. Also, since acceptance sampling involves less handling that 100% inspection, it involves less handling damage, which can be a significant advantage in some situations. In the case of items which can only be inspected by destructive testing methods, acceptance sampling is the only feasible method of quality assurance.

Acceptance sampling, like all quality control activities, is based on probability and statistics and, specifically, on some sort of probability distribution. In a sampling situation, the important variables are defined as follows:

N = the number of items in the batch
y = the number of defective items in the batch (unknown but hypothesized)
n = the number of items that to be inspected
x = the number of inspected items that allowed to be defective

In this sample plan definition, the decision is to accept or reject the entire batch of N items based on the result of our inspection of the n items in the sample. The n items are selected at random from the batch. Naturally, the greater the ratio of N to n, the more savings in comparison to 100% inspection.

If x or fewer of the n items inspected are defective, it is concluded that the batch is of sufficient quality to be accepted. If more than x of the n items are defective, the decision is to reject the entire batch. This means that the x/n ratio is used to infer information about the y/N ratio, based upon our knowledge of the probability and statistics of the situation.

The statistics of this particular situation is based on the Hypergeometric Distribution. This is the same distribution that governs selection of numbers in a lottery or the drawing of names from a hat. Its most interesting feature is that the probability of each successive item drawn from the batch having some characteristic (winning lottery number, defective item, etc.) is different than the probability of the previous item, since the proportion of items remaining in the batch has changed with the removal of the previous item.

9.3.5 Six-Sigma Fundamentals

There is much talk of "Six-Sigma Manufacturing" these days. What does this really mean? In a broader sense, the movement in favor of Six Sigma Manufacturing includes all manner of quality improvement techniques, in both technical and business processes. It also includes elements of organizational and educational improvements. But, at its core, Six Sigma is based on a mathematical concept that flows directly from the Process Capability Analysis discussed in Section 9.3.3.

Real manufacturing processes have some amount of variability in their output; the volume of soda dispensed into a nominally 12-ounce bottle will vary from bottle to bottle by a certain amount. The distribution of these values can be characterized mathematically with a Normal Distribution. This Normal Distribution will have some mean and some standard deviation (called sigma, or σ). How much of this distribution will fall within the acceptable tolerances of the bottle filling process? In many industries, plus-and-minus three sigma is considered acceptable. That is, if the upper tolerance limit is at least equal to the mean plus three standard deviations, and the lower tolerance limit is at most equal to the mean minus three standard deviations, then the majority of the "hump" of our Normal Distribution will be within tolerance. For a Normal Distribution that fits perfectly into this plus-and-minus three sigma range, this represents 99.73% of the process output, with only 0.27% falling outside of specifications. For a long time, for many industries, this was considered perfectly acceptable.

However, in modern, high-tech, high-component-count industries, this is sometimes not acceptable at all. If making a product a large number of components (e.g., a million components, which is not unusual in micro-manufacturing such as computer chips), then even a 0.27% defect rate could result in a product with 2,700 defective components! A higher standard needed to be developed.

That is where Six Sigma Manufacturing comes in. In this standard, it is desired for the tolerance range to include NOT merely three standard deviations above and below the mean, but six standard deviations above and below the mean. According to Normal Distribution, this range would include 99.9999998% of the process output. Or in other words, only 0.002 parts per million would be outside of the specifications range.

Of course, this extremely low level of defects is very difficult to achieve and even more difficult to maintain. Real processes tend to drift over time; even if their level of random variation can be maintained as constant, the mean tends to wander up and down over time, requiring periodic adjustments. For this reason, Six Sigma orthodoxy gives us a little wiggle-room. It assumes that the mean of the process will most likely drift now and again, and it allows for a drift of up to 1.5 standard deviations (1.5 sigma) before the drift is detected and corrected. This will of course decrease the percent of the output within spec. Allowing for this 1.5 sigma drift, a Six Sigma process would still have 99.999660% of its output within specifications and only 3.4 parts per million out of specification.

Many organizations claim to be Six Sigma companies. This usually means that they are *pursuing* Six Sigma levels of quality as a goal. Few companies have conquered this challenge to date. But, that doesn't invalidate the concept. The mere pursuit of such lofty goals is enough to bring organizations to world-class status, and it helps to invigorate their continuous improvement mentality. Millions of dollars in cost savings have been reported by many organizations which have adopted Six Sigma techniques and goals.

9.3.6   Cost of Quality

Does quality *cost* money? Or does it *save* money? The truth is that it does both. Expenditures in certain areas of quality will reap savings in other areas. The goal is to spend the right amount of money and effort in areas that will optimize overall costs associated with quality. To that end, one should look at both categories of quality costs: the good ones and the bad ones.

9.3.6.1   Good Costs of Quality

Money and effort spent deliberately and strategically in order to improve the quality of our products can be thought of as "good costs of quality." These are often thought of as Costs of Conformance; in other words, the costs associated with keeping our product quality as high as possible. It is further broken down into two sub-categories: Prevention Costs and Appraisal Costs.

> **Prevention Costs**: These include all the costed efforts for preventing quality problems from occurring in the first place. They are generally planning and design-type activities that are conducted far from the shop floor. They include such activities as quality planning, quality training, design for quality, design for reliability, design for manufacturability, design verification, and installation of quality data acquisition systems. All of them are efforts to increase the inherent quality of our designs and of our production processes.
>
> **Appraisal Costs**: These include all the costed efforts for detecting quality problems as soon as they occur. They are generally conducted on the shop floor during production. They include such activities as Acceptance Sampling of purchased materials, SPC and Control Charts, and testing and inspection of items produced. All of them are efforts at enforcing the inherent quality of our designs and our production processes.

9.3.6.2   Bad Costs of Quality

Costs that are incurred as a direct consequence of our lack of quality can be thought of as the "bad costs of quality." These are often thought of as Costs of Non-Conformance; in other words, the costs associated with NOT keeping our product quality as high as possible. These can be further broken down into two sub-categories: Costs of Internal Failure and Costs of External Failure.

> **Costs of Internal Failure**: These are the costs associated with discovery of bad quality product before it has shipped to the customer. The primary costs here are scrap and rework. In some industries, where much value has been put into a part that has then been found to not conform to specifications, an expensive review process might have to be conducted to decide if the part can be salvaged, repurposed, or if it must be scrapped. Other costs in this area include equipment downtime while problems are corrected, and loss of revenue when product has to be sold at a discount due to its reduced conformance level.
>
> **Costs of External Failure**: These are possibly the worst costs of all, as they are costs associated with defective product actually reaching a customer. Obvious costs in this category are warranty costs, replacement costs, and possibly costs of damage resulting from the use of defective product in the field. There are also liability issues if safety problems occur and possibly recall costs. There is also the intangible cost of damage to the organization's reputation for producing quality product, which can take years to recover from.

Obviously, the more money and effort that are deliberately spent on the good costs of quality, the less that will need to be spent on the bad costs of quality. The trick is to find the proper balance that optimizes overall costs, as well as the intangibles like reputation in the marketplace. Quality guru Philip Crosby famously published a book titled *Quality Is Free*, indicating that when quality activities are properly managed, good quality products cost less to produce, in the long run, than poor quality products.

9.3.7   TOTAL QUALITY MANAGEMENT

Total Quality Management (TQM) is a term that gets used a lot and represents some very good ideas, but unfortunately does not have enough details and specifics to really make a difference in the manufacturing world. It was a popular philosophy in the 1980s, and is still discussed today, but has largely been replaced in actual operations by more detailed and rigorous techniques such as Six Sigma and ISO 9000. Still, it is an interesting philosophical approach to quality improvement.

TQM began in the western world as a response to the growing competition from Japan and other eastern nations in the marketplace of manufactured goods. Its basic premise is a good one: that the highest quality products and operations can only be achieved when an entire organization embraces the devotion to quality, and that this devotion must be driven by top management all the way down through all levels of the organization. Few people would disagree with that attitude. However, TQM itself did little to supply operations means of achieving the results.

Some of the more specific directives of the TQM philosophy include:

- **Focus on the Customer**: If the customer doesn't consider products to be of high quality, then they are not of high quality.
- **Top Management Commitment**: All levels of management from top to bottom need to be involved in pushing the focus on quality. Strategic quality initiatives are originated and managed at the highest level possible.
- **Total Organization Involvement**: Specific tools are used to involve all employees in advancing quality: quality circles, suggestion boxes systems, cross-functional teams, etc.
- **Continuous Improvement**: Commitment to gradual but continuous growth in quality at all levels and in all processes.
- **Supplier Quality Initiatives**: Reaching out to suppliers to improve quality in the entire value chain.
- **Focus on Quality Writings of W. Edwards Deming and Joseph Juran**: Tools that long predate the TQM movement.

All of these elements of TQM are good ones, and all are used successfully in other quality initiatives. However, the TQM framework has largely been replaced by more successful quality management systems and international standards.

9.4 LEAN MANUFACTURING FUNDAMENTALS

Lean manufacturing is a broad term with many subtleties, but essentially it means manufacturing with minimal waste. The details of lean manufacturing come down to how one defines, identifies, and minimizes that waste. It evolved over a number of decades, starting with Toyota, and eventually becoming a world-wide movement. Lean manufacturing is also closely associated with Just-In-Time manufacturing.

There are four guiding principles of lean manufacturing:

1. Minimize waste
2. Perfect first-time quality
3. Flexible production lines
4. Continuous improvement

9.4.1 Minimize Waste

There are many types of waste in a factory, and each contributes to excess cost and time and reduces quality. Although waste can never be completely eliminated, it is the goal of lean manufacturing to reduce it to an absolute minimum. The Japanese term for waste is *muda*, which is used extensively in the literature.

One of many categorization schemes lists seven types of waste, which is convenient enough for discussion purposes:

1. **Production of Defective Parts**: Obviously this is a waste of material, labor, and time and is to be avoided.
2. **Production of More Than the Required Number of Items**: This is also a waste of material, labor, and time, although it is a phenomenon which often occurs despite our best efforts. There is a natural tendency to overproduce product, with a "just-in-case" mentality. For example, a worker, needing to produce 100 units of product, may fear that a few might go bad during the process. To avoid this, he starts with 105 blanks and hopes for no more than five defectives. Whether or not any become defective, it still ends up leading to a waste of resources.
3. **Unnecessary Inventories**: Any inventories on hand, be it raw material, work in process, or finished goods, represent a monetary value that is tied up and unavailable for other purposes. Some inventory is necessary for smooth production of course, but any more than necessary represents waste.
4. **Unnecessary Processing Steps**: Putting parts through steps that do not increase their value is a waste of labor and of time and possibly of consumable resources. Why would anyone perform unnecessary processing steps? Maybe they were necessary at one time, but changes elsewhere have rendered them unnecessary today. This is an insidious source of waste and requires constant re-evaluation of ones processes.
5. **Unnecessary Movement of People**: Moving people around a factory certainly wastes time and might possibly consume significant resources. Operations should be arranged to minimize the amount of reposition of labor that is needed.
6. **Unnecessary Movement of Materials**: Similar to the previous category, never move materials unnecessarily. Of course, if multiple people work on each unit of production, something must move some amount, either the people or the parts. These two categories are examples of "waste" that cannot be completely eliminated but must be studied for minimization.
7. **Workers Waiting**: Whenever a worker is ready and willing to perform a job, but parts, tools, or other resources are not available, that worker is waiting. The worker is still being paid, but the company is getting no return on that cost. Obviously, a lean system must be arranged such that all workers can remain productive as much of the time as possible.

These categories of waste are highly interrelated, and often improving one will make another one worse. For example, reducing movement of workers often necessitates an increase in movement of materials. Reducing movement of materials may result in more time workers spend waiting for them. On the other hand, some of the categories are symbiotic, such as not producing more than the required number of items will also reduce inventories. In the end, production controllers and other managers must use judgment in balancing these goals in their waste reduction efforts.

9.4.2 Perfect First-Time Quality

Obviously, perfect quality is desired at all times, but there is an attitude in many industries that first-time quality is not something to be expected. Losing a few "set-up pieces" at the start of a production run is considered an acceptable loss. Indeed, sometimes it is even considered a necessary step. Lean manufacturing does not allow for this type of waste, and for good reason, it makes short production runs unfeasible.

This concept puts lean manufacturing in conflict with some other time-honored quality techniques. For example, statistical process control (SPC) regularly uses concepts such as Acceptable Quality Level (an acceptable proportion of defective units out of a batch) and Lower Control Limits. If one were to be a purist, one might have trouble reconciling the two philosophies. A better approach, however, is to see them both as valid and useful within specific situations.

One of the ultimate goals of lean manufacturing is the ability to economically produce smaller and smaller batches of product. Indeed, "batch size one" is the holy grail of lean manufacturing. Obviously, by expecting to waste four or five blanks before getting the first good unit of production, a batch size of one becomes wildly uneconomical.

9.4.3 FLEXIBLE PRODUCTION LINES

If a production facility were to make only one product, all day, every day, for years on end, it would only need production equipment that can do the same thing all the time. A lean production system requires the ability to make multiple, often very different, products on the same production equipment. This advances us towards our goals of smaller batches and smaller production runs but without decreasing our machine utilization. Some of the largest investments in becoming lean are investments in more flexible production equipment. But, these investments pay off in the long run.

Quick changeovers are a related concept. Production equipment must not only be capable of shifting production to a new product, but it must be capable of doing it quickly. Looking at the formula for unit cost during batch production, it is seen that cost goes down with increasing batch size, as the setup time is distributed over more units. In order to mitigate this expense for smaller batches, it is necessary to reduce the unproductive setup time to a bare minimum.

Workers, as well, must be more flexible for lean manufacturing and capable of switching between products just as easily as the production equipment. This often requires a large investment in training. Indeed, lean organizations tend to have far fewer job classifications than non-lean organizations, since so many of the employees are trained and qualified in many areas.

9.4.4 CONTINUOUS IMPROVEMENT

The final guiding principle of lean manufacturing is a culture of continuous improvement. This requires considerable management support, because it only works if all employees are involved in finding ways of always improving operations.

These improvements could come from many areas: cost reduction, quality improvement, shorter setups, quicker processes, reduced waste, etc. Employees at all levels must be empowered and encouraged to see continuous improvement as part of their job.

There are many techniques used to implement continuous improvement. Suggestion Box systems allow employees to write down their proposals and turn them in to management, which will periodically review them and select the most promising for study. Some organizations will financially reward employees for their suggestions if implemented, based on the savings to the company. Quality Circles are regular meetings of employees who have similar jobs, to discuss the best way to perform common tasks. MBWA, management by walking around, is an activity conducted by managers taking them out of their offices and down to the production process where they can see what is really going on, what the real problems are, and what is needed to solve them.

The term *Kaizen* is the Japanese word for continuous improvement, and it is used extensively in the literature of this topic. A *Kaizen Event* or *Kaisen Blitz* is a specific, targeted effort to effect improvements in areas that have been identified as needing them urgently. However, these short-term events should not be considered a substitute for an overall cultural climate of slow but continuous improvement in all areas of operations.

As a final note on continuous improvement, it must be stressed that the lack of continuous improvement does not mean that you maintain the status quo. Indeed, lack of continuously moving forward leads to continuously moving backward, or degrading. Technology advances, industry advances, and most of all business competitors are advancing. An organization that is not continuously improving, is getting farther and farther behind.

9.5 MANUFACTURING SYSTEM EFFECTIVENESS

Manufacturing System Effectiveness (Se) is a method to monitor manufacturing systems and provide key information for making business decisions. The decisions of interest are those addressing the "readiness" of a manufacturing system, process, or product to support operational demands. It is a simple performance measure reflecting manufacturing system's output quality, throughput rate, and ability to be in a working (not broken state).

As a performance measure, Se gives a quick picture of a system's ability to operate as expected to maximize profitability. The Se value is directed to either the overall system or components therein. The "system" can be defined as a product, the overall manufacturing system or an equipment therein, or a distinct "business cell." How well the system is doing becomes obvious; the closer the Se demonstrated value is to 1.0, the better the system is doing. As a performance measure, Se reflects the ability that the system, whether a product or equipment, will successfully meet operational demand under specified conditions. It takes into account quality, reliability (or availability), and efficiency (or throughput) [26].

These are key indicators of system performance because of their impact on profit margin. Very simply, if the system is making good product in a working state at the rate expected, then profit margins will be at or below the standard product cost expected. Being at or below the standard cost maximizes profitability potential.

Note that Se = 1.0 does not guarantee profitability. It merely indicates that the system is operating at the level expected to maximize profitability. There are many operational costs beyond the manufacturing system that can reduce profitability of the product priced at a saleable level for the market place.

9.5.1 COMPONENTS OF SE

By fundamental definition, Se is the ability of the item to successfully meet operational demand when used under specified conditions. It is the product of three independent, not mutually exclusive probabilities.

$$Se = Cq \times Eo \times Aa$$

Se = quality capability *and* operational efficiency *and* achieved availability
   where:

   Se is System Effectiveness
   Cq is quality capability (the proportion of good stuff made)
   Eo is operational efficiency (the ratio of achieved throughput to expected throughput)
   Aa is achieved availability (probability of being in a working state)

As a probability, Se is always a value between (or equal to) 0.0 and 1.0. This makes it easy to chart to display ongoing progress (y-axis is Se value....x-axis is time). Charting all the manufacturing system component Se values makes it clear to see the greatest opportunities for improvement in the overall manufacturing system.

The Se variables each have much relevancy to ensuring system reliability, product quality, and system throughput capability. Together, each of these equation variables provides ongoing assessment of whether the manufacturing system, or equipment, or process, is performing as designed to deliver a good product at volumes expected.

9.5.1.1 Quality Capability

Quality capability, Cq, is the probability that the item, or an item output, is capable relative to a design tolerance(s) for a key characteristic(s). This simply highlights the interest in the proportion of data for a defined key characteristic(s) that is within the design requirement or tolerance. Capability is commonly indicated by a C_{pk}. Typical target C_{pk} values for manufacturing processes are:

$$\text{New Processes: } C_{pk} = 1.33 \left(\text{gives a Cq} = 0.999934 \right)$$

$$\text{Existing Processes: } C_{pk} = 1.0 \left(\text{gives a Cq} = 0.9973 \right).$$

$$\left(\text{Cq is also calculated as 1-\% defective.} \right)$$

9.5.1.2 Operational Efficiency

Operational efficiency, Eo, is the measure (pseudo-probability) that the item works as efficiently as design-intended. It is the ratio of achieved operating rate to specified operating rate. Operating rate is often described in terms of parts per minute or cycle time and is included in the cost justification.

$$Eo = \frac{\text{Achieved Throughput Rate}}{\text{Specified Throughput Rate}}$$

or

$$Eo = \frac{ActualProduction\ Qty}{Expected\ Production\ Qty}$$

If achieved rate is greater than specified *or* Eo is not applicable to subject item, use $Eo = 1.0$. Expected rate or quantity is determined based on the time the system is running.

9.5.1.3 Achieved Availability

Achieved availability, Aa, is the probability that the item will function as required when called upon at any point in time.

$$Aa = (Uptime) / (Uptime + Downtime)$$

The "downtime" includes corrective maintenance time and preventive maintenance time. It does not include delay time in performing the maintenance action (i.e., time to schedule task, waiting time for spare parts or equipment, lunch breaks). Chapter 10 provides a detailed discussion of availability and its relevance for predicting and determining the reliability of systems that are repairable.

Note that our downtime focus is limited to that caused by equipment maintenance. This is because the manufacturing system equipment is what ultimately constrains manufacturing capacity. However, availability can be calculated using downtime comprised of all contributors such as meetings, training, and maintenance; this value is Operational Availability. Operational Availability will be a smaller value than Achieved Availability and includes issues that are less a function of system capacity than of time/schedule management.

9.5.2 PERFORMANCE BENCHMARK

The benchmark Se value simply comes from multiplying the minimum target values for each of the Se components.
Requirement Level: Se > 0.9

The Se minimum requirement is derived from the multiplication of the three measure components.

$$Se = Cq \times Eo \times Aa$$

$$= 0.9$$

With a benchmark defined, data collection is the key to getting a good demonstrated Se value to compare actual performance with that expected.

9.5.2.1 Quality Capability Data

Variable or attribute samples are collected at defined intervals (e.g., every hour). (Use your appropriate corporate or divisional form to collect data.) Sequential collection of all sample data must be noted (time, date, collector) and maintained. Note all system, process, or product adjustments made during the demonstration.

For manufacturing "short-run" situations, a sample of at least 30 is desirable. If the system output is less than 30 individual samples, then all the samples are included in the statistical analysis. (Inferences based on confidence intervals are best for this scenario.) Note that capability is based on a "stable" or "in-control" process.

For variable data, the underlying population is assumed to be "normally" distributed. The sample data's ability to exhibit this distribution should be verified and noted within the analysis report. Use of the Normal Distribution facilitates calculation of the probability of the variable characteristic being within design specifications.

Depending on the scenario, attribute data may be collected. For attribute data, capability is the complement of the probability of the defective event occurring.

Typically, process capability is based on one to three key critical quality output characteristics. If more than one characteristic is looked at, then use the probability multiplication rule for all the independent, not mutually exclusive events occurring simultaneously.

9.5.2.2   Operational Efficiency Data

Capture the throughput information accurately. The actual quantity of material or product received is counted and recorded as required for closure of any manufacturing order. The "expected" quantity of output is derived from the item's design-specified operational rate.

The expected value is based on the actual operational run time, not the total time. That is, run downtime is not included in determining expected value. Also, note that the operational rate may change during an item's functional design. Make sure to take expected rate changes into account when calculating the output expected value. For example, bowl feeders are typically designed to slow down as they become empty; that is an expected rate change.

9.5.2.3   Operational Availability Data

It is important to maintain a summary record, which chronologically lists all failures and stoppage events, their description, their time (i.e., time or cycles), and their categorization as either relevant or nonrelevant.

Data and information include (as applicable):

1. Cycles or time to failure/stoppage event.
2. Reason for each failure/stoppage event (e.g., preventive maintenance).
3. Corrective action taken for each failure or stoppage event.
4. Identification of and the time used for various activities contributing to the downtime for any one downing event (e.g., 1 hour in tool and die, 1 hour for lunch break, etc.).
5. Classification of each failure/stoppage event as relevant or nonrelevant.

Relevant events include:

a. Design defects.
b. Manufacturing defects.

c. Functional degradation below design specifications.
d. Intermittent failures.
e. Failures of known limited life parts prior to specified life.
f. Failures not attributable to a specific cause.
g. Failure of built-in-test features.

Nonrelevant events include

a. Failures resulting from improper handling or installation.
b. Failures of instrumentation or monitoring devices external to the assessment item.
c. Failures due to stress beyond design specifications.
d. Failures resulting from procedural/human error.
e. Failures induced by maintenance actions.
f. A secondary failure which is the direct result of a failure of another part within the equipment.

Note the following for availability/reliability assessment:

1. Intermittent failures are relevant. If several such failures are due to a specific defect and the defect is eliminated during the test, then only one failure is counted.
2. Failures occurring during test setup, troubleshooting, or maintenance are not considered as reliability failures.
3. Only operational cycles or time occurring during normal "power on" portion of the assessment are considered. Operating cycles or time accumulated as a result of "checkout" or "verification" activities are not considered part of the assessment.
4. Assessment time restarts from time zero after implementation of a *major* design change.
5. Assessment time may be interrupted to make a *minor* design change. Assessment time resumes from the point of stoppage, and the design change has no effect on the classification of previous failures.

9.5.3 Business Decision-Making

With weekly Se reporting, each system's operation can be reviewed relative to the following questions.

- Does it functionally do what it is supposed to do?
- Is the reliable, or available?
- Does it provide a high-quality output?
- Does it run at the speed it is supposed to?
- Is immediate corrective action needed?…Near-term corrective action?…No corrective action?

The answers to these questions can easily come from the people working within the system. The demonstrated Se value is compared to the specified Se enabling everyone to visually see and understand if the system is in good shape. This enables the workers to take ownership of the responsibility for quality, efficiency, and/or reliability.

TMA Case Study: Stress-Strength Product Analysis

OVERVIEW

A new fastener product has been developed for marine applications, as well as for use on dry land in humid ambient environments. The product's fastening capability provides a competitive edge over all other competing products. Ensuring that the consumer experiences the desired holding force advertised, the stresses impacting the design application must be eliminated or be greatly reduced. If potential stresses impacting a design are minimized or eliminated, then the potential failure(s) caused by that stress are eliminated. There are four basic ways to deal with stress: (1) increase the product strength, (2) decrease the average stress, (3) decrease the stress variation, or (4) decrease strength variation.

ISSUE

The product experience in anticipated harsh environments seen in new global markets is minimal. Industry-wide research data was used to support and guide the use of adjustment or degradations factors to account for differences between laboratory specimens and manufactured parts.

STRATEGIC OBJECTIVE

The strategic objective is to launch the product in global markets as a unique product that works in harsh environments. The company wants to ensure that the product applications meet all holding strength capabilities defined in the product specification sheets and catalog in order to avoid exposure from product warranty costs.

CASE BACKGROUND

The probability distributions for strength can be approached in two ways. First, if it is assumed that the strength of components is determined by the weakest point, then the distribution will be determined by the lowest value of all samples taken from a distribution that describes the strength of all points in the material, which is the Extreme Value Distribution. Second is the assumption that the weaker points receive support from the stronger points surrounding them. That is, an averaging process occurs. Accordingly, the distribution of the strength is related to the mean value of samples from the distribution of strength for all points, which is the Normal Distribution.

The industrial-wide available research data supports the second assumption, as probability distributions of ultimate tensile, yield, and endurance strengths are found to be normally distributed. Accordingly, the subject new product is made of steel and the Normal Distribution for strength and stress was assumed valid.

Engineering Management endorsed the approach that the inherent reliability of the metal locking fastener was to be determined based on the assumption that strength and stress are random variables having Normal Density functions. The reliability of the steel locking fastener was to be derived from the equation:

$$R = 1 - \phi\left(-\frac{U_{ST} - U_{SS}}{\sqrt{\sigma_{ST}^2 + \sigma_{SS}^2}}\right)$$

where
R is reliability
U_{ST} is the mean strength
U_{SS} is the mean stress
σ_{ST}^2 is the variance of strength
σ_{SS}^2 is the variance of the stress

The reliability equation illustrates that the reliability depends on the lower limit of the data distribution, which is denoted by:

$$z = \phi\left(-\frac{U_{ST} - U_{SS}}{\sqrt{\sigma_{ST}^2 + \sigma_{SS}^2}}\right)$$

The random variable z is the standard Normal variable. From this, reliability is found by referring to a table of the cumulative probability distribution function values for the standard Normal Distribution.

RESULT/CONCLUSION

The following steel fastener strength data was derived from tensile test. The product catalog value for maximum fastener application strength is 100 lbs, which assumes 0 variance. This is the maximum stress to be seen by the product under defined product application parameters. The following tensile test data (in pounds) was collected for product failure from a laboratory environment:

123, 112, 175, 148, 159, 193, 176, 124, 197, 108, 186, 195, 123, 103, 105, 154, 155, 150, 143, 110, 153, 174, 135, 108, 190, 161, 118, 137, 114, 187

The test engineer calculated the following for predicting product reliability.

$U_{ST} = 147.2$
$U_{SS} = 100$
$\sigma_{ST}^2 = 31.04$
$\sigma_{SS}^2 = 0$

Using this data to calculate the standard Normal variable, z, gives:

$$z = -(147.2 - 100) / (31.04 + 0) = -1.52$$

Using a Standard Normal Distribution table, the z-value $= -1.52$ equates to 0.0643. The reliability for the metal fastener is derived to be:

$$R = 1 - 0.0643 = 0.936$$

Case Study Question: *What factors should be taken into consideration regarding basing a global product technology launch on laboratory test data? Do these factors provide an impact on perceived and real risk of commercial success?*

9.6   SUMMARY

This chapter addressed maintaining designed-in levels of product reliability, safety, and quality. The focus is on evaluating and monitoring the manufacturing system (and equipment therein) to ensure that it is capable of providing the optimal product from a benefit-cost perspective. That is, while still maintaining and providing the market required levels of reliability, safety, and quality.

Reliability control focuses on maintaining the inherent level of product reliability designed in and predicted. The objective is to integrate into the manufacturing system and process design specialized screens and supporting tests and inspections that eliminate, or minimize, product latent defects moving to the market place. Latent defects are not readily observable defects (or partial defects) that exposed themselves upon a non-overstress situation. The removal of these defects enables a product to quickly move out of the infant mortality phase and into the useful life phase of a product's characteristic life cycle.

Safety control addresses both the product and the manufacturing system design. Designs must exhibit safety features that illustrate their benefits in both paper analyses and special tests. Analyses such as FMECA and FTA are excellent methods for assessing product and system safety in a cost-effective manner. For the manufacturing system, sound design must be supported by a well-founded occupational safety program that focuses on human factors.

Quality control focuses primarily on patent defects. That is, defects that are readily observable. Once a product is designed with the customers' quality requirements in mind, it is up to the system to deliver consistently the corresponding product. Many statistical techniques are used to monitor product quality, including control charting as part of SPC.

QUESTIONS

1. Discuss the relationship between design of experiment and reducing the product development cycle.
2. How are "life" and "MTBF" different?
3. How does durability fit into the concept of the reliability bathtub curve?
4. Derive the reciprocal relationship between failure rate and MTBF.
5. Discuss the difference between random failure and latent defects.

6. How did reliability remain an issue with space satellites even with the maintenance capability provided by the U.S. Shuttle aircraft?

7. When is it best to use inductive versus deductive failure analysis methods?

8. Discuss the legal liabilities associated with a lack of human factor design consideration.

9. What is the difference between process monitoring and acceptance sampling? Suggest several situations where one would be more appropriate than the other.

10. List the advantages and disadvantages of single sampling plans versus double or multiple sampling plans. Think of a situation where each would be most appropriate.

11. Think of situations in your organization which might cause each of the following trends in a control chart: cyclic variation, gradual drift in the mean, a sudden jump in the mean, or two distinct populations.

12. A design has the choice of two competing designs. Alternative 1 is a redundant design of two components, each having a $MTBF = 1,000$ hours. Alternative 2 is a serial design of one component, having a $MTBF = 3,000$ hours. Each design will complete the same mission over 500 hours. If each design costs the same which design should be chosen and why?

13. Calculate the reliability for a series-parallel system. Blocks A1 and A2 are in parallel. Blocks B1 and B2 are in parallel. Reliability values are $A1 = 0.9$, $A2 = 0.8$, $B1 = 0.7$, $B2 = 0.6$.

14. Using the reliability data in Problem 13, calculate the reliability for a parallel-series system. Blocks A1 and B1 are in series. Blocks A2 and B2 are in series. Comparing the reliability calculated in Problem 13, which system design provides a higher predicted reliability level?

REFERENCES

1. Anderson, R. & Lakner A. (1986). *Reliability Engineering for Nuclear and Other High Technology Equipment: A Practical Guide.* London: Elsevier.

2. Brauer, D. (1986). *Materials & Reliability.* Chicago, IL: Report for University of Illinois at Chicago.

3. Brown, D. (1976). *Systems Analysis & Design for Safety.* Englewood Cliffs, NJ: International Series in Industrial and Systems Engineering. Prentice-Hall.

4. Kapur, K. & Lamberson, L. (1977). *Reliability In Engineering Design.* New York: Wiley.

5. U.S. Department of Defense. (1984). MIL-HDBK-781. *Reliability Test Methods, Plans, and Environments for Engineering Development, Qualification, and Production.* Philadelphia, PA: Military and Government Specs & Standards (Naval Publications and Form Center) (NPFC).

6. O'Connor, P. (1985). *Practical Reliability Engineering.* New York: Wiley.

7. Abernethy, R., Breneman, J., Medlin, C. & Reinman, G. (1983). *Weibull Analysis Handbook.* AFWAL-TR-83-2079. Wright Patterson AFB, OH: USA Air Force Systems Command.

8. Bazovsky, I. (1961). *Reliability Theory and Practice.* Englewood Cliffs, NJ: International Series in Engineering: Space Technology Series. Prentice-Hall.

9. U.S. Department of Defense. (1986). DoD-HDBK-344. *Environmental Stress Screening of Electronic Equipment.* Philadelphia, PA: Military and Government Specs & Standards (Naval Publications and Form Center) (NPFC).

10. U.S. Naval Avionics Center. (1986). R&M-STD-R00217. *Human Reliability Analysis.* Philadelphia, PA: Military and Government Specs & Standards (Naval Publications and Form Center) (NPFC).

11. U.S. Nuclear Regulatory Commission. (1981). NUREG 0492. *Fault Tree Analysis Handbook.* Springfield, VA: National Technical Information Service.

12. U.S. Department of Defense. (1980). MIL-STD-1629A. *Procedures for Performing a Failure Mode, Effects, And Criticality Analysis.* Philadelphia, PA: Military and Government Specs & Standards (Naval Publications and Form Center) (NPFC).

13. Nicholson, A. & Ridd, J. (Ed.). (1987). *Health, Safety, and Ergonomics.* London: Butterworths.

14. U.S. Department of Defense. (1989). *Human Engineering Design Criteria for Military Systems, Equipment & Facilities.* Philadelphia, PA: Military and Government Specs & Standards (Naval Publications and Form Center) (NPFC).

15. Hammer, W. (1980). *Product Safety Management and Engineering.* Englewood Cliffs, NJ: Prentice-Hall.

16. Hicks, C. (1982). *Fundamental Concepts in The Design of Experiments.* New York: CBS College Publishing.

17. Box, G., Hunter, W. & Hunter, J. (1978). *Statistics for Experimenters.* New York: John Wiley and Sons.

18. Miller, I. & Freund, J. (1985). *Probability And Statistics for Engineers.* Englewood Cliffs, NJ: Prentice-Hall.

19. Bendell, A., Disney, J. & Pridmore, W. (Ed.). (1989). *Taguchi Methods: Applications in World Industry.* London: IFS Publications.

20. Schmidt, S. & Launsby, R. (1992). *Understanding Industrial Designed Experiments.* Colorado Springs, CO: Air Academy Press.

21. Montgomery, D. (2001). *Design and Analysis of Experiments* (5th ed.). New York: Wiley. ISBN: 978-0-471-31649-7.

22. Lapin, L. (1983). *Probability & Statistics for Modern Engineering.* Monterey, CA: Brooks/Cole.

23. Banks, J. (1989). *Principles of Quality Control.* New York: John Wiley & Sons.

24. Besterfield, D. (1986). *Quality Control.* Englewood Cliffs, NJ: Prentice-Hall.

25. Taguchi, G., Elsayed, A. & Hsiang, T. (1989). *Quality Engineering in Production Systems.* New York: McGraw-Hill.

26. Brauer, D. (2004). *Manufacturing System Effectiveness.* Design Assurance Sciences.

10 System Maintenance

A key element of all manufacturing systems is maintenance. Obviously, if a system received no maintenance it would run until failure. In a perfect world, there would be no failure, and manufacturing systems would run for ever. However, a perfect world does not exist, and manufacturing systems, and products, are destined to experience failures.

In advancing towards Total Manufacturing Assurance (TMA), system maintenance is a very important area not to be overlooked. Since, when failures do occur, their cost can be very high. This is in regard to lost manufacturing time and in equipment repair or replacement. To minimize these costs, it is necessary to have in place a comprehensive and cost-effective maintenance program.

The key to a successful program is to minimize total maintenance life cycle costs. This is achieved by performing maintenance at the optimal time between maintenance interval for each major system component. The curves presented in Figure 10.1 illustrate this concept. The optimal maintenance interval is achieved by balancing the performance of corrective maintenance (CM) and preventive maintenance (PM).

The figure illustrates that as time increases, the cost of CM increases. This is due to increasing cost with the severity of the problem. By performing PM, costly corrective repairs can be avoided. But, PM can be expensive too, and it is desirable to extend the time between its performance. Not only for cost reasons but also to avoid introducing system problems from too much hands-on interaction.

This chapter provides insight into establishing a cost-effective maintenance plan for the manufacturing system. Special attention is given to establishing a maintenance program which cost-effectively enhances system "availability." This includes addressing the fundamental ideas of maintenance planning. Also, emphasis is placed on establishing a maintainability program and deriving quantitative data as part of enhancing the cost-effective achievement of maintenance.

10.1 RELIABILITY, MAINTAINABILITY, AND AVAILABILITY

In Chapter 9, the concept of reliability engineering was discussed, which focused on the item (e.g., product or manufacturing system) that works, fails, and is replaced. But, often, an item that fails can be repaired and used again (e.g., product or manufacturing system). This situation expands the discussion of reliability and Mean-Time-Between-Failure (MTBF). The term "reliability" really just focuses on an item failing and then being replaced. But, what about items that can be fixed and used again...and are cost justified to fix and use again (e.g., an automobile or manufacturing system equipment)? The subject of system reliability has now necessarily broadened to address maintenance and reparable items.

This leads to two additional design and probabilistic concepts: maintainability and availability. These become important topics because once something breaks, it is generally advantageous to get it fixed as quickly as possible (if it is not a throw-away

DOI: 10.1201/9781003208051-13

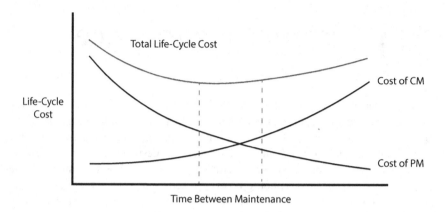

FIGURE 10.1   Maintenance life-cycle cost curve.

item). When something is not repairable, then the focus is on reliability, and when something is repairable, then the focus is on maintainability and availability. As with reliability, design, modeling and prediction, and test and demonstration activities are performed and managed.

Like reliability, maintainability and availability are figures-of-merit that describe the ability of an item to be maintained and remain in active service performing as intended. All three of these engineering design figures-of-merit can be specified, predicted, and assessed both on a qualitative and quantitative basis.

The maintainability of an item is represented by a Mean-Time-To-Repair (MTTR) value. The failure rate and repair time are combined to arrive at a CM action rate. This process is repeated for each replaceable part/assembly in the system. From the maintenance action rates derived for each replaceable item, the MTTR is determined, which is merely a weighted average of maintenance task times.

MTTR is that time associated with CM. It is the time required to localize and isolate a fault to a replaceable module or subassembly; disassemble the equipment to the extent necessary to gain access to the failed item, replace the defective item, and reassemble, align, and checkout the repaired system.

$$\text{MTTR} = \frac{\sum (\lambda_i)(M_{cti})}{\sum (\lambda_I)}$$

where: λ_I is the ith individual failure rate.
M_{cti} is the ith individual CM time.

Designing for ease of maintenance requires providing special means for identifying failure and/or potential or marginal failures and facilitating fault diagnostics, access to failed units, and removal and replacement of failed units.

As depicted in Figure 10.2, the total maintenance downtime is divided into several groups. The principal ones being active maintenance downtime and CM downtime, of which active repair time is a principal quantifiable element. CM is the action

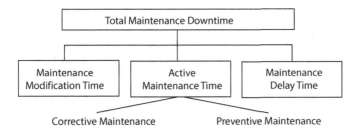

FIGURE 10.2 Maintenance time elements.

performed, as the result of a failure, to restore an item to a specified condition. The element of CM, which is important in terms of design, is the active repair time.

Availability is the other engineering topic that rounds out the trio in conjunction with reliability and maintainability. It is the probability of an item being ready to function as intended when required to do so. Availability provides a single combined measure of the reliable operation of an item and its ability to be effectively maintained.

Availability is based on the continuous duty-cycle of an item having both a constant failure rate and repair rate. When downtime is zero (i.e., when MTBF approaches infinity or when MTTR nears zero), the probability of being "available" (i.e., availability) equals 1.0. Zero downtime becomes an engineering objective, and MTBF and MTTR are the parameters which directly affect the percentage of time that a system is available for use.

Availability has a similar meaning for repairable equipment as reliability has for non-repairable equipment. The difference is that reliability only accounts for the single event failure and availability accounts for both failure and repair events.

Instantaneous availability function, A(t), is defined as the probability that an equipment performs a specified function under given conditions at a *prescribed time*. The instantaneous availability is bounded such that

$$R(t) < A(t) < 1$$

since $A(t) = R(t)$ for an item that does not undergo repair. An important difference between A(t) and R(t) is their behavior for large times (see Figure 10.3). As "t" (i.e. time, cycles, or events) becomes large, R(t) approaches zero, whereas availability functions reach some steady-state value.

The most practical way to achieve high availability is to supplement the design for reliability with a design for efficient, rapid repair, and high degree of maintainability. That is, increase item uptime and decrease item downtime. Increasing uptime and decreasing downtime is achieved by increasing MTBF and reducing MTTR.

MTBF improvement techniques include:

a. Better parts and materials
b. Derating
c. Production reliability screens or burn-in tests
d. Environmental control (e.g., cooling)
e. Environmental hardening (e.g., shock, vibration)

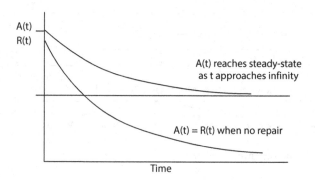

FIGURE 10.3   Reliability versus availability.

MTTR improvement techniques include:

   a. Modularity
   b. Diagnostics (bite)
   c. Other maintainability improvement techniques which do not directly impact system MTTR include:
      i.   Performance monitoring
     ii.   Personnel training
    iii.   Levels of support (online/base/depot)
    iv.   Spares support

In addition to the above MTBF and MTTR improvement techniques, engineering practices and controls applied during item acquisition to improve availability are:

   a. Warranty provisions
   b. R&M testing
   c. R&M control
   d. R&M growth
   e. Critical parts control
   f. Failure modes, effects, and criticality analysis (FMECA)
   g. Vendor control program

Increasing reliability, maintainability, and availability comes down to cost. As is often the case, best intentions are shrouded with the need to keep costs under control. This requires making life cycle cost trade-off decisions regarding how reliable and maintainable a design should be. Let it be shouted from the mountain tops that ultimately what dictates design reliability is the customer requirement, which is reflected in the design specification. Be that as it may, a quick overview of life cycle cost (LCC) is necessary.

As with any item, it is often advantageous to optimize the total LCC. As depicted in Figure 10.4, the optimal level is the point on the total LCC curve that provides the lowest LCC. LCC has two principal components:

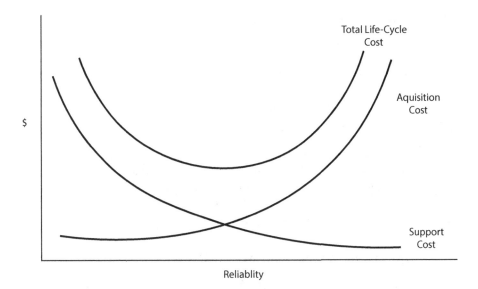

FIGURE 10.4   Life cycle cost curve.

1. The cost incurred before an item becomes operational (acquisition cost, AC)
2. The cost incurred after the item becomes operational (maintenance support cost, MSC).

$$LCC = AC + MSC$$

Figure 10.3 illustrates the trade-off decisions to be made. Designing in high reliability level decreases the support costs (e.g., maintenance personnel or spare part costs) but increases the acquisition costs (e.g., engineering design or initial part cost). The concept is pay more upfront to pay less down the road. On the other hand, designing in lower reliability increases the support cost and decreases the acquisition cost. The concept is pay less upfront and to pay more down the road.

In general, there is always an AC, but there may not be any MSC. The shape and existence of the LCC curves depend on the item and its application. Every item has its own designed-in life cycle that defines the expected LCC.

AC, Acquisition Costs include:

- Engineering Design
- Test and Evaluation
- Experimental Tooling
- Program Management
- Manufacturing & Quality Engineering
- Production

MSC, Maintenance Support Costs include:

- Spares
- Maintenance Performance
- Support Equipment
- Training
- Technical Data and Documentation
- Logistics Management

It ultimately comes down to meeting customer expectations. If the customer pays a premium to get the best brand on the market, then there is likely no expectation for any product failures. However, buying the cheapest brand on the market, there is likely no expectation for the product to last as long as the most expensive comparable item. Looking at LCC is the heart of "getting what you pay for."

10.2   MAINTENANCE PROGRAM PLANNING

10.2.1   CORRECTIVE AND PREVENTIVE MAINTENANCE

The performance of maintenance serves two functions: (1) correct failures or (2) delay the onset of failures. PM provides the delaying action and CM provides the corrective action. For either category of maintenance, the focus is on wear-out failures which occur as a result of material wear, fatigue, corrosion, etc.

Underlying maintenance theory is the concept of "force-of- mortality." It is represented by the hazard rate or instantaneous failure rate. The force of mortality causes the hazard rate to increase. It is through our maintenance program that some control can be levied against the force-of-mortality.

In Chapter 9, it was stated that it is important to quickly move products away from infant mortality and have them operate in chance mortality, or useful life. In this chapter, the focus is on the other end of the reliability "bathtub" (i.e., wear-out). It is important to keep manufacturing systems away from wear-out mortality and have them operate in their useful-life mortality phase.

In essence, manufacturing systems are products themselves and the same mortality concepts apply. The goal for both is to remain in the useful-life phase of the life cycle.

In discussing maintenance, the curves illustrated in Figure 10.5 are key. As shown, over time the hazard rate increases. Ultimately, all systems and equipment therein move into wear-out if no special attention is received. The objective of maintenance is to pull an item out of the wear-out phase or, in the case of PM, prevent or delay its entrance.

As a system ages, the probability increases that it, which has survived t increments of time, will fail during the next increment. This probability represents the force of mortality. Thus, it becomes clear that maintenance concerns itself with phenomena that occur as a result of aging.

It is desirable to avoid wear-out failures and to only experience chance (or useful-life) failures. With chance failures, the age of a system is of no consequence.

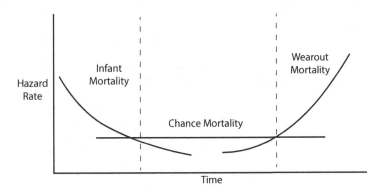

FIGURE 10.5   Force-of-mortality curves.

Therefore, a system is considered like new as long as it is still operating and has not reached its wear-out phase. During this period, the force of mortality is constant.

Through a comprehensive PM program, it is possible to keep a system operating in a state where the force of mortality is constant. In other words, the manufacturing system remains in its useful-life period and exhibits only chance, or random, failures.

The significance of this maintenance approach is illustrated in Figure 10.5. Consider the hypothetical case where "perfect" CM is possible. The hazard rate curve for perfect maintenance serves as a dividing line between PM and CM. Note that the hazard rate becomes constant with perfect CM. This indicates that wear-out failures can also occur by "chance." [1]. This chance characteristic is caused by the unequal age of the components. As a result, wear-out failures cannot be distinguished from pure chance failures.

Perfect CM implies that replaced or repaired items work properly. Also, no damage or error occurs during the repair task. However, perfect CM does not exist. In the case of faulty CM, the hazard rate lies above that which would result from perfect CM. In essence, the item is being pulled back into its useful-life phase.

Therefore, PM plays a critical role in maintaining a low hazard rate. In PM, items are replaced or repaired prior to the time when a malfunction might occur. Consequently, the hazard rate will lie below that of the hazard rate resulting from perfect CM.

Also depicted at the bottom of the Figure 10.6 is the constant hazard rate resulting from pure chance failures. It is important to realize that these failures cannot be reduced by maintenance. However, they can be increased by careless or faulty maintenance.

The maintenance strategy developed for the manufacturing system and its components identifies the type of maintenance to be performed. A key issue in developing the maintenance strategy is cost. Reliability-centered maintenance, RCM, (to be discussed later in this chapter) is an engineering method for determining the most cost-effective type of maintenance.

Two cost factors exist in maintenance: (1) the actual expenditure for maintenance action and (2) the gain obtained by maintenance action. These two factors must be balanced in order to make optimal maintenance decisions. For example, low

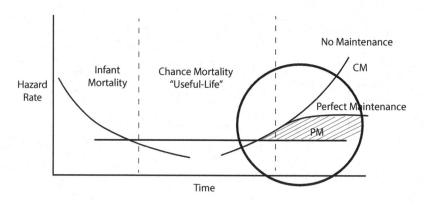

FIGURE 10.6   PM versus CM.

replacement cost of a failed item may rule out the use of PM. On the other hand, high replacement cost may necessitate the need for PM.

10.2.2   MAINTENANCE PLANNING PROCESS

Implementation of PM and CM requires the development of a detailed maintenance plan. A good maintenance plan provides an organized and disciplined approached to ensuring a high level of manufacturing system availability. It also ensures that the system operates in an efficient and safe manner, as intended.

The scope of a sound maintenance program plan addresses the overall manufacturing system and considers its interaction with other ongoing activities; particularly, production scheduling. However, it must provide the details necessary to properly maintain each key maintenance item (KMI), and parts thereof, comprising the system.

Figure 10.7 illustrates the process for developing a maintenance program plan [2]. The maintenance plan defines the requirements and tasks necessary to restore or sustain the operating capability of the system. It evolves from various analyses to identify the key elements of the program. These elements include:

1. Maintenance strategy
2. Reliability, maintainability, and availability requirements
3. Specific maintenance tasks
4. Maintenance organizations
5. Support and test equipment requirements
6. Maintenance standards
7. Supply support requirements
8. Facility requirements
9. Technical publications

The foundation of the overall program plan addresses management. The focus is on the assignment of maintenance responsibilities. With this defined, it becomes possible to prepare the detailed maintenance schedule for each KMI. These schedules

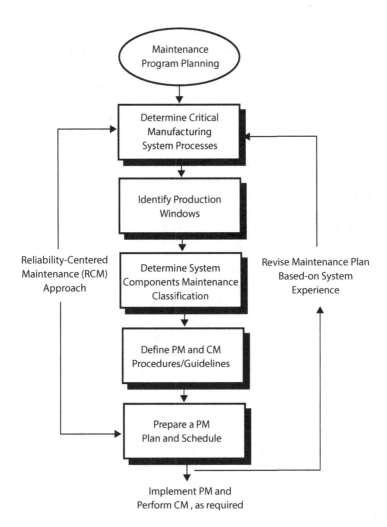

FIGURE 10.7 Maintenance program planning process.

control the performance of all "known" maintenance tasks in accordance with established manufacturing system and production priorities.

Maintenance schedules are relatively firm for PM activities accomplished on a periodic basis. Adjustments can readily be made to compensate for variations caused by operational requirements and current workloads. The PM schedules themselves are developed based on system reliability data, actual or estimated, as well as the best judgment of material maintenance specialists. Defining PM intervals will be addressed later in this chapter.

CM cannot be scheduled in the same sense as PM. The reason is that random failures occur in spite of the performance of PM. Once a failure occurs, the appropriate CM task must be scheduled into daily activities. Remember that the need for CM can arise at any moment. Therefore, the maintenance program plan must enable the responsible maintenance organization to respond in a timely manner.

10.2.2.1 Maintenance-Induced Problems

Previously discussed was how a broad range of factors contribute to the degradation of manufacturing system availability. This degradation is usually the result of the interaction of man, machine, and environment. This is particularly true during system maintenance.

Manufacturing systems are defenseless against poor maintenance practices. Excessive handling brought on by frequent and/or faulty maintenance often results in defects being introduced into the system. Typical examples include foreign objects left in an assembly, bolts improperly torqued, dirt injection, parts replaced improperly, wrong parts used, and improper lubricants.

Although it is important that a system be highly maintainable, where practical, maintenance tasks should be designed out. But, keep in mind that even with these efforts the effects of poorly trained, poorly supported, or poorly motivated maintenance technicians on system availability can be catastrophic.

10.2.2.2 Maintenance Documentation

A good means for offsetting maintenance-induced problems is the availability of sound maintenance documentation. This includes written, graphical, and other types of information to guide maintenance technicians in accomplishing both the PM and CM tasks identified in the maintenance plan for the system.

All maintenance documentation reflects the overall maintenance strategy and repair policies established for the system. A comprehensive documentation package provides clear-cut direction leading from failure detection and fault isolation to the actual repair task. Also, maintenance procedures must be clear and presented in a format which facilitates understanding and updating.

Typically, four types of information represent the minimum package for assuring that the manufacturing system is successfully operated and maintained. It is critical that these documents remain dynamic, clearly written, current, and up to date. The four information types include:

1. **Functional description and operating instructions for each major system component**: This includes:
   - A description of system capabilities and limitations
   - A technical description of the system operation
   - Step-by-step operating procedures
   - Confidence checks to verify satisfactory system performance
2. **Equipment and installation description**: This includes:
   - Flow diagrams
   - Schematics
   - Parts data in sufficient detail to permit reordering or fabrication
   - Instructions for installing and checking out installed or retrofitted components
3. **Maintenance aids (for troubleshooting)**: This includes:
   - Methods for system-level fault isolation when the system is "up" but operating in a degraded mode (use and interpretation of system readiness test results)

- Methods of systems level fault isolation when the system is totally down (use and interpretation of fault isolation test and monitor console displays)
- Procedures for functional equipment-level fault isolation (based on fault sensing indicators supplemented as required by test point measurements using built-in test equipment)
- Equipment-level isolation techniques to permit identification of the problem area to a single module or replaceable part (e.g., maintenance expert systems)
- Routine test, adjustments. alignment, and other preventive procedures performed at periodic intervals
- Reference to FMECA (failure modes, effects, and criticality analysis)

4. **Ready reference documentation**: This information is that routinely required by the maintenance technician. It is easily usable in the work area. It contains only routine checkout, alignment, and PM procedures; fault monitoring interpretation and replacement data; supplemental trouble shooting techniques required to complement the automatic fault detection and isolation system; and item and unit spare parts ordering data keyed to system identity codes.

10.3 MAINTENANCE STRATEGY

10.3.1 RELIABILITY-CENTERED MAINTENANCE

The cornerstone of an effective maintenance plan is the maintenance strategy. The maintenance strategy defines the application PM and CM for each major component, and parts therein, comprising the manufacturing system. By extending the overall strategy to the system's KMI indenture level, it is possible to develop an optimal maintenance plan, which allows us to restore or maintain the operational capability of the system most cost-effectively.

An engineering method for defining the KMI-level maintenance strategy is RCM. The RCM method utilizes a decision logic for systematic analysis of failure mode, rate, and criticality data. This method enables the most effective maintenance requirements for KMIs. Subsequently, the scheduled maintenance burden and support costs are reduced while sustaining system availability.

The RCM method focusses engineering attention to the major system or equipment in a corporate disciplined manner.

Benefits include:

1. Development of high-quality maintenance plans in less time and at lower cost.
2. Availability of a maintenance history for the system and KMIs therein.
3. Assurance that all KMIs and their critical failure modes are considered maintenance requirements.
4. Increased probability of optimal requirements.
5. Online information exchange among engineering and management staff.

The RCM method is initially applied during the design/development phase for the system. It is then reapplied, as appropriate, to sustain an optimal maintenance program based on actual operating experience.

RCM segregates KMIs into two distinct categories: (1) nonsafety-critical components and (2) safety-critical components. For nonsafety-critical components, scheduled maintenance is performed only when the task will reduce the life cycle cost of ownership. For safety-critical components, scheduled maintenance is performed only when the task will prevent a decrease in reliability and/or deterioration of safety to an unacceptable level, or when the task will reduce the life cycle cost of ownership.

The baseline premise of RCM is that reliability is a design characteristic to be realized and preserved during a manufacturing system's operational life. With this, an RCM logistic support program can be developed using a decision logic which focuses on the consequences of failure [3,4]. By applying the RCM method, it becomes easy and straightforward to develop a maintenance program providing the desired, or specified, levels of operational safety and reliability at the lowest possible overall cost.

This cost goal is achieved by targeting maintenance problem areas for the PM program. Fundamentally, PM consists of routine inspections and servicing. It is intended to correct and detect potential failure conditions and make corrections that will prevent major operating difficulties. It is most effective when service requirements are known or failures can be predicted with some accuracy.

PM is desirable when it increases the operating time of an item by reducing the severity and frequency of breakdowns. It typically includes cleaning, lubricating, inspection, calibration, testing, critical part replacement before failure, or complete overhauls.

When system failures occur, they idle men and machines, result in lost production time, delay schedules, and incur costly repairs. This argues that performing CM may not be the best maintenance strategy. On the other hand, PM should only be performed when it provides a cost-benefit over CM. A cost trade-off exists and PM can be carried too far. When immediate repair is not necessary and little harm is done by waiting, CM is advantageous.

Using the RCM method results in defining a cost-effective PM program where:

- Incipient failures are detected and corrected either before they occur or before they develop into major problems.
- The probability of failure is reduced.
- Hidden failures are detected.
- The cost-effectiveness of a maintenance program is improved.

10.3.1.1 The RCM Process

The RCM process aids in determining the specific maintenance tasks to be performed, as well as to influence system design maintainability and reliability. The process is based on a decision logic. Each KMI potential failure mode is evaluated using the decision logic to identify the maintenance tasks to be performed as part of the overall maintenance plan.

The process forces maintenance tasks to be classified into three areas:

1. **Hard-Time Maintenance**: for those failure modes that require scheduled maintenance at predetermined fixed intervals of age or usage.
2. **On-Condition Maintenance**: for those failure modes that require scheduled inspections or tests designed to measure deterioration of an item. Based on an item's deterioration, either CM is performed or the item remains in service.
3. **Condition Monitoring**: for those failure modes that require unscheduled tests or inspection on components where failure can be tolerated during operation of the system or where impending failure can be detected through routine monitoring during normal operations.

The following seven steps comprise the RCM method [5]:

STEP 1: DETERMINE KMIs: This is done by performing either a failure mode, effects, and criticality analysis (FMECA) and/or fault tree analysis (FTA) (as discussed in Chapter 6). These analyses identify KMIs and provide quantitative failure mode data to aid in answering the RCM decision logic questions. This includes determination of the impact of component malfunction, human error, and other potential causes on the system and their priority for improvement.

STEP 2: ACQUIRE FAILURE DATA: Data is compiled for each basic fault identified during Step 1. Part failure rate, operator error, and inspection efficiency data are necessary inputs for determining occurrence probabilities and assessing criticality.

Part failure rate data is typically available in either in-house or industrywide data systems. Operator error is represented by the probability that a failure caused by an operator takes place, whether intentionally or unintentionally. This data is developed through subjective techniques based on discussions with persons familiar with the system operating environment. Inspection efficiency is the probability that a given defect will be detected prior to its leading to a failure. It is treated similar to operator error probability.

STEP 3: DEVELOP CRITICALITY DATA: This involves computing the criticality probability of each potential failure mode. Sensitivity, or conditional probability, is the probability that a defined manufacturing incident will occur given that a basic system fault has occurred. The sensitivities are then used to compute the criticality of each basic fault. Criticality is the measure of the relative seriousness or impact of each fault on the top event. It involves both qualitative engineering (e.g., visual analysis of FTA (fault tree analysis) minimal cut sets) and quantitative analysis (e.g., exercising data through the fault tree). This provides a basis for ranking the faults in their order of severity. Criticality is defined by:

$$Cr = P\{X_i\} \times P\{F \mid X_i\}$$

where: $P\{F|X_i\}$ is the sensitivity

$P\{X_i\}$ is the occurrence probability of the basic fault.

STEP 4: APPLY DECISION LOGIC TO CRITICAL FAILURE MODES:
This involves using the RCM decision logic to define the most effective
PM task combinations for each KMI. As each failure mode is processed,
judgments are made as to the necessity of various maintenance tasks. The
selected tasks, together with the performance intervals deemed appropriate,
define the total scheduled PM program.

The decision logic is designed in two levels. Level 1 (Questions 1–4; see
Figure 10.8) requires evaluation of each failure mode for determination of
the consequence category. Level 2 (Questions 5–26; see Figures 10.9 and
10.10) takes the failure mode(s) into account for selecting the specific type
of task.

Level 1 questions lead to Level 2 questions via the applicable consequence
categories:

1. **Safety, Evident (Questions 5–10)**: PM tasks are required to assure safe
   operation. All questions must be asked. If no effective task(s) results
   from this category analysis, then component redesign is mandatory.
2. **Economic, Operational (Questions 11–15)**: A PM task is desirable if
   its cost is less than the combined cost of the operation loss and the cost
   of repair. Analysis of the failure causes through the logic requires the
   first question (Servicing) to be answered. Either a "Yes" or "No" answer
   of Question 11 still requires movement to the next level. From this point
   on, a "Yes" answer completes the analysis and the resultant task(s) will
   satisfy the requirements. If all answers are "No," then no task has been

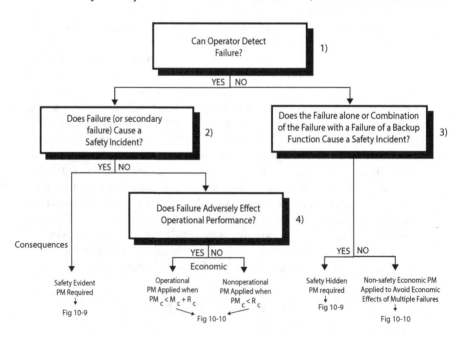

FIGURE 10.8   RCM decision logic level one.

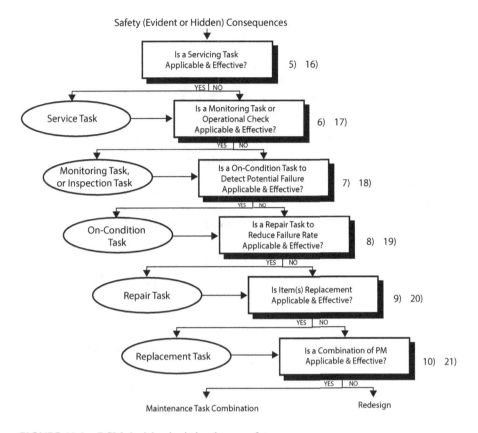

Safety (Evident or Hidden) Consequences

FIGURE 10.9 RCM decision logic level two, safety.

generated and if economic penalties are severe, a component redesign may be desirable.

3. **Economic, Non-operational (Questions 11–15)**: A PM task is desirable if its cost is less than the cost of repair. Analysis of failure causes is the same as Economic, Operational.

4. **Safety, Hidden (Questions 16–21)**: PM tasks are required to assure the availability necessary to avoid the safety effects of multiple failures. All questions must be asked. If there are no tasks found effective, then redesign is mandatory.

5. **Non-safety/Economic (Questions 22–26)**: PM tasks are desirable to assure the availability necessary to avoid the economic effects of multiple failures. Analysis of failure causes is the same as Economic, Operational.

During a decision logic application, at the user's option, advancement to subsequent questions is allowable after a "Yes" answer. But only until the cost of the last task being considered is equal to the cost of the failure prevented.

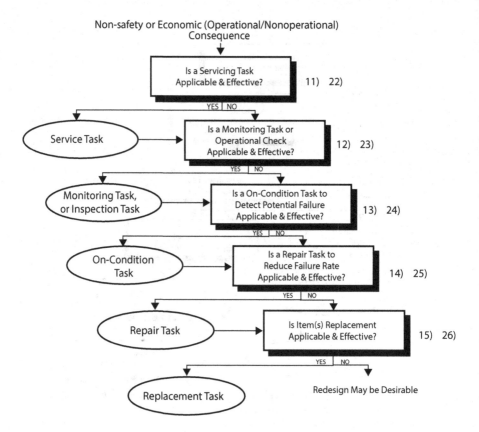

FIGURE 10.10   RCM decision logic level two, nonsafety.

Notice that default logic is reflected in the economic consequence categories by the arrangement of the logic questions. In the absence of adequate information to answer "Yes" or "No" to questions in the second level, default logic dictates that a "No" answer be given and the subsequent question be asked. As "No" answers are generated, the only choice available is the next question, which in most cases provides a more conservative and costly route.

Developing the most effective maintenance task strategy is handled similarly for each consequence category. To determine the most effective maintenance task, it is necessary to process each failure mode through Level 2 of the logic diagram.

For the Economic, Operational, and Nonoperational failure consequence categories, it is necessary to make a cost-effectiveness decision prior to moving into Level 2 of the logic. If the expected cost of system failure per period without PM is greater than the expected cost of system failure with PM, then PM is the best strategy.

The expected cost of system failure per period, if there is no PM, is the cost of system failure divided by the expected number of periods between system failures [6].

$$TC = C_b/E(n)$$

where

$C_b = Nc_b$ is the total cost of system failure

N is the total number of items in a group

c_b is the cost of a single failure

$E(n) = \Sigma n P_n$ is the expected number of periods between failures

$1/E(n)$ is the expected number of failures per period

n is the time period

P_n is the probability of a breakdown in period n

The expected cost of system failure per period with PM includes both the cost of the PM and the cost of those components that fail regardless of PM.

$$TC = \left[C_{pm} + (B_n \times c_b) \right]/n$$

where

$B_n = N(P_1 + P_2 + \ldots + P_n) + B_1 \times P_{n-1} + B_2 \times P_{n-2} + \ldots B_{n-1} \times P_n$

C_{pm} is the cost of PM

n is the number of time periods between PM

B_n is the expected number of failures with PM performed every n time periods

PM is more economical than CM if:

$$\left[C_{pm} + (B_n \times C_b) \right]/n < C_b/E(n)$$

Notice that if standby capacity exists in the system, then a single system failure may not be critical. (This assumes that the manufacturing process can be performed elsewhere in the overall manufacturing system.) Excess capacity favors CM over PM. When system utilization approaches capacity, PM is more desirable.

STEP 5: COMPILE/RECORD MAINTENANCE CLASSIFICATIONS: Applying the decision logic (in Step 4) segregates maintenance requirements into three general classifications [7] to define a maintenance task profile.

- On-condition – scheduled inspection or tests designed to measure deterioration of an item. Based on the deterioration of the item, either CM is performed or the item remains in service.
- Hard-time – scheduled removal at predetermined fixed intervals of age or usage.
- Condition-monitoring – unscheduled tests or inspection on components where failure can be tolerated during operation of the system or where impending failure can be detected through routine monitoring during normal operations.

The maintenance task profile indicates, by part number and failure mode, the PM strategy task selection for RCM logic questions answered YES.

STEP 6: IMPLEMENT RCM DECISIONS: With the maintenance task profile established, the task frequencies/intervals are set and integrated into the overall maintenance program plan. Setting the task frequencies/intervals requires first determining whether applicable data are available which suggest an effective interval for task accomplishment. Prior knowledge is used, as applicable, to define a scheduled maintenance task which will be effective and economically worthwhile.

If there is no prior knowledge, the task frequency/intervals are established initially by experienced engineering and maintenance personnel. Good judgment and actual operating experience are used in concert with accurate reliability data. If failures are adequately modeled by a Poisson process, then the Poisson Distribution can be used to predict the probability of occurrence of failures in a given period of time. With a target reliability and failure rate known, a PM interval can be calculated for each part.

STEP 7: APPLY SUSTAINING ENGINEERING BASED ON ACTUAL EXPERIENCE DATA: The RCM process has a life cycle perspective. The driving force is reduction of the scheduled maintenance burden and support cost while maintaining the necessary manufacturing readiness state. Therefore, it is prudent to review the RCM information, as available maintenance and reliability data move from a predicted state to actual operational experience values.

10.4   MAINTAINABILITY ANALYSIS

Maintainability analysis is performed to determine if a manufacturing system can be maintained in a cost-effective manner which satisfies its operational requirements. The analysis serves four main purposes: (1) to establish design criteria; (2) to allow for design evolution based on trade-off studies; (3) to contribute towards defining maintenance, repair, and servicing policies; and (4) to verify design compliance with the defined maintainability requirements.

Maintainability itself is a measure of the ease and speed with which the system can be restored to operational status following a failure. It is represented by the parameter MTTR which is a measurable and controllable parameter which is specified during design, measured during test, and sustained during operation. The achievement of a high level of maintainability is primarily a function of system design.

An integral part of a maintainability analysis is the performance of a maintainability prediction. Such a prediction supports design/development by deriving repair time, maintenance frequency per operating hour, PM time, and other parameters.

The calculation of MTTR is based on the active repair time associated with the four clock time elements of fault isolation, fault correction, calibration, and checkout (see Figure 10.11). Those time elements related to preparation and delay, while quantifiable, do not provide much insight into the maintainability design of manufacturing equipment.

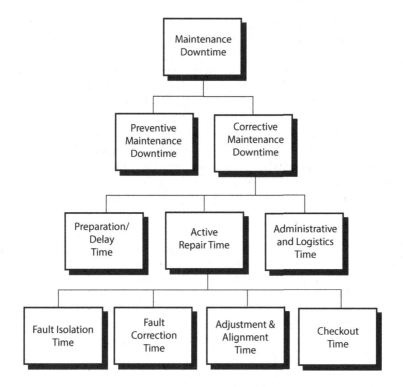

FIGURE 10.11   Elements of time comprising maintenance downtime.

10.4.1 Maintainability Improvement

The maintainability of a manufacturing system (and equipment therein) is highly dependent on its design. Therefore, it is essential that the system be designed for ease of maintenance. This involves considering design features addressing automatic detection, location, and diagnosis of failures, and the incorporation of easily accessible and interchangeable modules and subassemblies.

Additionally, the following design maintainability features are common:

1. Modular design techniques such as uniform size and shape, guide pins and keyed connectors, ease of test/checkout, quick disconnect, and minimum number of functions.
2. Built-in fault detection circuits.
3. Design for replacement at higher levels.
4. Design level fault diagnosis localization ability.
5. Built-in-test (BIT) capabilities.
6. Readily accessible and identifiable test points.
7. High-quality technical manuals and maintenance aids.
8. Limited access barriers to replaceable items.
9. Limited number of interconnections per replaceable item.
10. Use of plug-in elements.
11. Minimal requirements for special maintenance tools.

There also exist maintainability improvement techniques which do not directly impact quantitative parameters such as:

1. **Performance Monitoring**: The identification of performance degradation which may indicate impending system failure (thus preparing maintenance personnel for the type of repair task to be encountered) or identify an existing fault.
2. **Personnel Training**: The level and expertise of maintenance personnel trained for the proper performance of maintenance tasks.
3. **Levels of Support (Organizational/Intermediate/Depot)**: The maintenance philosophy prescribed for repair of a failed hardware.
4. **Spares Support**: The levels of spare support for failed units.

10.4.1.1 Automated Fault Isolation

In many cases, it is advantageous to incorporate some degree of automation in performance monitoring and fault isolation. This involves the integration of BIT features and automated test equipment (ATE) into the manufacturing system. ATE provides the following functions:

- Automatic control of test sequence and the selection of appropriate stimuli;
- Comparison of monitored responses with predetermined standards; and
- Display or recording of test results.

Recognize that automatic fault isolation functions are not a substitute for good maintainability design. Automatic testing is costly, and if it is not properly integrated into the system, it can induce more reliability and maintainability problems than it solves. When properly utilized, however, automatic testing reduces CM time and increases system availability.

In some cases, the automatic test features are used to detect (or predict) impending failure. This permits the correction of system degradation problems as a PM routine, thereby increasing reliability. Automatic fault isolation techniques also reduce both the number of maintenance personnel and the maintenance skill levels required for the system.

Selection of the test features involves consideration of the following requirements and constraints: (1) test function, (2) test modes, (3) level of detection, and (4) degree of fault isolation. System-level design requirements for automatic fault isolation features must be (to the extent possible) specified qualitatively and defined quantitatively in the overall manufacturing system specification. This includes defining the degree of failure detectability, false alarm rates, degree of fault isolation to be provided, fail-safe provisions, and reliability and maintainability of sensors, interface hardware, and ATE.

10.4.2 Maintainability Levels and Prediction

When looking at manufacturing system maintainability, it is advantageous to develop a maintainability functional-level diagram (Figure 10.12). This is done by dividing

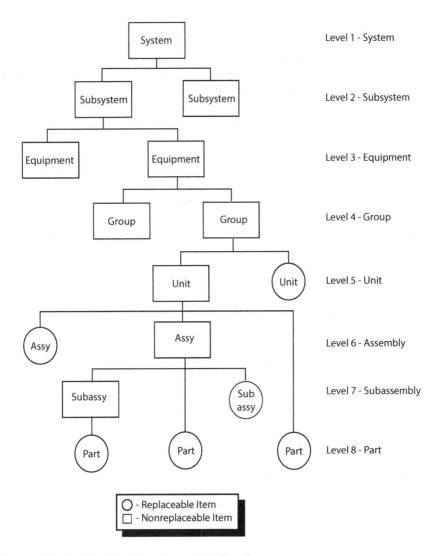

FIGURE 10.12 Functional level maintainability diagram.

the system into its various physical subdivisions beginning with the highest and continuing down to, ultimately, parts that may be replaced, repaired, or adjusted during CM. This facilitates the designation of levels at which repair is to be accomplished (either by modular replacement or by in-place repair), and the level to which fault isolation is to be extended.

As stated earlier, a maintainability prediction provides a quantitative evaluation of the manufacturing system (and equipment therein) in terms of MTTR, as well as other parameters. These parameters indicate the capability of the system to meet all specified quantitative maintainability requirements, including MTTR objectives and allocations.

The prediction method is applicable to any operational environment and type of system (i.e., electronic or mechanical). It enables one to monitor the overall system maintainability throughout design and development. Thus, if it appears that the maintainability requirements are in danger of not being met, system design changes can be made before they become prohibitively expensive.

The prediction method is initially applied by making use of "gross" design data. It is revised as the manufacturing system design evolves. The method is applicable to any system level and is based on the performance of CM.

CM actions consist of the followings: preparation, fault isolation, and fault correction. The latter is further reduced into disassembly, interchange, reassembly, alignment, and checkout. The time to perform each of these tasks is an element of MTTR.

The definitions for these elements are as follows:

Preparation: Time associated with those tasks required to be performed before fault isolation can be executed.

Isolation: Time associated with those tasks required to isolate the fault to the level at which fault correction begins.

Disassembly: Time associated with gaining access to the replaceable item or items identified during the fault isolation process.

Interchange: Time associated with the removal and replacement of a faulty replaceable item or suspected faulty item.

Reassembly: Time associated with closing up the equipment after interchange is performed.

Alignment: Time associated with aligning the system or replaceable item after a fault has been corrected.

Checkout: Time associated with the verification that a fault has been corrected and the system is operational.

The time dependency between the probability of repair and the time allocated for repair typically produces a probability density function in one of three common forms: (1) Normal, (2) Exponential, or (3) Lognormal. The Normal Distribution applies to the relatively straightforward maintenance tasks and repair actions which consistently require a fixed amount of time to complete. Typically, the use of automated fault isolation techniques makes this choice of distributions most appropriate.

Historically, the Lognormal Distribution was used to represent maintenance task performance due to the excessive amount of time required to isolated the problem. However, in recent years, electronic design capabilities have moved the time emphasis to the actual repair task.

There are several quantitative parameters of interest in describing a manufacturing system's maintainability. These are defined in the remainder of this subsection [8].

Mean CM Task Time, MCT: The mean time required to complete an individual maintenance task.

$$MCT = \left(\Sigma M_{cti} \right) / N$$

where
  M_{cti} is the CM time required to complete the ith individual maintenance task.
  N is the number of observations

Mean Time To Repair, MTTR: The mean time required to complete a maintenance action over a given period of time.

$$MTTR = \Sigma\left(\lambda_i \times M_{cti} \right)/\Sigma\lambda_i$$

where
  λ_i is the failure rate of the ith element of the item for which the prediction is being performed

Mean PM Task Time, MPT: The mean system downtime required to perform scheduled PM. This excludes any PM time used during system operation, and administrative and logistic downtime.

$$MPT = \Sigma\left(f_i \times M_{pti} \right)/\Sigma f_i$$

where
  f_i is the frequency of individual (ith) PM action in actions per operating hour adjusted for system duty cycle
  M_{pti} is the average time required for i_{th} PM action

Corrective Downtime Rate, MDT_{ct}, is the CM downtime per hour of operation.

$$MDT_{ct} = \Sigma\left(\lambda_i \times MCT \right)$$

Preventive Downtime Rate, MDT_{pt}, is the PM downtime per hour of operation.

$$MDT_{pt} = \Sigma\left(\lambda_i \times MPT \right)$$

Maintenance Downtime Rate, MDT, is the total downtime for CM and PM combined.

$$MDT = MDT_{ct} + MDT_{pt}$$

As indicated throughout this chapter, a key requirement for manufacturing systems is a high level of operational readiness or availability. "Availability" is the probability of being operationally ready at any point in time. It is similar to reliability except that it takes into account system maintenance.
  Operational availability is derived from:

$$A_o = MTBF/\left(MTBF + MDT \right)$$

The "inherent" availability of the system reflects the characteristic of design without consideration of administrative or logistic time. It is derived from:

$$A_i = MTBF/(MTBF + MCT)$$

or

$$A_i = MTBF/(MTBF + MTTR)$$

TMA Case Study: Technology Transfer with Maintenance Planning

OVERVIEW

A new connector assembly machine is being transferred to an existing manufacturing facility in the Pacific Rim. This technology transfer is a necessary activity to establish productive and efficient logistical pathways to customers. As companies grow and expand, their footprint and operational presence around the globe will necessarily increase to strategically support moving product and manufacturing technology between domestic and international facilities.

ISSUE

The product line has gained widespread customer acceptance and demand growing rapidly. The manufacturing system is struggling to keep up with global growth and is planning strategic locational moves to maintain customer preference. Strategic reasons for this product international technology transfer include:

1. Localize manufacturing in the market region
2. Reduce logistics cost and time
3. Reduce/avoid import tariffs and value-added tax issues
4. Access supply chain partners

STRATEGIC OBJECTIVE

The strategic objective is to put in place manufacturing system contingencies to carry on operations should a situation occur that hurts or impacts production (e.g., fire, political unrest, and governmental retaliatory actions).

CASE BACKGROUND

To better understand the manufacturing system and equipment maintenance support needs, a comprehensive Weibull analysis is to be performed. Weibull analysis application to failure analysis includes:

- Plotting the data
- Interpreting the plot
- Predicting future failures
- Evaluating various plans for corrective actions
- Substantiating engineering changes that correct failure modes.

A significant advantage of Weibull analysis is that it provides a simple graphical solution. The process consists of plotting a curve and analyzing it. The horizontal scale is some measure of life, such as start/stop cycles, operating time, or mission cycles. The vertical scale is the probability of the occurrence of the event. The slope of the line, β is particularly significant and provides a clue to the physics of the failure in question. This type of analysis relating the slope to possible failure modes can be expanded by inspecting libraries of past Weibull curves, which are maintained in the organization's big data system holding all manufacturing system failure recording, analysis, corrective action (FRACA) data.

Another advantage of Weibull analysis is that it may be useful even with inadequacies in the data. For example, the technique works with small samples, which is the case with the connector manufacturing equipment.

Most applications of Weibull analysis are based on a single failure class or mode from a single part or component. An ideal application would consist of a sample of 20–30 failures. Except for material characterization laboratory tests, ideal data are rare; usually the analysis is started with a few failures embedded in a large number of successful, nonfailed, or censored units. The connector assembly equipment of interest has seven data points to evaluate via the Weibull analysis.

The first use of the Weibull plot to be developed will be to determine the parameter β, which is known as the slope or shape parameter. Beta (β) determines which member of the family of Weibull failure distributions best fits or describes the data. This provides insight to the life cycle phase that failure mode is occurring in. That is, the analysis will reveal where the failure mode is occurring in the reliability bathtub curve. The Weibull plot reveals failure mode infant mortality with slopes less than one, random with slopes of one, and wear out with slopes greater than one. The Weibull plot will also be inspected to determine the onset of the failure.

As the concern is with equipment failure occurring during the production time, the responsible project engineer is interested in predicting the number of failures that might be expected into the future (e.g., over the next 3, 6 months, a year, or 2 years). This process is anticipated to provide information on whether or not the failure mode applies to all similar manufacturing system equipment or to only one specific equipment. After the responsible engineer develops alternative plans for corrective action, including production rates and retrofit dates, the risk predictions will he repeated. The engineering manager has indicated that she will require these risk predictions in order to select the best course of action for effective technology transfer.

Going forward in support of global technology transfer, Weibull analysis will routinely be used to evaluate engineering changes for corrective action as to their effect on the same and similar system equipment. The implemented maintenance schedules will continually be evaluated using the Weibull analysis. It has been decided that for all future failure mode assessments, a baseline Weibull analysis will be conducted prior to the engineering change or maintenance change. The study will then be repeated with the estimated effect of the change modifying the Weibull curve. The difference in the two risk predictions represents the net effect of the change, whether design related or maintenance related. The risk parameters will be the predicted number of failures, LCC, maintenance downtime, spare parts usage, hazard rate, or equipment availability.

TABLE 10.1
System Equipment Failure Mode Data

Event #	Failure Time (minutes)	Failure Mode Comment
1	90	Track jam failure
2	96	Track jam failure
3	100	Flipper loose
4	30	Track jam failure
5	49	Flipper loose
6	45	Track jam failure
7	82	Track jam failure

TABLE 10.2
Screened System Failure Mode Data

Rank Order	Event #	Failure Time (minutes)	Failure Mode Comment	Median Rank
1	4	30	Track jam failure	12.9
2	6	45	Track jam failure	31.3
3	7	82	Track jam failure	50.0
4	1	90	Track jam failure	68.6
5	2	96	Track jam failure	87.0

RESULT/CONCLUSION

The system equipment data of interest for the Weibull analysis was focused on a single failure mode associated with a connector housing jam at the track flipper mechanism. Table 10.1 provides the data collected.

In looking at the data it was determined that events three and five are not representative of the failure mode of interest. In order to plot the data, the rank order and median rank is needed [9]. Failure event median rank simply identifies the percentage of the total population failing before it. The updated data are provided in Table 10.2.

The Weibull analysis revealed a beta value, β of 2.1 and a characteristic life of 81 minutes for this failure mode.

Case Study Question: *What actions are necessary to justify the technology transfer as exhibiting an acceptable risk for successful system equipment operation? What is more important for this system equipment; reliability or availability? Why?*

10.5 SUMMARY

This chapter addresses the importance of maintenance to the manufacturing system. The whole push for TMA is dependent on the ability of the manufacturing system to be up and running in a mode which makes products adhering to the customers' performance requirements. The objective is to maximize system "availability." That is, make sure that the system's uptime is much greater than its downtime. Maintenance plays a key role in maximizing manufacturing system availability.

One of the first things that must be done in regard to maintenance is development of a maintenance program plan. There are two fundamental types of maintenance performed: (1) PM and (2) CM. The program plan defines the maintenance strategy which takes into account PM and CM activities. The main difference between the two is that PM is scheduled prior to its performance; whereas, CM cannot be scheduled ahead of time.

The reliability-centered maintenance (RCM) approach to defining specific PM tasks for KMIs enables establishment of the most cost-effective PM program. The RCM approach uses a decision logic to lead the user to optimal PM task for the KMI. All tasks are classified as hard-time, on-condition monitoring, or condition monitoring.

To quantitatively assess the system design and operational performance level, maintainability formulae are used. These formulae are used to assess competing design trade-offs intended to enhance the ability to perform maintenance, as well as to track improvements in the overall maintenance program.

QUESTIONS

1. Explain the maintenance life cycle curve. What role does the time between maintenance play with accounting?
2. How does maintenance fit in a discussion about the reliability "bathtub" curve?
3. What is the primary difference between PM and CM?
4. What does PM do in relation to the reliability "bathtub" curve?
5. Discuss the probability of "perfect maintenance" occurring.
6. What are the steps of the maintenance program planning process?
7. What is RCM? What is its objective?
8. Define the three classifications of maintenance tasks?
9. Exercise the RCM decision logic for a helicopter transmission.
10. What is the inherent availability for product consisting of three parts each having a random failure rate of 10 failures/million hours and a repair time of 30 minutes?
11. What are the components of MTTR?
12. Discuss the advantages of automated monitoring capability.
13. Discuss the trade-offs that are present in the "availability" equation.

REFERENCES

1. Pieruschka, E. (1963). *Principles of Reliability*. Englewood Cliffs, NJ: Prentice-Hall.
2. Kelly, A. (1987). *Maintenance Planning & Control*. London: Butterworths.
3. U.S. Department of Defense. (1985). MIL-STD-1843. *Reliability-Centered Maintenance for Aircraft, Engines, and Equipment*. Philadelphia, PA: Military and Government Specs & Standards (Naval Publications and Form Center) (NPFC).

4. U.S. Department of Defense. (1993). MIL-STD-1388-1. *Logistic Support Analysis.* Philadelphia, PA: Military and Government Specs & Standards (Naval Publications and Form Center) (NPFC).

5. Brauer, D. and Brauer, G. (1987). "Reliability-Centered Maintenance." *IEEE Transactions on Reliability,* Vol. R-36, pp. 17–24.

6. Tersine, R. (1981). *Production/Operations Management: Concepts, Structure, and Analysis.* New York: North Holland.

7. Brauer, D. (1988). NTIAC-85-1. *Depot Maintenance Handbook.* San Antonio, TX: Southwest Research Institute.

8. U.S. Naval Air Systems Command. (1977). NAVAIR 01-1A-33. *Maintainability Engineering Handbook.* Philadelphia, PA: Military and Government Specs & Standards (Naval Publications and Form Center) (NPFC).

9. Abernerthy, R., Breneman, J., Medlin, C. & Reinman, G. (1983). *Weibull Analysis Handbook.* AFWAL-TR-83-2079. Wright Patterson AFB, OH: USA Air Force Systems Command.

Section IV

System Improvement Monitoring

THE GOAL OF YESTERDAY WILL BE THE STARTING POINT OF TOMORROW.

—CARLYLE

DOI: 10.1201/9781003208051-14

11 Big Data System Planning

Assessing one's position in regard to attaining and maintaining Total Manufacturing Assurance (TMA) requires the establishment of a "useful" data system. That is, a data system for compiling pertinent management and manufacturing system and product reliability, safety, and quality performance. It is this type of data that significantly enhances a company's ability to make strategically sound manufacturing management and engineering decisions.

Obviously, it is necessary to perform detailed planning in defining and structuring a data system; particularly, in regard to the types and amount of data. Ideally, the organization wants to collect as much information that makes sense (i.e., data that can and will be used intelligently). Not too much data, but not too little. The breadth of the data system is a function of the costs and difficulty of gathering data, as well as the costs associated with not collecting specific data.

Prior to actually implementing a TMA-focused data system, a few items need defining. These include (1) its purpose, scope, and usage; (2) its structure in terms of user responsiveness both in the near term and upon future evolution; (3) the fundamental steps involved in its operation, as well as input sources; and (4) the various outputs and benefits to be derived. With this accomplished, it is then possible to establish a working, real-time accessible data system.

This chapter provides an overview of the fundamentals of a TMA-focused data system. Emphasized is the data system's role as an integral part of product engineering, manufacturing, and service activities. Obviously, overall management data needs can be analogously addressed.

11.1 DATA SYSTEM ORGANIZATION

Throughout this textbook the availability and importance of data collection in assessing and monitoring organizational performance, including the performance and effectiveness of product development and manufacturing projects, involves gathering and organizing data and information. This data comes from a wide range of tests, experiments, and inspections being performed as an integral part of the effort. Also, supplementary (i.e., generic) information deemed applicable to the case at hand may prove useful.

In general, much data is routinely generated through the evaluation and engineering of a specific product, as well as subsequent related product development efforts. However, unless a means exists whereby the data can be captured, its usefulness is limited and valuable lessons learned are lost over time. It is in this light that the need for, and benefits of, a comprehensive data system is most easily recognized.

A good data system facilitates the early definition of essential parameter values (for design, manufacture, and support purposes) by its experience-based content. The logical organization associated with a properly planned data system encourages a

DOI: 10.1201/9781003208051-15

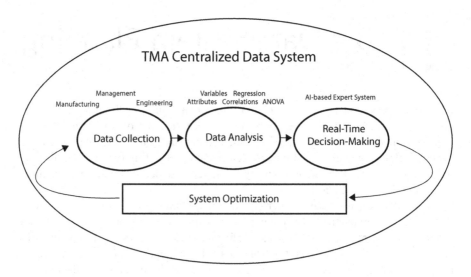

FIGURE 11.1   TMA data system.

rapid and orderly design evolution towards a product meeting its reliability, safety, and quality assurance (RSQA) requirements.

A TMA-focused data system (conceptualized in Figure 11.1) provides a means whereby essential cause and effect relationships can be determined. It represents a primary experience pool for planning new development projects and for improving existing manufacturing systems and products with respect to RSQA.

To support the data system, a data recording and feedback program (to be discussed in Chapter 12) must be in place. This program is built around a formalized communication link between engineering, manufacturing, and product service. This enables the efficient transmission of data and information inputs, as well as requests and inquiries. The key to making this work effectively is to remove inhibiting departmental walls and communication barriers.

In conjunction with the data recording and feedback program, a data system typically performs the following functions:

1. Provides data resulting from ongoing activities for analyzing, evaluating, and assessing product manufacturability, performance, effectiveness, operational suitability, and logistic support requirements.
2. Provides historical data that are applicable to the design and development of new products, processes, and experimental routines based on similar function, structure, or operational horizon.

Realization of both functions requires a capability that is responsive in a clear and timely manner (i.e., it must satisfy management's need for current assessment information). Also, it must incorporate the mechanics necessary for generalized data collection, storage, organization, processing, purging, and retrieval [1].

For each project undertaken, the data recording and feedback program identifies the specific data elements which are required. In addition to the types of data, the appropriate data recording format and media is determined. Early consideration of these factors not only helps to enhance the basic data system architecture, but they also establish the system's reporting burden and characteristics.

Obviously, a primary focus of a TMA data system is on those data elements necessary to evaluate and verify the functional and operational characteristics of specific products and manufacturing systems and processes. But, don't forget key data pertaining to the day-to-day operational management activities. The number and scope of potential TMA data elements can rapidly become overwhelming. This is particularly true with the advent of automated documentation systems.

Table 11.1 provides a list of sample TMA data elements (not exhaustive). Obviously, the scope of a company's data system should reflect its specific needs. This includes a robust range of information and data to ensure effective and continuous movement towards TMA.

As much of the basic data addressed in Table 11.1 is, undoubtedly, routinely generated, it is just as routinely not collected (due to either lack of initiative or management support) and, consequently, is lost as a resource. Having a working data system eliminates this problem by providing relevant data for current and future activities (e.g., product development). It does this by providing insight and answers to questions which have been addressed previously, thereby eliminating the duplication of effort (e.g., laboratory environmental testing). This inherent characteristic enables a data system to acquire the confidence of all parties actively involved with the product throughout its life cycle.

11.2 DATA SYSTEM STRUCTURE

A centralized, comprehensive data system to collect and organize information such as inspection, test, failure, and repair data (and other related data) is a key. A data system provides maximum insight into design and operational deficiencies and enables tracking and measuring the effectiveness of the various management and engineering activities. For example, activities addressing design approaches, material and configuration alternatives, and improvements.

Perhaps the most difficult aspect of establishing a data system is implementing the mechanism (i.e., the data recording and feedback program) whereby the data and information can be captured; *not* generating data and information. For example, a product development and manufacturing effort is actually developing a technology base on relevant materials, system and component designs, performance characteristics, operational factors, and manufacturing techniques. A working data system formally classifies and consolidates this engineering knowledge (including lessons learned) and then augments it with other similar product reliability, quality, maintenance, and safety information. This results in a significantly enhanced knowledge base which can materially reduce the time and cost of developing, manufacturing, and validating future products.

Consider the case where a company is involved in developing complex medical products where patient safety is of primary concern, a key element of the data system

TABLE 11.1
TMA Data Elements

Management

- Operational requirements and performance measures
- System utilization (modes of operation and operating hours)
- System costs and effectiveness
- Operational availability, dependability, reliability, maintainability, safety, and quality
- Maintenance level and location
- Work-in-process tracking
- Payroll processing
- Personnel information
- Scheduling
- Line balancing
- Shop loading
- Equipment requirements
- Maintenance of engineering standards
- Piece rate maintenance
- Integration with standard data systems
- Tracking operator responsibility
- Purchasing orders
- Supplier performance
- Sales literature
- Product data sheets
- Maintenance data sheets
- Parts catalogs
- Product schematics
- Technical documentation
- Product costs

Engineering

- Design reviews
- Value/cost analyses
- Modification orders
- Engineering releases
- Work order requests
- Material data sheets
- Setup sketches
- Operating instructions
- Tooling specifications
- NC operator sheets
- NC tape prove-cuts

Product Reliability

- Quantitative requirements
- Demonstration methods and results
- Program milestones
- Procedures for evaluating and controlling design corrective action
- Reliability contributions to total design
- Supplier responsiveness
- Design review results

(Continued)

TABLE 11.1 (*Continued*)
TMA Data Elements

Product Safety

- Program monitoring
- Review, direction, and close-out of safety reports
- Procedure for initiating design corrective action
- Failure mode and effects analysis results

Quality Assurance

- Quality assessment: destructive or nondestructive tests,
- 100% inspection or statistical methods
- Frequency of inspection and defect levels
- Inspection records
- Metrology techniques and standards used
- Burn-in procedures
- Frequency of calibration of inspection equipment
- Procedures for configuration control

Product/Process Maintenance

- Spare/repair part types and quantities by location
- Supply responsiveness
- Item replacement rates

must be RSQA information. The data system should be structured to highlight the effects of particular design features and interface stresses. This will provide valuable insight for product RSQA enhancement, as well as degradation control.

The real advantage of a TMA dedicated, centralized data system is its ability to provide detailed and specific data for use in various analytical exercises. As a matter of course, the data system catalogs and classifies all data and information in a form that makes it readily identifiable and accessible and amenable to further analysis or evaluation.

Figure 11.2 illustrates a data system's structure consisting of a technical information base (TIB) and a quantitative database (QDB) for each resident data bank.

The TIB portion of the data bank accumulates product research and development reports, test reports, failure analysis and corrective action reports, design specifications, incident and field service reports, and other textual information concerning components and products. Aside from using this information for engineering purposes, it is frequently used for classifying and interpreting quantitative data.

The QDB portion of a data bank reflects relevant engineering parameters. This includes functional performance, test results, and maintenance addressed during product design, through manufacture, and into field operation.

Its access could be a function of automated search and retrieval routines which key on selected descriptors and content identifiers. The selection of meaningful descriptors is essential in the development of a useful data system. Descriptors identify component reliability, quality, maintenance, and/or safety characteristics to enable later correlation between these characteristics and specific component and product design configurations, manufacturing efficiencies, and, ultimately, commercial applications.

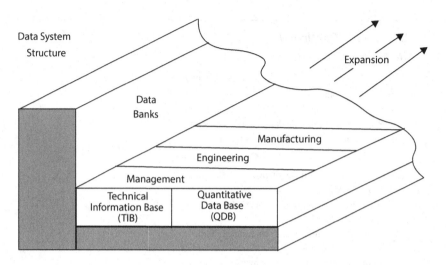

FIGURE 11.2   Data system structure.

An evolutionary approach is typically in establishing data systems [2]. Ultimately, an automated system featuring realtime data entry and user access is desirable in light of Industry 4.0. Initially, a more modest approach may be appropriate where the system relies on manual entry and retrieval of data.

Because of the availability of personal computers and commercial database software, automation is very practical. At first, output data and information might be transmitted via hard copy or electronic reports. However, networking is an economical median and will likely be adopted after the data system becomes fully operational. As the data system grows, a more powerful central distributed processing capability may be practical.

As stated previously, the data system must be organized as a centralized entity providing one point of data management (i.e., submittal and distribution). However, the data system must be readily accessible to all potential users, that is, engineering, manufacturing, and product service. Also, the data must be accurate. Use of the system should be encouraged via special alerts, narrative reports, and other unique outputs. These outputs notify involved parties about specific problems, trends, and corrective actions.

In general, it is wise to publicize all uses of the data. Additionally, in so far as is practical, the collection of data and information should be compatible with other data systems accessible by all intended users, whether industry-wide or internal to themselves.

11.3   DATA SYSTEM OPERATION

The general operational procedure for compiling and reducing applicable engineering, manufacturing, and field (i.e., service) experience data consists of three tasks: (1) data collection, (2) data processing, and (3) data reporting. As part of the data

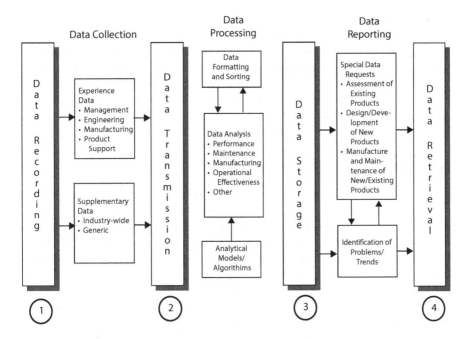

FIGURE 11.3   Data system operation.

system's operational protocol, each task must be flexible to adapt to the specific input material and properly process and output a desired purview of information.

Figure 11.3 depicts the operational cycle of the data system. As part of this cycle, four distinct milestones are defined: (1) data recording, (2) data transmission, (3) data storage, and (4) data retrieval. Initially, all appropriate data is recorded (Milestone 1) in accordance with the overall data recording and feedback program in place. Data and information are recorded via mechanisms designed for efficient completion, yet comprehensive capture of pertinent facts.

A key source of data is a uniform and coordinated mechanism for reporting, analyzing, and implementing corrective actions for manufacturing system and product failures. This includes those failures that occur during laboratory qualification and reliability tests, manufacturing tests, and field operation and repair activities. Obviously, capturing this data requires that failures are accurately reported and thoroughly analyzed (and that corrective actions are taken on timely basis to reduce or prevent recurrence).

Once recorded, all data and information is transmitted (Milestone 2) to the data system. At this point, the material is formatted and sorted into the applicable data bank(s). Once in the data bank(s), it undergoes analysis via appropriate analytical methods which combine the new data with that already resident in the data bank and the overall data system. Upon the completed running of analytical routines, the data is stored (Milestone 3) as part of permanent memory for future reference.

The data and information resident in the data system is retrievable (Milestone 4) on an "as needed" basis (i.e., special data requests). Or, upon identification of problems/ trends, pertinent material can automatically be sent to the appropriate organizations. Special data requests are a primary information transfer catalyst and are typically made in regard to a specific need. Overall, data and information can be retrieved to support topical areas such as reliability, maintainability, and safety analyses; maintenance planning; design improvement; manufacturing planning; and procurement decisions. Many potential applications exist.

11.3.1   DATA COLLECTION

Both product-specific experience and industry-wide generic data on similar parts should be collected. Experience data represents that obtained (e.g., from developmental testing; manufacturing inspections, tests, and screens; field experience; and design analyses). Industry-wide generic data represents that obtained from existing data systems. Initially, a data system can be based on industry-wide data. However, when actual system and product experience data becomes available, it is incorporated into the data system as the primary resource.

The following are examples of hardware-oriented industry-wide data systems [3]. These data systems may provide a good source of initial material for input into the TMA data system, as well as support "similar" system and product analyses. Note that this type of information does not replace the need for acquiring actual product or process experience data.

IEEE Reliability Data Manual (IEEE STD 500-1985): A source of reliability data for electrical, electronic, and mechanical components. The data includes failure modes, failure rate ranges, and environmental factor information on generic components. Reliability data appear in the form of hourly and cyclic failure rates and failure mode information. Low, recommended, high, and maximum failure rates are given for each individual mode.

Government-Industry Data Exchange Program (GIDEP): A cooperative program between Government and Industry for exchanging data to reduce time and cost for researching relevant areas. The program provides a means of exchanging certain types of technical data used in research, design development, prediction, and operation of systems and equipment used mainly in electronic or electromechanical application. Participants in GIDEP are provided with access to several major data interchanges: Engineering Data, Metrology Data, Reliability & Maintainability Data, and Failure Experience Data. The information is applicable for qualitative studies such as failure modes and effects analysis, decision tree analysis, and event tree analysis, as well as for quantitative studies such as reliability prediction, test interval calculation, or spare part studies.

Nonelectronic Parts Reliability Data Notebook (NPRD-3): A publication providing test and reliability data primarily from military and space applications. The information is presented in four sections: (1)

generic-level failure rate data, (2) detailed part failure rate data, (3) part data from commercial applications, and (4) failure modes and mechanisms. The information is suitable for qualitative studies such as failure modes and effects analysis and decision tree analysis, as well as quantitative studies such as a reliability prediction of systems composed of nonelectronic parts.

MIL-HDBK217, "Reliability Prediction of Electronic Equipment,": A compilation of data and failure rate models for nearly all electronic components along with qualifying factors which design or reliability engineers can use to perform reliability analysis on electronic systems in specific applications. It provides a basis for performing, comparing, and evaluating reliability predictions of related or competing designs.

11.3.2 DATA PROCESSING

The objective of data processing is twofold. First, the system does all primary calculating, sorting, formatting, and organizing of the data and information to make it available for use as quickly and accurately as possible. Second, the data banks (and files therein) are defined to aid potential users in readily gleaming out pertinent and essential data.

This can be achieved via three fundamental data processing task elements: (1) preprocessing of data by participating organizations, (2) transfer of data to the centralized data system, and (3) specialized analysis of trends and significant problems highlighted by the data system.

It is common for user descriptors to define with the input and output data. Examples include the following [4]:

1. **Item Descriptors**: A characterization of the components and products for which engineering data is being collected. Unique data elements are necessary for each product type to properly characterize it. For example, the data elements or parameters used to characterize and classify a crankshaft certainly will differ from those for an electronic microcircuit device. Item descriptors should be chosen for their known or postulated influence on reliability, quality, and/or safety.

2. **Application Descriptors**: A characterization of the possible RSQA influences relative to the manner in which the item being tracked is applied in the component and product. They help to classify the nature and level of application stresses that are present during product operation.

3. **Operational Descriptors**: A characterization of the environmental and operational stresses impacting a unit. Since individual product units are likely to be used in different geographic locations under differing load conditions, the unique operational conditions and environments to be experienced are recorded. For controlled tests, the specific test conditions are logged.

4. **Operating Data**: The actual operating record in appropriate units (hours, cycles, actuations) experienced by a unit during test or operation. Typically,

the data is recorded over defined calendar periods or test duration. Where cyclic operation is involved, actual cycling rates and/or onandoff periods are logged as many units are susceptible to transient stresses, thermal expansion rate differences, nonuniform lubrication, and other detrimental influences that are not present during continuous operation.

5. **Discrepancy and Failure Event Data**: The discrepant and/or failure event. The event record should indicate the time of failure (and, if possible, the time to failure), an indication of how the event affected product function, a description of the event, and where possible, the failure mode and/or failure mechanisms. Also, the corrective action taken to alleviate the discrepant event is important.

11.3.3 DATA REPORTING

There are three general ways in which data and information are retrieved from a data system. First is by written or oral request. Second is through periodic standard reports. Third is by direct access (via computer terminal). Also, an internet of things (IoT) retrieval mechanism could be established at some point after the data system has matured.

Once the mechanism for data retrieval is in place, a broad range of analyses are possible using the basic data elements resident in the data system. Analyses could produce information for tracking RSQA status and progress for a specific product development/manufacturing program. Or, general engineering information could be provided for application to future design and manufacturing improvement efforts.

Examples of output information include:

- Identification of data recording periods and all discrepancies with reference to tests
- Failure rates (product, component, and part) and modes
- Reject rates by inspection and screen tests
- Inspection, test, and screen effectiveness factors
- Responsible failure mechanisms
- Recommended or accomplished corrective actions
- Failure analysis reports of hardware discrepancies, including accumulated operating hours to time of failure, failure modes, and cause and type of failure modes
- Cumulative plots of failure events versus time
- General reliability analysis that correlates design analyses with test results and field experience

11.4 DATA SYSTEM IMPLEMENTATION

The TMA data system will undoubtedly be globally accessible for ease of use and efficient data management. This provides an effortless method for inputting and extracting data and information to assist in evaluating product performance throughout its life cycle. This also enables design, manufacturing, and industrial engineers; management; and other potential users to readily audit manufacturing system and product reliability, safety, and quality.

The availability of digital data collection, simulation software, and cloud-based archival systems provide unique features for data collection and use, such as:

1. Uniformity and standardization
2. Increased prediction/analysis speed
3. Reduced prediction/analysis costs
4. Automated compilation of output data
5. Compatibility with other corporate data systems.

An IoT-based, real-time system enables management and engineers to develop an acute awareness of reliability, safety, and quality issues as they impact product and manufacturing improvement efforts. With a strong hold on these issues, the ability to attain TMA is enhanced through optimization of management and engineering activities.

In addition, a well-structured TMA data system supported by failure analysis and corrective action provides the means to maintain visibility over the effectiveness of the overall engineering and manufacturing programs. Product or system adjustments can then be made as necessary to minimize cost and maximize effectiveness.

As the data system evolves, other useful input data and information, aside from RSQA will be identified. Likewise, more sophisticated statistical analysis methods can be incorporated, enabling a broader range of useful output intelligence to be produced in response to the changing needs resulting from technological advances and increased user awareness. In essence, the data system remains dynamic to accommodate advances in technology and processing capabilities.

TMA Case Study: International Supplier Visit

OVERVIEW

Big data collection and analysis encompasses company's "soft" data, as well as "hard" data. The "hard" data includes all the automated data collection capabilities that is possible with the digital age. This includes all Industry 4.0 technologies that connect "operations" to "analytics" to facilitate real-time advanced decision-making using "big data," which simply means accelerated system optimization to meet productivity and profitability requirements. The global institutionalization of the internet and computing power has led to routine applications of artificial intelligence, virtual/augmented reality, machine learning, etc., to manufacturing optimization.

However, there remains a need for "soft" data. "Soft" data are those pieces of qualitative and quantitative data that typically come from face-to-face contact. While the accelerated use of virtual/online technologies and products has re-defined meeting and travel routines globally, there will always be a need for direct human contact and the recording and retention of key communications and interfaces.

ISSUE

The MTI company is evolving its global strategic plan regarding manufacturing locations and market accessibility. This has brought profitability objectives to forefront regarding ongoing supplier performance evaluation regarding cost, capabilities, and long-term partnering rationalization. A decision to expand facilities or to contract manufacturing in India is needed as part of the development of the MTI global strategic plan.

STRATEGIC OBJECTIVE

The supplier base in India is being evaluated regarding: (1) manufacturing capabilities to meet quality mandates to satisfy global consumers and (2) breadth of historical products and volume. The strategic objective is to determine long-term partnering priority.

CASE BACKGROUND

The supplier data in Table 11.2 provides a summary of current Indian supplier activity for a variety of products. An MTI executive traveled to India to meet with the current supplier base to assess the relationship from a face-to-face perspective. This was deemed necessary to make a very personal decision regarding investing time and resources over the long term to support global expansion. Various topics come into play, such as manufacturing capabilities; ability to expand to meet increased order demand; workforce training initiatives; protection of proprietary product and business information, patents, and distribution channels; and relationship trust.

TABLE 11.2
Supplier Information

Vendor Part				Invoiced Amount	Invoiced Quantity (EA)	Avg. Invoice Unit Price (EA)
1	High-M	A1	INSTRUCTION SHEET	$7,486	155,100.000	$0.05
1	High-M	A2	INSTRUCTION BULLETIN	$2,472	144,400.000	$0.02
1	High-M	A3	INSTRUCTION SHEET	$1,503	88,100.000	$0.02
1	High-M	A4	INSTRUCTION SHEET	$155	9,000.000	$0.02
1	High-M	A5	INSTRUCTION SHEET	$88	5,000.000	$0.02
1	High-M	**Result**		**$11,704**	**401,600.000**	**$0.03**
2	Jup T & M	B1	FEED THRU PLATE	$188,081	15,500.000	$12.13
2	Jup T & M	B2	FEED THRU PLATE	$4,361	215.000	$20.28
2	Jup T & M	B3	*** HANGER	$1,393	6,000.000	$0.23
2	Jup T & M	B4	WAVEGUIDE HANGER SUPPORT	$432	5,000.000	$0.09
2	Jup T & M	**Result**		**$194,267**	**26,715.000**	**$7.27**

(Continued)

TABLE 11.2 (*Continued*)
Supplier Information

	Vendor Part			Invoiced Amount	Invoiced Quantity (EA)	Avg. Invoice Unit Price (EA)
3	S-Tech Industrial	C1	*** TAPE 24" LENGTH	$188,686	183,000.000	$1.03
3	S-Tech Industrial	C2	TAPE 24" LENGTH	$139,913	136,000.000	$1.03
3	S-Tech Industrial	C3	CUSHION	$28,247	40,550.000	$0.70
3	S-Tech Industrial	C4	CUSHION	$11,892	16,150.000	$0.74
3	S-Tech Industrial	C5	CUSHION	$1,313	1,850.000	$0.71
3	S-Tech Industrial	C6	CUSHION	$1,174	1,650.000	$0.71
3	S-Tech Industrial	C7	PLUG FOR ENTRY BOOT	$137	1,700.000	$0.08
3	S-Tech Industrial	C8	CUSHION	$92	100.000	$0.92
3	S-Tech Industrial	C9	PLUG FOR ENTRY BOOT	$23	700.000	$0.03
3	S-Tech Industrial	**Result**		**$371,477**	**381,700.000**	**$0.97**
4	S-Industrial	D1	HARDWARE KIT	$64,744	114,000.000	$0.57
4	S-Industrial	D2	STUD, THREADED, 3/8-	$47,910	135,100.000	$0.35
4	S-Industrial	D3	HARDWARE KIT	$34,615	13,613.000	$2.54
4	S-Industrial	D4	COMPACT ANGLE ADAPTOR	$28,371	12,900.000	$2.20
4	S-Industrial	D5	STUD ADAPTOR	$24,678	136,000.000	$0.18
4	S-Industrial	D6	1.5" SST PHILLIP OVAL HEAD	$16,706	121,000.000	$0.14
4	S-Industrial	D7	STUD, THREADED, 3/8-	$11,870	48,300.000	$0.25
4	S-Industrial	D8	HARDWARE KIT	$5,532	27,226.000	$0.20
4	S-Industrial	D9	STUD, THREADED, 3/8-	$4,212	18,726.000	$0.22
4	S-Industrial	D10	SCR, HCS, 5/16-18, 1.25, SST, PASS	$3,904	38,300.000	$0.10
4	S-Industrial	D11	#14 SST FINISHING WASHERS	$3,899	121,000.000	$0.03
4	S-Industrial	D12	STUD, THREADED, 3/8-	$1,793	3,036.000	$0.59
4	S-Industrial	D13	#14 PLASTIC SCREW ANCHORS	$1,115	121,000.000	$0.01
4	S-Industrial	D14	WSHR, FLT, 3/8, SST, PASS, 0.375×1	$479	24,000.000	$0.02
4	S-Industrial	D15	HARDWARE KIT	$346	90.000	$3.85

(Continued)

TABLE 11.2 (*Continued*)
Supplier Information

Vendor Part			Invoiced Amount	Invoiced Quantity (EA)	Avg. Invoice Unit Price (EA)
4 S-Industrial	D16	STUD, THREADED, 3/8-	$295	1,000.000	$0.30
4 S-Industrial	D17	HARDWARE KIT	$79	180.000	$0.44
4 S-Industrial	D18	STUD, THREADED, 3/8-	$5	12.000	$0.40
4 S-Industrial	**Result**		**$250,554**	**935,483.000**	**$0.27**
5 SKY Polymers	E1	CLICK-ON HANGER HALF	$239,283	1,305,360.000	$0.18
5 SKY Polymers	E2	SINGLE CLAMP PLASTIC	$35,487	340,000.000	$0.10
5 SKY Polymers	E3	CLICK-ON HANGER HALF	$17,934	61,200.000	$0.29
5 SKY Polymers	E4	CLICK-ON HANGER HALF	$13,581	45,300.000	$0.30
5 SKY Polymers	**Result**		**$306,285**	**1,751,860.000**	**$0.17**
6 Polymer Wires Ltd.	F1	16 MM WIRE, 7 STRAND, BLACK	$35,203	166,594 FT	$0.21/FT
6 Polymer Wires Ltd.	F2	GREEN 16MM GROUNDING WIRE	$2,964	4,000 M	$0.74/M
6 Polymer Wires Ltd.	**Result**		**$38,167**	*	*
7 B-Stat Industrial	G1	PLATE CRIMP LUG	$9,051	68,300.000	$0.13
7 B-Stat Industrial	**Result**		**$9,051**	**68,300.000**	**$0.13**
8 Panad Packaging	H1	CARTON	$18,579	16,380.000	$1.13
8 Panad Packaging	H2	PACKING BOX	$3,483	24,960.000	$0.14
8 Panad Packaging	**Result**		**$22,062**	**41,340.000**	**$0.53**

*wire in feet and meters

RESULT/CONCLUSION

The following notes were submitted to the MTI data system upon returning from India:

Realities:
1. Local suppliers of hardware, adapters, and grounding assemblies are abundant
2. MTI products intellectual property is openly copied
3. MTI overheads will always be greater than these local suppliers

4. The local suppliers want desperately to be a supplier to MTI so that they can sell that they sell to MTI
5. MTI competitive advantages include ability to pay suppliers, logistics capability, design leadership, and global reputation
6. There is a broad spectrum of manufacturing capabilities...MTI will never be the lowest cost supplier, but MTI has corporate infrastructure that can be leveraged (e.g., global, product development, and logistical lanes to move product)

Recommendations:
1. Have S-Industrial copy the wire hanger design...look at time to copy, performance characteristics, and implications for MTI's reputation when copy-cat products fail in field...is it worth chasing every low-cost option for this market (that will never be the lowest cost design).
2. Look at "integral" hardware design for customer added value
3. Look at "enviro-coated" hardware at 202 modified and 304
4. Look at potential global supply opportunities for hardware and cushions/boots
5. Bring up double riveter resident at India
6. Bring up stamping die via India (trade off cost of product with China supply)
7. Look at cost difference with between producing grounding assemblies to current design specifications and new grounding assembly design
8. Introduce grounding assembly manufacturing by July 2030
9. Get regional suppliers centered on MTI products
10. Focus on products that have a huge market leverage on...for example, molded components
11. Focus on selling MTI without sacrificing the company's reputation...sell the brand

Case Study Question: *What opportunities exist for supplier consolidation and/or increasing supplier diversity? How does purchasing magnitude leverage buying power? What are the trade-off considerations for global logistical supply substitution?*

11.5 SUMMARY

This chapter addresses the establishment of a TMA-focused data system. The objective of such a data system is to provide a centralized depository for pertinent management, product, and manufacturing system data and information that is routinely generated. Availability of data and information reports enable smarter management decisions based on specific status and trends and alerts concerning major problems. One of the biggest benefits comes from the ability to capture lessons learned for later reference.

The data system needs to be structured to easily capture key data and information for a vast array of topical areas including reliability, safety, and quality. Typically, two fundamental database categories are included in a comprehensive system. These are the TIB and the QDB.

There are three key tasks in the operation of a data system: (1) data collection, (2) data processing, and (3) data reporting. These tasks enable the data and information to be recorded, transmitted, stored, and ultimately retrieved upon demand.

Commercially available data systems exist for all levels of computer technology and IoT global implementation. The data system is dynamic in its sophistication and contents. As the system matures, the scope of data captured should increase as makes sense. Also, the technology used to support the systems operational requirements should continually keep pace with user demands for simplicity in data input and retrieval.

QUESTIONS

1. Discuss the dangers of collecting all the data possible.
2. Discuss the need to maintain a centralized data system.
3. What type of program is necessary to support the data system?
4. Review Table 11.1: What additional data might a corporation's president require?
5. What are the two categories of databases?
6. Identify and discuss various factors that would impact data system evolution.
7. What are the three major tasks involved in a data system operational cycle? What are the major milestones?
8. What are the purposes of generic data systems?
9. What are user descriptors? How do they benefit data system users?
10. How does the data system interface with failure analysis and corrective action?

REFERENCES

1. Kelly, A. (1987). *Maintenance Planning & Control*. London: Butterworths.
2. Cosenza, R. & Duane, D. (1988). *Business Research for Decision Making*. Boston: PWS-Kent Publishing.
3. Brauer, D. & Bass, S. (1986). *Reliability Analysis of Gas-Engine Heat Pump Systems: Methodology, Model, and Data System*. Chicago, IL: Gas Research Institute.
4. Cleveland, E. & Duphily, R. (1978). *Power Plant Data Systems*. Palo Alto, CA: Electric Power Research Institute.

12 Data Recording and Feedback

The key to manufacturing control, as well as organizational control, resides in the ability to know what is happening. This is why a Total Manufacturing Assurance (TMA)-focused big data system is established, as discussed in Chapter 11. To actually collect data which gives us knowledge about the overall business operation requires a data recording and feedback system (DRFS). Ideally, this is a "real-time" activity enabling immediate and knowledgeable management, engineering, and operator responses, to immediate situations.

This real-time scenario eliminates the poor practice of addressing problems a week or two later after they occur. It does little good to have a DRFS in place if it is difficult to collect, digest, interpret, and respond to manufacturing and product information; assuming that such a system is in place.

The need for a DRFS is inevitable and necessary. The vast amounts of information generated by a company on a daily basis provide valuable insight and guidance for attaining TMA. The key is to maintain an efficient corporate database, that is, collect only information that is "useful." This usefulness may be either a short-term or long-term basis.

However, it must be recognized that it is impossible to collect all the information that is generated daily. Nor should one try. The focus must be on database efficiency. This does not imply passing over information to limit database size. It is always easier to reduce the amount of data collected and maintained than it is to add collection and maintenance requirements down the road.

A DRFS typically takes on one of two main orientation options. The first option dictates that information (e.g., a document) is not replaced, but only managed. This is analogous to a card catalog system. Important information is collected and maintained but not every possible piece of information.

The second option dictates that the system contain as much information as possible. However, this requires having a comprehensive management practice and capabilities as the information and the DRFS become highly complex and difficult to use.

Regardless of which orientation the system assumes, it is necessary to efficiently record and ultimately return the data back to the potential user. The best way to do this is to implement a real-time, automated data collection system. Many commercially available systems of this type are now commonly used. Such systems are commonly referred to as "factory information systems or enterprise planning systems."

Factory information systems enable real-time data collection and processing, which obviously provides great benefits in the manufacturing environment. In addition, applications of real-time DRFS are commonly extended to overall company management, product development, and product deployment, as well as other activities.

DOI: 10.1201/9781003208051-16

A key benefit derived from factory information systems is increased management awareness, engineering efficiency, and operator productivity. This is achieved via a reduction in strain and frustration.

Chapter 11 addressed the fundamentals of data system planning. This chapter focusses on effectively capturing data and using it to continuously improve the manufacturing system and its products, and in general to improve organizational health. An overview is provided for several data recording media. Also, special attention is given to need for a failure recording, analysis, and corrective action (FRACA) process as a means for feeding product information back into the continuous manufacturing system improvement initiative.

12.1 ORGANIZATIONAL DATA COMMUNICATIONS

Perhaps the most important aspect of establishing a DRFS is identifying the type of data that is worth collecting and desirable to review. Information is generated in four primary areas: (1) product, (2) manufacturing system, (3) company business structure, and (4) product operational environment.

As stated earlier, the potential amounts of acquired information are enormous. It is probably fair to say that data collection is one of the most time-consuming functions within a corporation. However, a critical element of company success is the ability to acquire accurate and timely data and ensure its feedback to all concerned business areas as soon as possible to enhance decision-making.

Historically, DRFSs had one thing in common, that is, paper. Information for the most part was recorded and fed back via paper. Getting rid of all the reams of paper eliminated a major bottleneck in using manufacturing and product information. Today, the availability of data digital collection, storage, and transmission facilitates data effective uses.

The next logical step is to embrace the concept of the "paperless" business operating environment. The technology exists to move in this direction and the money saved in manual labor costs alone would no doubt justify the hardware costs for the change.

This technology is commonly referred to as electronic shop documentation (ESD). Basically, computers are used to replace paper documents with electronic ones. A typical ESD system consists of a network of computers, operator terminals, and software. For the most part, any kind of document is adaptable to the electronic format.

Organizational success is facilitated by having a data communications system which simultaneously handles product design, manufacturing, testing, and business area information. With fast and accurate communication between management, engineering, and the shop floor, the ability to attain and maintain TMA is enhanced. For example, flexible manufacturing cells can be quickly reconfigured to manufacture an alternative product and TMA is not lost. Good communication enables this continuity to be possible since TMA lessons learned are recognized and available as a fundamental resource.

Data communications are possible based on three types of DRFSs: (1) manual, (2) semi-automated, and (3) real-time, automated. In the manual system, all information is manually recorded and stored on a paper. This type of system is inefficient since mounds of paper rapidly develop. Such a system is inherently cumbersome and expensive.

The semi-automated system is similar to the manual system except that it takes advantage of computer technology. The computer generates paper documents and can maintain a data system. Computational errors are reduced and the speed and diversity of the reporting mechanism is improved. However, this type of system is not real-time oriented.

The third type of DRFS is the real-time, automated. This system permits information to be collected, verified, and processed as it is generated. This creates direct two-way communications between the computer and the user(s) of the information.

A real-time, fully automated DRFS consists of an engineering and database management software and a hardware network. The system integrates nicely into the corporate routine and facilitates simultaneous engineering.

As discussed previously, simultaneous engineering is a very cost-effective approach to developing a product. Essentially, it works by getting contributions from applicable corporate business areas early during the development cycle. This results in shortening the overall product introduction time and a more robust product design.

With this in mind, databases must be flexible to accommodate all possible data inputs and requests. The database must be designed to accommodate the real world, not just some abstract data. This includes providing a means for moving and manipulating data as it makes sense. Additionally, a database must be accessible to all potential users via an online network capability. Poor accessibility does nobody any good and only forces redundant work. If the same information reports are generated for several departments, the benefits of simultaneous engineering are lost.

Automating the DRFS requires the use of information management programs, search programs, and/or form generation programs. Advances in state-of-the-art workstation technology increase the potential uses of DRFSs. Workstations create forms, call up and run subroutines or outside programs, access other systems, search and find data based on ambiguous instructions.

12.1.1 TYPES OF DATABASES

An important element of an automated DRFS is its supporting database structure for storing information. Numerous database warehousing management systems are commercially available which provide a means for quickly developing custom databases. This enables the development of databases for managing large amounts of distinct information. The key to a successful database is its ability to provide the right information for each unique request.

Two database structure approaches are commonly used. The first approach is the "relational" database. Relational databases treat data as an enormous table where each record is represented as a row and the attributes are seen as columns. This provides two major advantages: (1) it is easy to ask for data sorts and (2) sorts can be made to address several distinct data files. This provides great flexibility in defining data sorts that can identify all information available.

The other common database approach is the "object-oriented." In this type of database, every record is treated as a separate object with a collection of attributes. These objects are automatically incorporated into a larger data assembly.

12.1.1.1   Automatic Identification Technology

Recording information is an integral element of the DRFS. Doing this efficiently is desirable. A principal means for achieving this is by using some form of automatic identification (AID) technology [1].

AID has a broad application horizon. It serves as a supporting technology to computer-integrated manufacturing (CIM) (discussed in Chapter 5). It also is a valuable tool in controlling inventory processes, keeping track of personnel information, and ensuring corporate security. Benefits include lower operating costs through higher productivity, better inventory control, and improved customer service. This technology assures accurate and efficient collection of data in response to corporate needs.

A company's use of AID technology enables it to monitor it itself in a real-time fashion and more closely. For example, the effectiveness of material flow control can be greatly enhanced, which fulfills a baseline requirement of CIM systems.

Tracking ability is the inherent feature of AID technology. Everything can be tracked and monitored, from work in process to finished production units. In some cases, people are tracked for human resource purposes. Obviously, AID technology is applicable to a diverse number of environments within the corporation.

The fact that AID can generally be applied wherever automatic recording of information is appropriate, means greater productivity in performing the information recording activity. Of course, AID has eliminated the need for manual recording system. By eliminating the human element, information accuracy increases since fewer errors will occur during the recording activity.

Many specific technology types comprise the AID family. Each has the goal of rapid and accurate data collection and recording but differs in its approach to capturing and processing data. Bar coding is the most popular form of AID in terms of size and growth rate [2]. However, other technologies exist such as optical character recognition, voice entry, vision, magnetic stripe, and radio frequency (RF). Each provides advantages based on the specific application.

12.1.1.2   Bar Coding

Bar coding refers to the use of a symbol which corresponds to combinations of digits, letters, or other punctuation symbols. It consists of a pattern of narrow and wide bars and spaces to represent information. Each pattern represents a different alphanumeric character. The technology integrates four independent processes: (1) printing, (2) reading, (3) transforming the code to a time phase (which is then decoded), and (4) a coding scheme. There are many commercial systems are available; however, each is typically unique making it difficult to compare system specifications.

The most common bar code systems use scanners which read bar code information by shining a light at the symbol and monitoring the amount of light reflected back. The light source is an incandescent light bulb or a light-emitting diode (infrared or laser). The reflected patterns of light are captured by a photodetector and ultimately converted into in digital signals and transferred to the data system.

A second type of scanner uses a solid-state light-sensitive element, or charge-coupled device. With this, the reflected image of the bar code information passes through an ordinary camera lens and is then projected onto the charge-coupled device.

Bar codes provide a rapid means for recording data. With a single scanner pass, data is read and input to the database at a speed many times faster than that possible via good manual methods. Convenience is also a key benefit of bar codes. They are readable at significant distances, they provide unique sensing capabilities for timing purposes, and they are very flexible from a permanency standpoint in monitoring manufacturing systems.

The potential for data recording errors is significantly reduced with bar codes from both a design reliability and user-friendliness standpoint. They work well with shop personnel who may have a low skill level or understanding of the corporation's manufacturing system.

12.1.1.3   Optical Character Recognition

Optical character recognition (OCR) is similar to bar code technology in that it uses a scanning technique to record data. OCR reads alphanumeric characters, not coded information. The two types of OCR systems used are document carriers and page readers. The latter reads a full-size page, while the first only reads a few lines of information at a time.

The use of this technology is reduced in manufacturing systems where rapid data recording is required and data accuracy is required. Its use is expected to increase with increased use of vision systems.

12.1.1.4   Voice Entry

Voice entry technology is based on pattern recognition as is bar coding. Instead of images, words are pre-programmed vocabulary are stored and recognized. The user speaks into a microphone and the phrase is recognized by the machine. The information is then transformed into electrical signals, which are subsequently sent to the data system.

Depending on the sophistication of the system, the recorded information is interpreted and related information or instructions are relayed back to the user. A key part of the system is for users to teach the system to recognize their voices in the environment in which they will be working. This enables background noise to essentially wash out. The advantage of this technology resides in the user's hands and eyes being free to perform a manufacturing activity.

12.1.1.5   Radio Frequency

Radio frequency (RF) technology is typically used in dirty and harsh manufacturing environments. This includes layout situations where physical obstructions or manufacturing system characteristics deter the use of other data recording media. It frequently replaces bar codes or magnetic stripes where they are rendered useless due to reading difficulties.

The technology uses bi-directional radio signals to transmit data between an RF reader and a transponder located on, in, or near the object to be detected. The transponder operates on a specified frequency and transmits its message when it is addressed by a reader.

RF is commonly used in automation and material handling applications where there is no line-of-sight between scanner an identification tag or where read/write capability is required. A transponder on the object being tracked sends a unique signature or data stream in when addressed by the transmitter/reader. The antenna picks up the signal and the reader decodes and validates the signal for transmission to the computer containing the database. This technology also enables other AID technologies to transmit information to the data system via a remote radio link.

12.1.1.6   Vision

Machine vision is a very complex and consists of a television camera, a monitor, a keyboard, and a processor. A camera picks up the image and sends it to the processor as an analog signal. The processor converts the signal to a digital matrix and compares the picture with information stored in its memory. Upon a match, the image is recognized and an output is generated. This may consist of a series of identification numbers, a numeric representation of the image, a list of flaws, or a go/no go command.

In general, any legible (to a person) characters provided on a label are readable and interpretable to a vision system. Characters or objects are recognizable at all angles, including upside down and different lenses provide changes in the observable depth of field. The speeds at which characters and objects are recognized varies. Generally, character recognition ranges from 30 to 60 items per second. The speed of recognizing objects changes based on the number of features to be recognized or inspected, as well as the environmental conditions.

12.1.1.7   Magnetic Stripe

Magnetic stripe involves encoding information onto a special material. A decoder reads the magnetic stripe and passes the information onto a computer. The advantage of this technology is that large amounts of data are easily stored on a single stripe. Therefore, it is widely used for personnel identification badges and bank credit cards.

The amount of information recordable on magnetic stripes is significantly greater than that possible in bar codes. Magnetic stripes also hold up to wear and tear. Even if a magnetic stripe is mutilated or subjected to heat, dirt, grease, and other factory conditions, it is generally still readable with a high degree of reliability.

12.1.2   Programmable Logic Controller

The previous subsection described numerous neat technologies, which are essentially used in capturing and transferring data a step or two away from a process. They do not directly influence any ongoing processes by making system operational adjustments. To do this, the programmable logic controller (PLC) is used.

PLCs have been around for a long time and are, in recent years, witnessing rapid growth in their extended for monitoring and controlling, on a real-time basis, manufacturing system (and processes supported thereby). They are well known for their efficient and flexible industrial control capabilities, and current trends have them increasingly exploited for their data gathering ability.

State-of-the-art PLCs are built with advanced instruction sets, memory, data processing, and communication capabilities. Consequently, PLCs are used to perform data acquisition and analysis in addition to other traditional control duties.

The PLC technology consists of microprocessor-based central processing unit, user memory, inputs and outputs, and a power supply. They control machines and processes by accepting data from field input devices, using the data to solve user-written control programs and creating outputs. The nature of the PLC's memory enables quick changes to the control program.

The maturity of networking and communication capabilities allows PLCs to interface with host computers, terminals, displays, other PLCs, and other devices such as expert systems [3]. This gives operators, as well as the DRFS, immediate access to vital process information needed for analysis and decision-making.

There are two primary reasons for using PLCs for data acquisition. First, PLCs are widely used in an automation role. The expansion of the PLC's role to include data acquisition, the need to purchase additional, or dedicated data acquisition devices is reduced. Second, PLCs have direct access to process and systems information.

In a data acquisition role, PLCs monitor, collect, and store manufacturing data. This aids in improving the overall productivity and efficiency of computer-controlled manufacturing systems. In addition, material waste is reduced and product quality improves.

An advantageous characteristic of the PLC is its ability to monitor a broad range of manufacturing variables and immediately manipulate the data for analysis. Examples include calculating material misfeed rates, the number of "good" versus "bad" parts over time, and system running efficiency. Plus, this information can be displayed in real time for immediate interpretation by the operator.

An additional benefit of the PLC mathematical capabilities is the need for shop personnel to worry about statistics, etc. The fact that the PLC collects vast amounts of data by itself makes it a significant contributor to increased efficiency and productivity.

12.2   FAILURE RECORDING, ANALYSIS AND CORRECTIVE ACTION

If the DRFS does nothing else, it must provide a mechanism for FRACA [4]. A formal FRACA effort effects the early elimination of failure causes. Consequently, this forces the reliability of manufacturing systems and products to grow. A natural extension from this is the attainment of high levels of operational availability.

Although FRACA was frequently highlighted as an activity employed for hardware items, it need not be limited to that. Non-hardware items such as late shipments and faulty invoices are also serious problems or failures. Keep in mind that wherever a word referring to hardware pops up, it can most likely be exchanged for any product outgoing from a staff function.

To get the full benefit of FRACA, it is implemented early during the design/development phase of the manufacturing system or product. This allows failure causes to be effectively identified at the point where it is most cost-effective to implement corrective action. As designs, documentation, and hardware mature, corrective action is still identified, but its implementation becomes more difficult and costly. This underlines the importance of a having a "closed-loop" FRACA activity in place at the initiation of a project.

The term closed loop refers to an activity which begins with the source of the reported problem and ends with reporting the results of the failure analysis and corrective action back to that originating source. In order to make this closed loop work in a timely manner, FRACA integrated into the existing TMA data system and DRFS.

The overall effectiveness of a FRACA process depends on the accuracy/thoroughness of the input data (i.e., reports documenting failure/malfunctions and failure analysis). The essential data inputs are made by a failure reporting activity. The scope of failure data includes all information pertinent to the failure in order to facilitate determining the failure cause.

A prerequisite of a healthy FRACA process is the use of a standardized failure report. The report itself must be designed to simplify the documentation of failure data and enhance the effectiveness of the FRACA process. This means having a form which is easy to fill out and capture only the essential data.

The FRACA report merely initiates the closed-loop process. As shown in Figure 12.1, the report follows a rigorous path. This includes performing a correlation/trend analysis in the data system, performing a detailed failure analysis, determining a corrective action, receiving concurrence from the failure review board, and finally, implementing the corrective action.

Keep in mind that FRACA is only one piece of the whole DRFS pie, but a very important and significant piece. Also, the data system itself must be designed to support the DRFS. As discussed in previous chapters, the data system needs to be robust in its content, as well as functionally able to perform various statistical, correlation, and trend analyses.

A key objective of the FRACA process is to routinely identify failure trends and correlations as they become evident. If the data system indicates the appearance of a trend, a failure analysis is performed. This serves as a means for determining the failure cause and the appropriate corrective action needed to eliminate or minimize its recurrence.

Failure analysis may be as simple as technical dialogue between design and reliability engineers to identify the failure cause(s). However, in some cases formal, detailed laboratory failure analysis may be required in order to reveal the failure mechanisms and provide a strategy for deriving an effective corrective action. Note that corrective action generally focuses on design changes, process changes, or maintenance changes. These options exist for either manufacturing system problems or product.

The FRACA effort needs to be flexible enough to accommodate all failure occurrences during a project including those that occur during actual operation. Along with this flexibility, a key FRACA output is a summary report which delineates information about failures and evidence of trends, as well as the extent of contemplated corrective action and its estimated impact (cost and effectiveness).

12.2.1 ELEMENTS OF THE FRACA SYSTEM

As stated earlier, FRACA is a closed-loop, real-time activity for identifying failure/problem areas. This naturally leads to the initiation of aggressive follow on activities to effect and verify corrective action. Effectively correcting observed problems

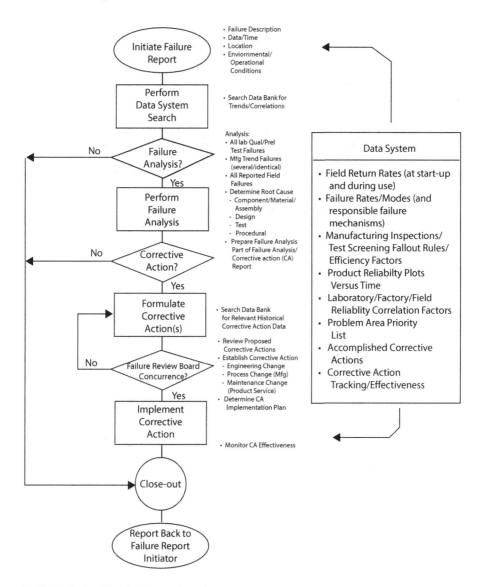

FIGURE 12.1 FRACA interaction with the data system.

requires rigorous and persistent data tracking and coordination of all the necessary actions performed by different organizations and disciplines.

In order to have a cohesive effort, an FRACA plan is developed. The plan identifies all the provisions necessary to assure that proper data analysis is performed and that effective corrective actions are taken on a timely basis. Additionally, the various levels of assembly are identified, definitions for failure cause categories are provided, logistic support requirements are identified, and the FRACA data items required are identified.

12.2.1.1 Responsibility

In most cases, the Reliability Engineering Department (or group performing this function) is responsible for instituting and managing the FRACA activity. It establishes policy, provides direction, and monitors the status of a failure analysis investigations. Specific responsibilities include:

1. Assigning identification numbers to reports received, completing the reports, and determining the need for failure analysis and corrective action.
2. Performing trend/correlation searches in the data system.
3. Conducting failure analyses and corrective action investigations.
4. Monitoring the effectiveness of corrective actions implemented.
5. Informing the appropriate corporate departments, as well as the report originator, of problems closed-out including the results of the investigation and action taken.
6. Maintaining an FRACA experience base as an integral part of the overall data system.

12.2.1.2 Activity Architecture

Figure 12.2 depicts the FRACA activity as a closed-loop operation. This configuration consists of several fundamental steps which must exist in order for the activity

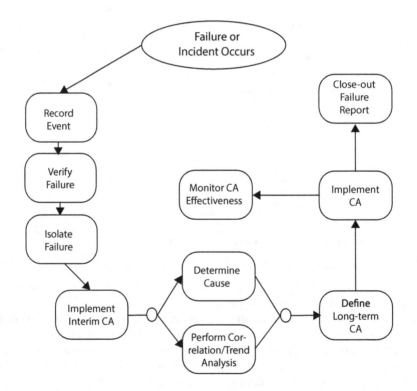

FIGURE 12.2   Closed-loop FRACA activity.

to be complete and effective. The closed-loop activity in its leanest form consists of the following steps:

1. Observe and document the problem
2. Verify that the problem actually occurred
3. Isolate the problem to lowest level possible
4. Implement interim corrective action
   a. Determine problem cause (failure analysis)
   b. Search the data system for trends/correlations
5. Determine and approve the necessary corrective action encompassing design change, process change, or procedure change
6. Implement the long-term corrective action and monitor its effectiveness
7. Close out the problem report
8. Feedback the results of investigation to the originator of problem report

12.2.1.3 Reporting

Obviously, to make this activity work, it is imperative that failure reporting and resulting corrective actions be effectively documented. FRACA forms must be designed to meet the needs of the individual manufacturing system or product, as well as departmental responsibilities, requirements, and constraints.

Key information to capture includes part identification data, conditions under which problem occurred, operating parameters indicating degradation, replacement part(s), repair times, references to applicable plans and procedures, and complete details leading up to or surrounding the incident. Failure analysis information includes a complete description of the failure parts, including manufacturer and lot production code. Once the failure analysis is complete and recorded, the decision is made whether or not to proceed with corrective action. Finally, the problem cause is described in detail, and recommendations are made for corrective action.

TMA Case Study: Creating a Reliability Specification

OVERVIEW

For manufacturing system components and equipment, the determination of the appropriate levels of reliability for incorporation into the design specification involves the evaluation of a large number of factors and the use of trade-off analysis to arrive at an optimum level. The process is driven by the overall objective to meet reliability/availability requirements at minimum life cycle cost. Reliability requirements are formulated covering both quantitative and qualitative design and development program provisions, which are ultimately reflected in a design specification to govern the technology acquisition and operational process.

All design, development, and deployment projects and activities generate enormous amounts of data (e.g., from a rigorous testing program). Used properly, this data enables real-time feedback to guide design adjustments and finalization.

ISSUE

A new research and development project is to be initiated, which will lead to the system technology being applied by the manufacturing organization for the first time. As this technology will ultimately be used globally, it is imperative that the design for manufacture process be guided by a well-vetted reliability specification.

STRATEGIC OBJECTIVE

The strategic objective is to develop and implement a new manufacturing system in an accelerated manner while maximizing the reliability and maintainability to minimize overall life cycle costs (i.e., development and operational support costs).

CASE BACKGROUND

The engineering team was given the task to create a reliability specification for a revolutionary molding machine product transfer and packaging development project. The system is the next generation of existing technology used in existing global manufacturing system operations. Key design features include the following: (1) the molded product runner trimmer component is integrated into the packaging equipment, (2) all system components have artificial intelligence-based "built-in test" and "diagnostic" capability, (3) enhanced product counting capability, and (4) accelerated product transfer into the packaging equipment.

The new system components are an integral part of the overall manufacturing system and needs to operate effectively with a <10 second cycle time. This project does not include changes to the molding process. Acquisition cost is reduced due to historical system design and operational data, as well as from disciplined FRACA activity retrieved from the organization's big data system. Additionally, corrective and preventative maintenance are to be performed similarly to that performed currently for existing system equipment. The use of preventative maintenance will be expanded.

In creating the reliability specification, the design team was to answer the following questions: (1) what is the definition of failure, (2) what are the reliability and maintainability design requirements, and (3) what level of reliability program is appropriate?

RESULT/CONCLUSION

The design team created the following reliability specification to guide the new technology development project.

Next-Generation Product Packager Reliability Specification

The following specifications are the next-generation product packager development project.

1. *Definition of Failure*

   Next-Generation Product Packager failure is the inability to:

   a. *Trim the product off the mold runner per product design requirements*

   b. *Correctly count and orient product in the packager*

   c. *Completely dispense the product into package staging*

2. *Reliability and Maintainability Design Requirements*

The *Next-Generation Product Packager shall meet the following performance requirements:*

   a. Mean Time between Failure: 2400 cycles
    b. Mean Time To Repair: 25 minutes
    c. Inherent Availability: 0.95
3. Reliability Program
     The reliability program shall be in accordance with the following:
    a. Conduct a program per Level B as determined by the Optimized Strategic Testing AI-base Expert System
    b. Develop a design testing program specification tailored to the Program Level B
    c. Conduct design reviews (preliminary and critical)...with all production team members
    d. Design for reliability and manufacture
       i. Mitigate equipment performance degradation due to stringing caused by the molding process
      ii. Define preventative maintenance tasks and the corresponding maintenance intervals
     iii. Determine sparing needs for the equipment
     iv. Perform a human factors analysis to identify and mitigate specific man–machine interactions degrading equipment and system performance
    e. Perform a FMECA (failure mode, effects, and criticality analysis)
       i. Preliminary FMECA to be presented at the preliminary design review
      ii. Final FMECA to be presented at the critical design review
    f. Develop a test plan and perform testing
       i. Test plan to include design approval testing (DAT) and design qualification testing (DQT)
      ii. DAT to consist of "dry cycling" and subsequent transition to manufacturing area (product layout) for integration with existing system equipment...both segments of DAT shall be failure or time truncated as defined in the test plan
     iii. DQT to consist of integrating the test units into the anticipated manufacturing system configuration with the molding machine and accessories...test shall time truncated as defined in the test plan
    g. Perform FRACA in conjunction with all system equipment testing

Case Study Question: *What role does the data recording and feedback play in impacting design for manufacture methods? How is quantitative and qualitative data information the driver of tactical strategies across all levels of the manufacturing organization?*

12.3 SUMMARY

This chapter addresses establishing and maintaining a organization-wide DRFS. The DRFS enables all persons involved in the manufacturing effort to feed the TMA data system with timely and accurate data. The ideal situation is to have a common database, a resource shared in real-time, online, and updated immediately when new information is generated.

Numerous automated technologies are available to collect data in real time. These are referred to as AID technologies and include bar code, optical character recognition, voice entry, RF, vision, magnetic stripe. Also, PLCs are widely used and are very capable data collectors.

An essential part of the DRFS is a working closed-loop FRACA activity. This activity assures that failures are accurately reported and thoroughly analyzed, and that effective corrective actions are taken on a timely basis to eliminate, or minimize, recurrence of the failures. The FRACA activity is most cost-effective when it is open to all failure and problem data regarding both hardware and nonhardware output products and systems.

QUESTIONS

1. What is meant by "real-time" data collection?
2. Discuss the various types of data that different departments might require in a real-time manner.
3. What are major benefits of moving to electronic shop documentation.
4. Discuss the disadvantages of a manual DRFS.
5. What are the two types of databases? Determine what type of database you access to.
6. What are the major technologies used for automated identification? Identify applications of each in the manufacturing system.
7. Discuss the relationship between FRACA and the TMA data system.
8. What is meant by a closed-loop FRACA process?
9. In the FRACA process, what is the difference between short-term and long-term corrective action?
10. What type of information is necessary to close out a failure report?

REFERENCES

1. Allen-Bradley Company. (1989). *Barcode Basics*. Milwaukee, WI: Industrial Control Group.
2. Attaran, M. (1989). "Strategic Issues in the Automated Factory." *Industrial Management & Data Systems*, Vol. 89, No. 4, pp. 14–19. https://doi.org/10.1108/02635578910132819
3. Hayes-Roth, F. (Ed.). (1983). *Building Expert System*. London: Addison-Wesley Publishing Company, Inc.
4. Reliability Analysis Center. (1988). *RADC Reliability Engineer's Toolkit: An Application Oriented Guide for the Practicing Reliability Engineer*. Rome, New York: Rome Air Development Center.

13 Performance Analytics

Performance analytics describes the use of metrics, or key performance measures, to provide insight to the manufacturing operation, as well as the overall business operation. Their purpose is to provide a timely, quick, and focused overview of how the operation is doing relative to strategic goals, objectives, and tactical activities. The specific metrics selected are identified and evolve out of the organization's desire to implement an overall strategic direction leading to optimal operational environment and performance by "aligning the operational stakeholders."

As discussed in Chapter 3, strategic modeling consists of three fundamental levels: (1) strategic goals, (2) strategic objectives, and (3) tactical strategies. Very simply, "aligning the stakeholders" [1] means that the strategies implemented and controlled at each operational level are (1) intimately related to each other, (2) universally understood and accepted, and (3) their outcomes are monitored via measures agreed to and reviewed by all stakeholders. Ultimately, the fundamental purpose of performance metrics is to provide quantitative, and qualitative, visibility to drive continuous improvement based on accountability by the "aligned stakeholders." (Figure 13.1).

13.1 KEY PERFORMANCE INDICATORS

Performance metrics are only useful if they monitor how operations are going in meeting expected outcomes the business is serving as a mechanism to record and facilitate continuous improvement relative to the overall strategic objectives, tactical strategies, and operational strategies [2–4].

Operational performance measures, or metrics, provide insight to operational performance relative to defined strategic goals, objectives, and tactical strategies [5]. This performance snapshot enables quick identification of problems (i.e., opportunities) and the implementation of appropriate action. It also facilitates inter- and intra-divisional communication and the transplanting of "best practices" from one operation to another

Examples of key manufacturing performance measures include:

1. Outgoing acceptable quality level
2. Yield/scrap
3. On-time delivery… (e.g., service level for A/B/C items)
4. Demand/output ratio for domestic and international operations
5. Demand/forecast ratio for domestic and international operations
6. Inventory value for raw/work in process and finished goods
7. Average days sales on hand
8. Inventory turnover rate
9. Cycle count accuracy
10. Fill rate

DOI: 10.1201/9781003208051-17

Alignment

Executive   -- Collect The Right Data

Management   -- Feedback Loop to all Levels

Production Team   -- Do Something With It
Members

FIGURE 13.1   Organizational stakeholder levels.

11. Commitment performance
12. Weekly system effectiveness (Se) reporting
13. Capacity utilization
14. Standard manufacturing times
15. Productivity
16. Efficiency
17. Manufacturing variance
18. Product Std cost
19. Sales forecasts
20. Customer complaints
21. Monthly employee turnover
22. Headcount
23. Oldest order
24. Manufacturing lead time

In the following sections, some key metrics are presented along with highlighted key features. Give special attention to the "Ability to Control" and "How the Business Reacts" information. "Control" is indicated by one to three stars and identifies the amount of control the manufacturing operation has to directly influence the metric's value. The more the stars, the more the organization controls its own destiny relative to the performance issue reflected by the metric.

☆ No control

☆☆ Some control

☆☆☆ Absolute control

The "business reaction" of interest is the impact on cost of goods sold (COGS) and gross profit since these are controllable by the manufacturing operation. Shown via arrow is their general change as a percentage of net sales. In some cases, it is uncertain what happens to COGS and/or gross profit. For example, an increase in COGS may be the result of increased production volume in response to increased product demand. In this case, the gross profit actually may not change. The best approach is to look at the actual value for percentage of net sales to identify variances.

To get the best overall understanding of operational performance, look at all the metrics in parallel. Often, what is happening in one metric is linked to what is happening in another.

13.1.1   PRODUCTION ANALYTICS

This section provides metrics relevant for manufacturing operations monitoring.

SYSTEM EFFECTIVENESS

Ability to Control: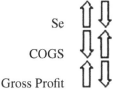

Source:	Manufacturing & Quality Assurance
Frequency:	Weekly/Monthly
Volatility:	High
Goal:	Se >0.9

HOW THE BUSINESS REACTS

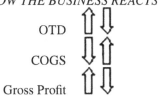

Se

COGS

Gross Profit

Definition: Indicates ability to meet operational demand and achieve defined standard cost. Se = Eo × Aa × Cq; where Eo is operational efficiency, Aa is achieved availability, Cq is quality capability

Addresses: Is the process running at the speed expected? What is the proportion of good product produced? Is system in working condition/status?

MFG ORDER ON TIME DELIVERY

Ability to Control:

Source:	Materials
Frequency:	Monthly/Annual
Volatility:	Moderate
Goal:	OTD% = 100%

HOW THE BUSINESS REACTS

OTD

COGS

Gross Profit

Definition: Indicates ability to get manufacturing orders done on time per MRP date. OTD = (# orders complete on time/ # total orders due) × 100

Addresses: What is ability to complete production and transfer product into an on-hand, available finished good status?

PRODUCTIVITY

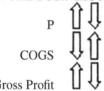

Ability to Control:
Source: Accounting
Frequency: Monthly/Annual
Volatility: Moderate
Goal: P > 100%

HOW THE BUSINESS REACTS

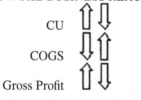

P

COGS

Gross Profit

Definition: Indicates the total direct and indirect labor time spent on a product(s) relative to the standard hours allocated. P = std hours/(direct hours + indirect hours) × 100

Addresses: Is in the actual labor time out of line with the defined standard time? Is there too much indirect time spent?

CAPACITY UTILIZATION

Ability to Control:
Source: Manufacturing
Frequency: Weekly/Monthly
Volatility: Low
Goal: CU = 90%–95%

HOW THE BUSINESS REACTS

CU

COGS

Gross Profit

Definition: Indicates use of current mfg time available; both scheduled and possible. CU = [(Mfg output achieved or time used)/(Mfg output possible or time available)] x 100

Addresses: Are equipment/facilities use time maximized? Is the need for manufacturing time or equipment expansion looming in near future? Is poor system throughput extending manufacturing time required?

DEMAND/OUTPUT RATIO

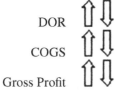

Ability to Control:
Source: Materials
Frequency: Weekly/Monthly
Volatility: Moderate
Goal: DOR >1.0

HOW THE BUSINESS REACTS

DOR ⇧⇩

COGS ⇧⇩

Gross Profit ⇧⇩

Definition: Indicates ability to meet market demand and to over produce. DOR = Order Qty/Production Qty

Addresses: Producing too much?...too little? Has demand changed drastically?

RAW & WIP INVENTORY LEVEL

Ability to Control:
Source: Materials and Accounting
Frequency: Monthly/Annual
Volatility: Moderate
Goal: RWIL = 2 weeks $ max... on-hand

HOW THE BUSINESS REACTS

RWIL ⇧⇩

COGS ⇧⇩

Gross Profit (uncertain)

Definition: Indicates $ value of materials carried. Look at the % of net sales. RWIL = $ of material on hand

Addresses: Experiencing low inventory turns? Is the investment in materials on hand increasing?

STD MFGING TIMES

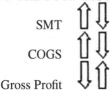

Ability to Control:
Source: Manufacturing
Frequency: Annual
Volatility: Low
Goal: SMT = 100% accuracy

HOW THE BUSINESS REACTS

SMT   ⇧ ⇩

COGS   ⇧ ⇩

Gross Profit ⇩ ⇧

Definition: Indicates the time to produce a product. SMT = production time per product #

Addresses: Is Mfg time excessive? What is relevance of product with lengthy Mfg time to total product offering? Is Mfg time reduction possible?

YIELD

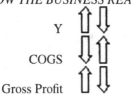

Ability to Control:
Source: Mfg & QA
Frequency: Weekly/Monthly
Volatility: High
Goal: Y% = 100%

HOW THE BUSINESS REACTS

Y   ⇧ ⇩

COGS   ⇩ ⇧

Gross Profit ⇧ ⇩

Definition: Indicates the ability of the manufacturing system to produce product of acceptable quality. Y% = (qty accepted items/qty in production run) × 100

Addresses: Is manufacturing scrap (fallout) increasing? Too high? Is it taking more time to produce a given quantity of good product? Is fallout impairing the ability to meet market demand?

OUTGOING QUALITY

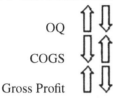

Ability to Control:
Source: QA
Frequency: Weekly/Monthly/Annual
Volatility: Low
Goal: OQ% = 100%

HOW THE BUSINESS REACTS

OQ

COGS

Gross Profit

Definition: Indicates the customer's rejection of product due to quality assurance issues. $OQ\% = (1 - (\text{qty customer rejects/qty product shipped})) \times 100$

Addresses: Shipping acceptable product? Are finished good quality standards correct?…adhered too?

EFFICIENCY

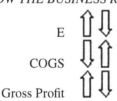

Ability to Control:
Source: Accounting
Frequency: Monthly/Annual
Volatility: Low
Goal: E > 100%

HOW THE BUSINESS REACTS

E

COGS

Gross Profit

Definition: Indicates the direct labor effort expended relative to that required per the standard time. $E\% = (\text{std hours/direct hours}) \times 100$

Addresses: Is it taking more labor to produce than that expected? Are labor standards incorrect?

DEMAND/FORECAST RATIO

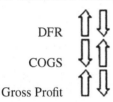

Ability to Control:
Source: Materials
Frequency: Monthly/Annual
Volatility: Moderate
Goal: DFR > 1.0

HOW THE BUSINESS REACTS

DFR

COGS

Gross Profit

Definition: Indicates customer demand relative to the forecast used for production planning. DFR = order qty/forecast qty

Addresses: Is the forecast too low?...too high? Has sales/marketing done something to stimulate sales? Is appropriate manpower in place to match production needs?

WORKER TURN-OVER

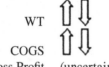

Ability to Control:
Source: Human Resources
Frequency: Monthly/Annual
Volatility: Low
Goal: WT% <1%

HOW THE BUSINESS REACTS

WT

COGS

Gross Profit   (uncertain)

Definition: Indicates loyalty of workforce to company. Look at the % of net sales. WTR% = (qty departures/qty employees) × 100

Addresses: Are wages and benefits competitive with other employers in the area? Are there manager – workforce relationship problems? Is investment in employee development adequate?

HEADCOUNT

Ability to Control:
     Source: Human Resources
  Frequency: Monthly/Annual
   Volatility: Low
      Goal: HC = production requirement

HOW THE BUSINESS REACTS

HC ⇧⇩
COGS ⇧⇩
Gross Profit (uncertain)

Definition: Indicates the number of people employed by the facility. Look at the % of net sales. HC = qty employed

Addresses: Is an educated workforce available? Is the location attractive to candidate workers? Is each position value-added to the business?

13.1.2 FINANCIAL ANALYTICS

This section provides metrics relevant to financial monitoring.

PROFIT – LOSS STATEMENT

Ability to Control:
     Source: Accounting
  Frequency: Monthly/Annual
   Volatility: Moderate
      Goal: PLS to show positive Net Income

HOW THE BUSINESS REACTS

PLS ⇧⇩
COGS (uncertain)
Gross Profit (uncertain)

Definition: Indicates financial health of operation. Look at the % of net sales.

Addresses: Are net sales low? Are expenditures high? Are operating expenses where they should be as a % of net sales? Is there a quality problem?

FINISHED GOODS INVENTORY VALUE

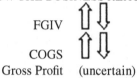

Ability to Control:
Source: Accounting & Materials
Frequency: Monthly
Volatility: Moderate
Goal: FGIL = 2 weeks $ max... on-hand

HOW THE BUSINESS REACTS

FGIV ⇧⇩

COGS ⇧⇩

Gross Profit   (uncertain)

Definition: Indicates $ value of materials carried. Look at the % of net sales.

Addresses: Experiencing low inventory turns? Is the investment in finished goods on hand increasing? Has product demand decreased? Too much product being produced?

PRODUCT STANDARD COST

Ability to Control:
Source: Accounting
Frequency: Annual
Volatility: Moderate
Goal: PSC decreasing

HOW THE BUSINESS REACTS

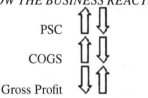

PSC ⇧⇩

COGS ⇧⇩

Gross Profit ⇩⇧

Definition: Indicates actual Mfg costs attained. PSC = (std Mfg time × VOH) + (std Mfg time × FOH)

Addresses: Is variable overhead increasing?...decreasing? Is fixed overhead increasing?...decreasing? Are warranty costs increasing? ...decreasing?

AVERAGE DAYS SALES ON HAND

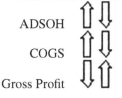

Ability to Control:
Source: Materials & Accounting
Frequency: Monthly/Annual
Volatility: Moderate
Goal: ADSOH = Mfg Lead Time

HOW THE BUSINESS REACTS

ADSOH ⇧ ⇩

COGS ⇧ ⇩

Gross Profit ⇩ ⇧

Definition: Indicates level of excess finished goods inventory on hand relative to manufacturing lead time. ADSOH = (period inventory $ value/avg daily COGS) − Mfg lead time

Addresses: Too much inventory on hand? How does days-worth of sales compare to manufacturing lead time?

SALES FORECAST VARIANCE

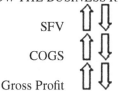

Ability to Control:
Source: Materials & Product Management
Frequency: Monthly/Annual
Volatility: High
Goal: SFV% <10%

HOW THE BUSINESS REACTS

SFV ⇧ ⇩

COGS ⇧ ⇩

Gross Profit ⇧ ⇩

Definition: Indicates the quantity of sales expected and reflected in the master production schedule. SFV% = ((qty achieved − qty expected)/qty expected) × 100

Addresses: Is the sales growth rate suspiciously high?…low? Does the forecast reflect any cyclical market patterns known to exist?

FINISHED GOODS INVENTORY TURNS

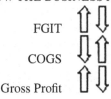

Ability to Control:
        Source: Accounting & Materials
     Frequency: Monthly/Annual
     Volatility: High
          Goal: FGIT >10/year

HOW THE BUSINESS REACTS

FGIT   ⇑⇓

COGS   ⇓⇑

Gross Profit ⇑⇓

Definition: Indicates the movement pace of finished goods to the customer. FGIT = (annual COGS)/(avg inventory $ on hand)

Addresses: Producing more than what the market is absorbing? Are inventory levels balanced with customer expectations of receipt time after ordering? Has something happened in the market to make current inventory levels unacceptable?

CYCLE COUNT VARIANCE

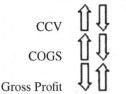

Ability to Control:
        Source: Materials
     Frequency: Weekly/Monthly
     Volatility: Low
          Goal: CCV% <10%

HOW THE BUSINESS REACTS

CCV   ⇑⇓

COGS   ⇑⇓

Gross Profit ⇓⇑

Definition: Indicates that actual inventory on hand is the same as that believed to be on hand. CCV% = ((qty counted-qty expected)/qty expected) × 100

Addresses: Is inventory management in control? Are inventory transactions correct? Is inventory disappearing?

13.1.3 CUSTOMER SERVICE ANALYTICS

This section provides metrics relevant for customer service monitoring.

SHIPPING FILL RATE

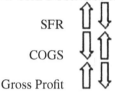

Ability to Control:
Source: Materials
Frequency: Monthly/Annual
Volatility: Low
Goal: SFR% = 100%

HOW THE BUSINESS REACTS

SFR ⇑ ⇓

COGS ⇓ ⇑

Gross Profit ⇑ ⇓

Definition: Indicates ability to ship complete customer order. SFR% = (ship qty/order qty) × 100

Addresses: Is inventory unavailable to fill orders? Are the wrong products, and quantities, carried in inventory? Does the customer receive shipments from multiple manufacturing locations to fill an order?

COMMITMENT VARIANCE

Ability to Control:
Source: Materials
Frequency: Monthly/Annual
Volatility: Moderate
Goal: CV = 0 days

HOW THE BUSINESS REACTS

CV ⇑ ⇓
COGS (uncertain)
Gross Profit (uncertain)

Definition: Indicates ability to get the product to the customer on the date promised. Look at the % of net sales. CV = arrival date − due date... ...(− days:early + days:late)

Addresses: Is product arriving too early?...too late? Are shipping lead times wrong? Is there a shipment carrier performance problem?

OLDEST MFGING ORDER

Ability to Control: ★★☆

Source: Manufacturing
Frequency: Weekly/Monthly
Volatility: Moderate
Goal: OMO <5 days

HOW THE BUSINESS REACTS

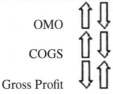

OMO

COGS

Gross Profit

Definition: Indicates ability to get a product manufactured in accordance with defined router time. OMO = current date − due date…(− days:early + days:late)

Addresses: Are processes running at the throughput speed expected? What is causing the production delay or slow down?

FINISHED GOODS INVENTORY LEVEL

Ability to Control: ★★☆

Source: Materials
Frequency: Weekly/Monthly
Volatility: Moderate
Goal: FGIL = A, B, C, D, service-level requirement

HOW THE BUSINESS REACTS

FGIL

COGS
Gross Profit   (uncertain)

Definition: Indicates ability to satisfy customer product availability expectations. Look at % of net sales.

Addresses: Producing the correct product? Producing the correct quantities of product? Do inventory levels match defined strategic service requirements?

CUSTOMER COMPLAINTS

Ability to Control: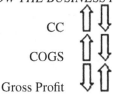
Source: Customer Service & Quality Assurance
Frequency: Monthly/Annual
Volatility: Moderate
Goal: CC=0

HOW THE BUSINESS REACTS

CC ⇑⇓

COGS ⇑⇓

Gross Profit ⇓⇑

Definition: Indicates ability to meet customer quality expectations in product and service. CC=qty product or service complaints

Addresses: Are there product quality or reliability problems? Are customer services/ processes poor and error-prone?

MFG LEAD TIME

Ability to Control: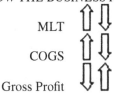
Source: Manufacturing
Frequency: Monthly/Annual
Volatility: Moderate
Goal: MLT <10 days

HOW THE BUSINESS REACTS

MLT ⇑⇓

COGS ⇑⇓

Gross Profit ⇓⇑

Definition: Indicates length of time for a product production cycle. MLT=Mfg start date − Mfg complete date

Addresses: Is the production process too long and inefficient? Are supplier capabilities in sync with customer expectations?

TMA Case Study: Manufacturing System Effectiveness Monitoring

OVERVIEW

Performance metrics are necessary to fulfill the adage of "what gets measured, gets done." The need to meet local markets and overall business performance improvements arising from increased manufacturing integration and global technology transfer continues to be a primary competitive issue. Effective manufacturing system deployment and integration results, in part, from effective deployment of business objectives down through the organization and the subsequent measurement of performance in critical areas as key elements of sustainable competitive advantage [6].

The performance measures serve as benchmark references to allow auditing of existing operational performance to effect more robust, flexible, and integrated manufacturing systems. Correctly structured and designed performance metrics provide the basis for a rigorous and effective performance management system, which is used as a management tool by strategic, tactical, and operational levels of management.

ISSUE

Manufacturing system technology is being transferred to additional locations to meet global demand. The originating location is North America with transfer locations being Europe and Asia. This provides logistical advantages to meet growing markets, as well as provide contingency manufacturing capabilities

STRATEGIC OBJECTIVE

The strategic objective is to ensure that the manufacturing system performs as good, or better, than the originating location operation. A standard company metric for manufacturing effectiveness will be used for locational benchmarking.

CASE BACKGROUND

The performance management activity is the process by which the organization manages its performance in line with its organizational and functional strategic goals, objectives, and tactics. The performance data is active with the organization's big data system providing a proactive closed-loop control system, where the organizational functional strategies are deployed to all business processes, activities, tasks and personnel, and feedback is obtained through the performance measurement system to enable optimal decision-making.

The organization's big data system collects all information relevant to managing the business. Typical information collected includes items, such as:

- Identification of business units and manufacturing capabilities.
- The market requirements in terms of qualifiers and differentiators.
- The development plans or objectives for the business, and each subunit.
- The performance measures used and reviewed by the executive leadership team and, as applicable, at all organizational levels (i.e., the strategic performance analytics).

- The specific performance metrics used and reviewed within each function of the organization.
- Personal objectives and any associated incentive scheme for the executives, managers, supervisors and operational personnel.
- Review, reporting, and performance responsibilities associated with measures used at all levels.

For this technology transfer to Europe and Asia, performance metrics will be used that continue consistent monitoring and analysis across the manufacturing operations. To maintain its competitive advantages regarding product quality, availability, and cost, it was decided to benchmark the performance of all similar manufacturing systems using the manufacturing system effectiveness metric. This performance metric addresses quality capability, operational efficiency, and achieved availability.

RESULT/CONCLUSION

The manufacturing system in North America was duplicated and transferred to additional new locations in Europe and Asia. The Se (manufacturing system effectiveness) was used to provide comparative benchmarking, as well as to monitor for ongoing continuous improvement of technology, methods, training, and maintenance planning. A full calendar year of data for each location is provided in Table 13.1.

Case Study Question: *What inferences are to be made regarding the system technology transfer success? What are the opportunities to be evaluated to facilitate continuous improvement at each global location?*

13.2 SUMMARY

A key element of improving manufacturing profitability is to identify and understand areas of improvement. This begins with taking a quantitative look at the operation. Establishing and implementing performance metrics begins that process. Each manufacturing operation and system is unique and requires strategically defining and implementing key performance measures that are of value. The measures, or metrics, provide insight to operational performance relative to defined strategic objectives and goals.

This performance snapshot enables quick identification of problems (i.e., opportunities) and the implementation of appropriate action. It also facilitates interdivisional communication and the transplanting of "best practices" from one operation to another. An essential feature of each metric is in the information it provides regarding the company's "ability to control" and "how the business reacts." "Control" is the amount of control the company has in directly influencing the metric's value. The greater the control, the greater the ability to impact the performance issue reflected by the metric.

The "business reaction" of interest is typically in the impact on COGS and gross profit since these are controllable by the business operation. In some cases, it is uncertain what happens to COGS and/or gross profit. For example, an increase in COGS may be the result of increased production volume in response to increased product

TABLE 13.1
Global Manufacturing System Effectiveness Data

Location: North America — CY25

Month	Eo	Aa	Cq	Se
January	0.87	0.98	0.99	0.84
February	0.88	0.95	0.99	0.83
March	0.86	0.9	0.95	0.74
April	0.89	0.97	0.99	0.85
May	0.86	0.9	0.98	0.76
June	0.89	0.95	0.97	0.82
July	0.85	0.87	0.99	0.73
August	0.93	0.96	0.99	0.88
September	0.91	0.93	0.99	0.84
October	0.89	0.96	0.98	0.84
November	0.9	0.96	0.99	0.86
December	0.92	0.98	0.99	0.89

Location: Europe — CY25

Month	Eo	Aa	Cq	Se
January	0.9	0.9	0.97	0.79
February	0.9	0.9	0.98	0.79
March	0.94	0.91	0.99	0.85
April	0.95	0.95	0.99	0.89
May	0.96	0.95	0.99	0.90
June	0.98	0.95	0.99	0.92
July	0.97	0.96	0.99	0.92
August	0.96	0.95	0.99	0.90
September	0.98	0.97	0.99	0.94
October	0.99	0.95	0.99	0.93
November	0.97	0.96	0.99	0.92
December	0.98	0.97	0.99	0.94

Location: Asia — CY25

Month	Eo	Aa	Cq	Se
January	0.8	0.85	0.9	0.61
February	0.82	0.9	0.9	0.66
March	0.84	0.87	0.95	0.69
April	0.83	0.88	0.93	0.68
May	0.85	0.86	0.95	0.69
June	0.87	0.87	0.96	0.73
July	0.88	0.88	0.98	0.76
August	0.9	0.93	0.99	0.83
September	0.89	0.93	0.98	0.81
October	0.93	0.95	0.99	0.87
November	0.95	0.95	0.99	0.89
December	0.94	0.96	0.99	0.89

demand. In this case, the gross profit may not actually change. The best approach is often to look at the actual value for percentage of net sales to identify variances.

Performance metrics need to be holistically aligned and reviewed to provide meaningful insights and guidance to maximizing manufacturing effectiveness. To get the best overall understanding of operational performance, it is critical to look at all the key metrics in parallel. Often, what is happening in one metric is linked to what is happening in another. Additionally, it is important to facilitate transparency of key metrics to all stakeholders for review and analysis. Looking at manufacturing performance from many different perspectives enables eliminating knee-jerk reactions and leads to making strategically correct decisions.

QUESTIONS

1. Why is COGS an important foundational measure?
2. How many key performance indicators are needed?
3. Discuss the analysis and review of performance analytics.
4. What is the guideline for the number of people reviewing the performance metrics?
5. Why is the ability to control a performance measure important?
6. Does metric volatility make it a better metric or a worse metric? Why?
7. How do all the metrics relate to financial considerations?
8. If the organization is a "nonprofit," what performance metrics apply?
9. What metrics are most important... production, financial, or customer service?
10. How does an organization address the cycling of adding or deleting performance metrics?

REFERENCES

1. Lynch, R. & Cross, K. (1991). *Measure Up! Basil.* Cambridge, MA: Blackwell, Inc.
2. Wheelwright, S. (Ed.) (1979). *Manufacturing Strategy.* Cambridge, MA: Harvard Business Review.
3. Institute for International Research (1990). Performance measurement for manufacturers. *International Conference,* Chicago, IL.
4. Chen, G. & McGarrah, R. (1982). *Productivity Management: Text & Cases.* Chicago, IL: Dryden Press.
5. Brauer, D. (1999). *The Atlas of Key Production Performance Measures.* Mokena, IL: Design Assurance Sciences.
6. Carrie, A. & Macintosh, S. (1992). *UK Research in Manufacturing Systems Integration, Integration in Production Management Systems.* London: Elsevier Publishers.

Appendix: Reactions to Group Situations Test (RGST)

RGST Questions (circle A or B)

1. When I wanted to work with Pat, I...
   A. felt we could do well together.
   B. asked if it would be all right with him.
2. When the group wanted his views about the task, Sam...
   A. wondered why they wanted his views.
   B. thought of what he might tell them.
3. When the leader made no comment, I...
   A. offered a suggestion of what to do.
   B. wondered what to do next.
4. When Taylor said he felt closest to me, I...
   A. was glad.
   B. was suspicious.
5. When I felt helpless, I...
   A. wished that the leader would help me.
   B. found a friend to tell how I felt.
6. When Sydney was annoyed, Alex...
   A. thought of a way to explain the situation to him.
   B. realized just how he felt.
7. When Andy felt eager to go to work, she...
   A. got mad at the late-comers.
   B. wanted to team up with Morgan.
8. When Mackenzie bawled me out, I...
   A. lost interest in what we were supposed to be doing.
   B. thought that some of his ideas would be useful.
9. When the leader lost interest, Riley...
   A. suggested a way to get everybody working.
   B. started talking with his neighbors.
10. When Avery felt warm and friendly, she...
    A. accomplished a lot more.
    B. liked just about everyone.
11. When the leader was unsure of himself, Jordan....
    A. wanted to leave the group.
    B. didn't know what to do.

12. When the group just couldn't seem to get ahead, I...
    A. felt like dozing off.
    B. became annoyed with them.
13. When the group wasn't interested, I...
    A. just didn't feel like working.
    B. thought that the leader should do something about it.
14. When the leader said he felt the same way I did, I...
    A. was glad that I had his approval.
    B. thought we would probably begin to make progress now.
15. When I became angry at Adrian, I...
    A. felt like dozing off.
    B. ridiculed his comments.
16. When the leader wanted me to tell the class about my plan, I...
    A. wished I could get out of it.
    B. wished that he would introduce it for me.
17. When Aiden criticized Sam, I...
    A. wished that the leader would help Sam.
    B. felt grateful to Aiden for really expressing what we both felt.
18. When Pat and Taylor enjoyed each other's company so much, I...
    A. thought that I'd like to leave the room.
    B. felt angry.
19. When the leader changed the subject, Sydney....
    A. suggested that they stick to the original topic.
    B. felt glad that the leader was finally taking over.
20. When the others became so keen on really working hard, I....
    A. made an effort to make really good suggestions.
    B. felt much more warmly towards them.
21. When I felt angry enough to boil, I...
    A. wanted to throw something.
    B. wished that the leader would do something about it.
22. When Lee was not paying attention, I...
    A. did not know what to do.
    B. wanted to tell her she was wasting our time.
23. When Alex thought that he needed a lot of help, Andy....
    A. warmly encouraged him to get it.
    B. helped him analyze the problem.
24. When Morgan reported her results so far, I...
    A. laughed at her.
    B. was bored.
25. When everyone felt angry, I...
    A. suggested that they stop and evaluate the situation.
    B. was glad that the leader stepped in.
26. When no one was sticking to the point, I...
    A. got bored with the whole thing.
    B. called for clarification of the topic.

27. When Mackenzie said he felt especially friendly towards me, I...
    A. wanted to escape.
    B. wanted to ask his advice.
28. When the group agreed that it needed more information about how members felt, I...
    A. described my feelings to the group.
    B. Wasn't sure I wanted to discuss my feelings.
29. When the leader offered to help Riley, Avery....
    A. wanted help too.
    B. resented the leader's offer.
30. When Jordan and Adrian argued, I...
    A. asked Aiden how she felt about them.
    B. hoped they would slug it out.
31. When Sam felt especially close to Lee, he...
    A. let him know it.
    B. hoped he could turn to him for assistance.
32. When several members dropped out of the discussion, Pat...
    A. thought it was time to find out where the group was going.
    B. got sore at what he thought was their discourtesy.
33. When Taylor told me she felt uncertain about what should be done, I...
    A. suggested that she wait awhile before making any decisions.
    B. suggested that she get more information.
34. When Sydney realized that quite a few people were taking digs at each other, he...
    A. wanted to call the group to order.
    B. got angry at the stupidity of their behavior.
35. When the group suggested a procedure, I...
    A. thought the leader ought to express approval or disapproval of it.
    B. thought we ought to decide whether to carry it out.
36. When Alex seemed to be daydreaming, Andy...
    A. winked at Morgan.
    B. felt freer to doodle.
37. When Mackenzie and Riley arrived 20 minutes late for the meeting, the group...
    A. went right on working.
    B. was very annoyed.
38. During the argument, Avery's opposition caused Jordan to...
    A. withdraw from the discussion.
    B. look to the leader for support.
39. When Jordan suggested we evaluate how well we were working us a group, I...
    A. was glad that the meeting was almost over.
    B. gladly backed him up.
40. When the group seemed to be losing interest, Pat...
    A. became angry with the other members.
    B. thought it might just as well adjourn.

41. Together Adrian and Aiden....
    A. wasted the group's time.
    B. supported one another's arguments.
42. When Sam offered to help me, I...
    A. said I was sorry, but I had something else to do.
    B. was pleased that we would be partners.
43. When the other group became so interested in their work, Pat...
    A. wanted to ask their leader if he could join them.
    B. felt resentful that his group was so dull.
44. When Taylor left the meeting early, Lee....
    A. and Sydney told each other what they felt about Taylor.
    B. was glad that he had gone.
45. When Alex turned to me, I...
    A. wished that she would mind her own business.
    B. asked her for help.
46. When Aiden felt hostile to the group, he...
    A. wished he would not have to come to the meeting.
    B. was glad that Andy felt the same way.
47. While Jordan was helping me, I...
    A. became annoyed with her superior attitude.
    B. felt good about being with her.
48. When I lost track of what Morgan was saying, I...
    A. asked the leader to explain Morgan's idea to me.
    B. was pleased that it was Sam who explained Morgan's idea to me.
49. While the group was expressing friendly feelings towards Mackenzie, Pat....
    A. thought that now Mackenzie would be able to work.
    B. opened a book and started to read.
50. When the leader offered to help her, Riley....
    A. said that she did not want any help.
    B. realized that she did need help from someone.

REACTIONS TO GROUP SITUATIONS TEST ANSWER KEY

Instructions: For each item answer, circle the letter corresponding to the answer on the RGST Answer Sheet. For example, if A was marked for item 1, circle the letter P on this answer key for that item. To obtain each of the five scores, count the number of times each letter was circled. (The letters denoted as "BA Modality" indicate the basic assumption preference designed into the question responded to.) The highest score indicates the dominant BA predisposition.

Question	BA Modality	Response A	B		Question	BA Modality	Response A	B
1	W	P	D		26	FL	FL	W
2	W	F	W		27	P	FL	D
3	D	W	D		28	W	W	FL
4	P	P	F		29	D	D	F
5	D	D	P		30	F	P	F
6	F	W	P		31	P	P	D
7	W	F	P		32	FL	W	F
8	F	FL	W		33	D	FL	W
9	FL	W	P		34	F	W	F
10	P	W	P		35	W	D	W
11	D	FL	D		36	FL	P	FL
12	D	FL	F		37	P	W	F
13	FL	FL	D		38	F	FL	D
14	P	D	W		39	W	FL	P
15	F	FL	F		40	FL	F	FL
16	W	FL	D		41	P	FL	P
17	F	D	P		42	D	FL	P
18	P	FL	F		43	W	D	F
19	FL	W	D		44	FL	P	F
20	W	W	P		45	P	F	D
21	F	F	D		46	F	FL	P
22	FL	D	F		47	D	F	P
23	D	P	W		48	FL	D	P
24	W	F	FL		49	P	W	FL
25	F	W	D		50	D	F	W

SCORE

W = Work:_____

F = Fight:_____

FL = Flight:_____

D = Dependency:____

P = Pairing:_____

Index

Printed in the United States
by Baker & Taylor Publisher Services